Environmental Changes and Human Occupation in East Asia during OIS3 and OIS2

Edited by

Akira Ono
Masami Izuho

BAR International Series 2352
2012

Published in 2016 by
BAR Publishing, Oxford

BAR International Series 2352

Environmental Changes and Human Occupation in East Asia during OIS3 and OIS2

ISBN 978 1 4073 0938 5

BAR Publishing is the trading name of British Archaeological Reports (Oxford) Ltd. British Archaeological Reports was first incorporated in 1974 to publish the BAR Series, International and British. In 1992 Hadrian Books Ltd became part of the BAR group. This volume was originally published by Archaeopress in conjunction with British Archaeological Reports (Oxford) Ltd / Hadrian Books Ltd, the Series principal publisher, in 2012. This present volume is published by BAR Publishing, 2016.

Printed in England

BAR
PUBLISHING

BAR titles are available from:

BAR Publishing
122 Banbury Rd, Oxford, OX2 7BP, UK
EMAIL info@barpublishing.com
PHONE +44 (0)1865 310431
FAX +44 (0)1865 316916
www.barpublishing.com

Table of Content

Preface

The Japan Association for Quaternary Research (JAQUA) and the Geological Survey of Japan (GSJ), National Institute of Advanced Industrial Science and Technology (AIST), celebrated their 50th and 125th anniversaries, respectively, with an international symposium entitled "Quaternary Environmental Changes and Humans in Asia and the Western Pacific" November 19-22, 2007, in Tsukuba, Japan.

We organized a session *Environmental Changes and Human Occupation in North and East Asia during OIS 3 and OIS 2*, focusing on the correlation between environmental changes and human activities among Palaeolithic sites in North and East Asia.

Assembled articles in this volume primarily originated from contributions to this symposium in both oral and poster sessions. Explicit interests on environmental changes and archaeology during OIS3 and OIS2 in East Asia have developed particularly in the last decade among research communities of various Quaternary disciplines in Japan, Korea, and the Russian Far East.

Chronologically, the middle of OIS3 corresponds broadly to the Eurasian Middle to Upper Palaeolithic transition, and this period also incorporates critical issues of human occupation in the Japanese islands from the Asian mainland. Although traces of Upper Palaeolithic people during OIS2 have been found at many locations in North and East Asia, human occupation in the context of environmental change is still open to discussion.

This volume consists from three focuses. First is climatic reconstruction and the temporal correspondence between archaeological and palaeoenvironmental records in the central part of the Japanese islands. Second is terrestrial mammal assemblages in Japanese discussing both the general features and specific aspects of the Okinawa islands. The last is archaelogical aspects of various lithic industries. The origins of microblade industries in northeast Asia, and reevaluation of the lithic chronology in the Imjin-Hantan River area in Korea provide controversial points of issue. New interpretations of the geochronological framework of Upper Palaeolithic Hokkaido, and early obsidian use of the Japanese islands in the Upper Palaeolihtic represent new aspects of recent studies.

This volume will provide diverse aspects of environmental changes and human occupation in East Asia during OIS3 and OIS2.

We express our gratitude to contributors and constructive reviewers for this volume. Also, we thank Akira Iwase for his help on editing. Special thanks are due to Dr. Ian Buvit for his intensive revision of all English texts.

Editors:

Akira Ono
Center for Obsidian and Lithic Studies, Meiji University
1-1, Kanda-Surugadai, Chiyoda-Ku, Tokyo, Japan 101-8301
Email: onoak@meiji.ac.jp

Masami Izuho
Archaeology Laboratory, Faculty of Social Sciences and Humanities, Tokyo Metropolitan University,
1-1, Minami Osawa, Hachioji City, Tokyo 192-0397, Japan.
Email: izuhom@tmu.ac.jp

High-Resolution Climate Reconstruction for the Past 72Ka from Pollen, Total Organic Carbon (Toc) and Total Nitrogen (Tn) Analyses of Cored Sediments From Lake Nojiri, Central Japan

Fujio Kumon

Department of Environmental Sciences, Faculty of Science, Shinshu University, Matsumoto 390-8621, Japan. Email: shkumon@shinshu-u.ac.jp

Sayuri Kawai

Institute of Mountain Science, Shinshu University, Matsumoto 390-8621, Japan. Email: skawai@shinshu-u.ac.jp

Yoshio Inouchi

Department of Human Behavior and Environmental Science, Faculty of Human Science, Waseda University, Mikajima, Tokorozawa 359-1192, Japan. Email: yinouchi@waseda.jp

Abstract: *Pollen analysis was performed on the upper three-fourths of a 44-m sediment core (NJ88) taken from the northern part of Lake Nojiri, central Japan in short segments of 2 to 6cm, which corresponded to intervals of 80 years on average. Total organic carbon (TOC) and total nitrogen (TN) analyses were also performed on the same core and on an additional 35-m sediment core (NJ95) for average intervals of 35 years. Lake Nojiri is located on the northwestern margin of the Japanese Alps, and has an altitude of 654m. The basin reaches a maximum depth of 38m deep with an average depth of 20.8m. The sediments consist mostly of homogeneous silty clay with tephra intercalations. The combined data on the both sediment cores cover the past 72ka, judging from sedimentation rates calculated from 14C dates and marker tephra ages.*

Pollen assemblages showed a cool-temperate period with deciduous broad-leaved trees spanning 0 to 12ka (marine isotope stage (MIS) 1) and a cold climate with subarctic conifer forests typically from 18.5 to 29ka (major MIS2) and 61 to 72ka (MIS4). Various intermediate stages exist between both forest types with frequent sudden peaks of cool-temperate deciduous broad-leaved trees from 29 to 59ka (MIS3). TOC and TN analysis on 2,064 samples showed three major modes; high TOC and TN from the present to 12ka, a constantly low TOC and TN from 12 to 29ka, and frequently fluctuating TOC and TN from 29 to 72ka. These climate changes correspond well with the isotope changes in ice cores from Greenland, and with MIS1 to MIS4 in both the variation profile and age. In addition, the short TOC peaks can be correlated with interstadial stages 1 to 18 identified in the Greenland ice cores.

This detailed reconstruction of the paleoclimate from the late Last Glacial age to the post-glacial offers direct information on the environments in which ancient people lived. The Upper Paleolithic record first appeared in the Kanto Plain of Japan at approximately 33ka. The lithic component changed quickly and eventually developed to an incipient pottery culture at 12ka. These cultural changes may have corresponded to the climate changes mentioned above.

***Keywords**: pollen composition, total organic carbon, total nitrogen, climate change, Last Glacial age, tephrochronology, Lake Nojiri*

Introduction

Global climate change has become a great concern recently because climate severely affects human lives in terms of food supply, health and safety for the present and in the future. Ancient people should also have been affected strongly by natural climate changes, especially during glacial times. Paleoclimate studies have progressed with data from ocean sediments and polar ice sheets. For example, $\delta^{18}O$ data of ice cores from the Greenland ice sheet are popular and excellent standards of past climate (e.g., Dansgaard *et al.*, 2003; NGRIP [North Greenland Ice Core Project members], 2004), but they may reflect local conditions in the north Atlantic. We need to know the local paleoclimate of Japan, as well as to understand the past global climate changes that may have strongly influenced the lives and cultures of people who lived on the Japanese islands.

Pollen analysis is an excellent and traditional way to reveal the paleoclimate of a land area, although the analysis requires mature skills and considerably intensive, lengthy labor for high-resolution study. Therefore, total organic carbon (TOC) and total nitrogen (TN) data were added in

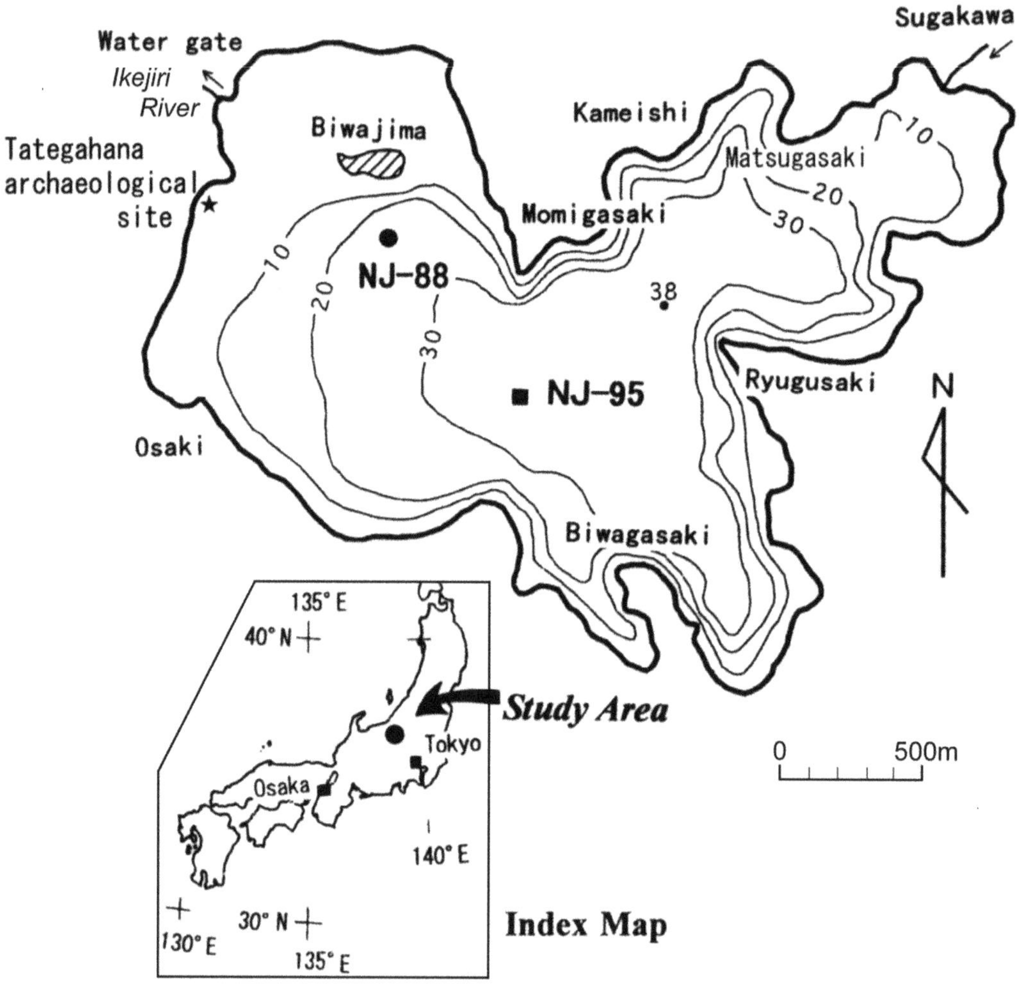

Figure 1 Locality of the study area showing the two drilled sites in Lake Nojiri, central Japan

this study to obtain higher resolution signals and to verify previouse climate information. TOC and TN contents have recently been used as useful paleoclimate proxies (Inouchi *et al.*, 1996; Meyers, 1997; Adhikari and Kumon, 2001; Adhikari *et al.*, 2002; Kumon, 2003, Tawara *et al.*, 2006). TOC and TN measurements require less labor and enable us to collect data in dense intervals.

In this paper, we reveal details on climate changes over the past 72ka on the basis of data obtained from analyses on a 44.16m sediment core together with another 35.4m core, both from Lake Nojiri, central Japan (Fig. 1). The data show continuous climate change from the middle Last Glacial to the present in 80-year intervals for the Japanese islands. In addition, the Palaeolithic archaeological site Tategahana is located on the western coast of Lake Nojiri associated with abundant mammal fossils (Utashiro, 1980; Nojiri-ko Excavation Research Group, 1984, etc.). The paleoclimate information can be correlated with the fossil and cultural remains at the site through marker tephra beds. Understanding past climate variation may be a key to elucidating the migration and developments of ancient people on the islands.

Geographical setting of Lake Nojiri

Lake Nojiri is located on the northwestern margin of the Japanese Alps and is at an altitude of 654.3m. The annual average temperature varies from 7.9°C to 10.1°C (average 9.0°C) and the annual precipitation is from 871 to 1,676mm (average 1,254mm) at the Shinanomachi meteorological station (36°32.9'N, 137°59.8'E, altitude 509m), a few km south of the lake, based on data from 1979 to 2004. At present, the natural vegetation around Lake Nojiri is a deciduous broad-leaved forest. The boundary between the temperate deciduous broad-leaved trees zone and subarctic conifer zone is located around an altitude of 1,800m at Myoko Volcano 10km northwest of the lake.

The lake surface has an area of 4.6km^2 and is surrounded by hills and low mountains. The drainage area of Lake Nojiri is only 2.8 times larger than the lake surface area and comprises a few small streams. The lake has maintained a quiet and constant condition for a long time. Lake Nojiri is an oligotrophic lake. Steady thermo-stratification occurs during the summer, but overturning of the whole lake takes place in spring and late autumn.

Table 1 List of major marker tephra beds of the NJ88 core from Lake Nojiri, central Japan
The key tephra beds used for an age model in Figure 3 are shown in Bold.

tephra name	*codes in Figs. 2 and 3*	*depth (cm)*	*calibrated depth (cm)*	*Age (ka)*	*reference/remarks*
Myoko-Otagirigawa	My-Ot	234.6	230.1	4.5	Machida and Arai (2003)
Myoko-Akakura	My-A	329	310.5	6	Machida and Arai (2003)
Kikai-Akahoya	K-Ah	387	368.3	7.3	Machida and Arai (2003)
Aira-Tanzawa	AT	1021	981.1	29	Oba *et al.* (1995)*
Joichi-Pink	Joichi	1394	1347	38	Sawada *et al.* (1992)*
Akasuko	Akasuko	1621	1569	43	Sawada *et al.* (1992)*
Dry Curry	Doraikare	1990	1848	46.2	Sawada *et al.* (1992)*
Breccia Zone	B.Z.	2093	1916	49	Sawada *et al.* (1992)*
Daisen-Kurayoshi	DKP	2765	2475	62	Nagahashi *et al.* (2007)

* These data are calibrated after Fairbanks *et al.* (2005).

The lake basin has a flat bottom plain with average and maximum depths of 20.8m and 38.5m, respectively. The sediment thickness of the lake varies from a few to several tens of meters, based on the results of an acoustic survey (Acoustic Research Subgroup for Nojiri-ko Excavation, 1987). The sediments at the deeper part of the lake are thicker than at the shallower part. The drill site for the NJ88 core was 28m deep at the northeastern flank (138°13.0'E, 36° 49.8'N), and that of 36m deep for the NJ95 core at the central part of the basin plain (138°13.3'E, 36°49.5'N) (Fig. 1).

Lake sediment stratigraphy and age

The Lake Nojiri core was drilled at its northeastern flank in 1988 and at the central part of its basin plain in 1995. The first drilling (NJ88) reached bedrock and a core of 44.16m was recovered as shown in Fig. 2 (Kumon and Inouchi, 1990). The lower one-fourth of the sediment is peat, while the upper portion is lake sediments composed of homogenous silty clay with many tephra layers. The core drilled in 1995 (NJ95) is 34.4 m long and also consists of homogenous silty clay. Both cores contain several of the same tephra beds and can be correlated precisely as shown in Fig. 2. As such, the NJ95 core corresponds with the upper half of the NJ88 core.

The marker tephras shown in Fig. 2 and are also listed in Table 1. The ages of well-known tephras such as K-Ah, AT and DKP follow Machida and Arai (2003), Oba *et al.* (1995), and Nagahashi *et al.* (2007), respectively.

The radiometric ages of local tephras such as Joichi-Pink, Akasuko, and Breccia Zone were determined as follows. Sawada *et al.* (1992) reported 31 ^{14}C dates on collagen extracted from faunal remains excavated on the western coast of Lake Nojiri where these tephras were identified using the AMS facilities at Nagoya University. Extracted collagen was used to date the animal bones. Therefore, we believe that the measured dates are reliable except for some erroneous dates measured on antlers of *Sinomegaceros yabei* . The dates in and around the maker tephra horizons

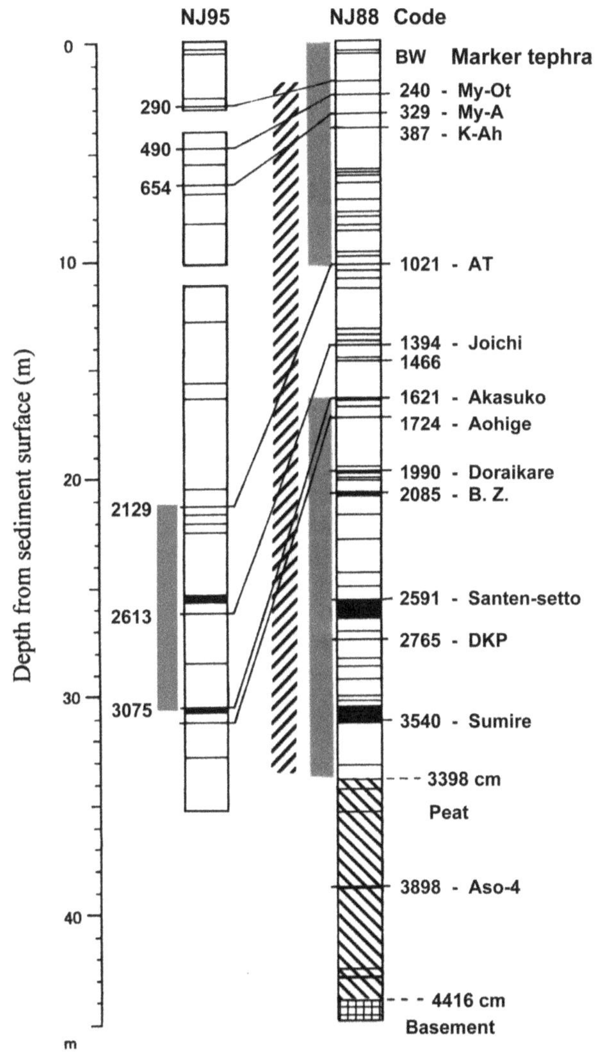

Figure 2 Columnar sections of the two drilled cores from Lake Nojiri showing correlation by marker tephra beds
Tephra codes are after Volcanic Ash Research Group for Nojiri-ko Excavation (1993) and Machida and Arai (2003). Shaded horizons were used for TOC and TN analyses in this study.

were compiled using only those on molars of *Paleoloxodon naumanni* and wood, and the averages of the conventional ages were then calibrated based on the work done by Fairbanks *et al.* (2005).

In addition to the tephra ages, four ^{14}C ages from the NJ88 core dating around 10ka were reported by Kumon *et al.* (2003), and calibrated with INTCAL98 (Stuiver *et al.*, 1998). As tephra accumulates in a very short time, compared with the age discussed here, we used calibrated depth for precise age estimation. This calibrated depth means the depth of a tephra bed excluded all tephra thicknesses above its horizon.

We chose several sets of depth and dates as key points; K-Ah (370.9cm, 7.3ka), one ^{14}C date (511cm, 10.3ka) (Kumon *et al.*, 2003), AT tephra (981.1cm, 29ka), Joichi-Pink (1,347cm, 38ka), Akasuko (1,569 cm, 43ka), Breccia Zone (1,916 cm, 49ka) and DKP tephras (2,475cm, 62ka) (Table 1). We then developed an age model to determine the depth-age relationship for the NJ88 core (Fig. 3), assuming that the sedimentation rate was constant between the key points. The depth-age relationship of the NJ95 core was also determined in the same way; AT, Joichi-Pink and Akasuko tephra beds were used as the key points. The depths of both sediment cores were translated into sedimentation ages.

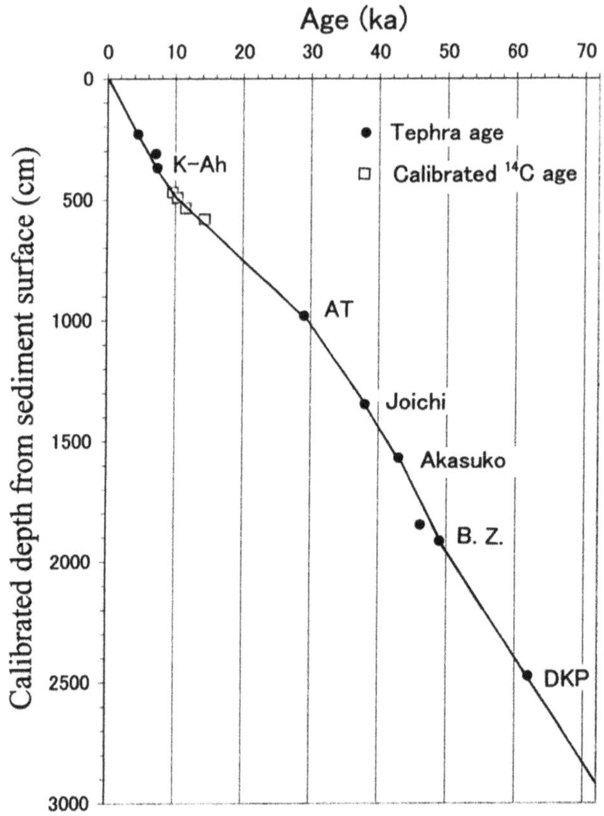

Figure 3 A depth-age model of the NJ88 sediment core from Lake Nojiri

The marker tephra beds used in this figure are listed in Table 1. Four ^{14}C dates are after Kumon *et al.* (2003). Six tephra beds with codes and one date of ^{14}C measurement were used as key points in this age model. After Kumon and Tawara (2009).

Methods

Pollen analysis was performed on 888 samples of 2 to 6cm segments from 171 to 3,398cm of the NJ88 core in the following procedure. A dried sample weighing a few grams was submerged in a 10% KOH solution at room temperature for a few days. After rinsing the sample with distilled water, it was submerged in a 10% KOH solution for 5 minutes and rinsed again. This was repeated until the rinse water was clear (20 to 30 times). A 47% HF solution was then added to dissolve fine-grained minerals and kept at room temperature for a day. Dense-media separation was carried out using 70% $ZnCl_2$ (specific gravity 1.75) for 10 min. using a centrifuge at 2,000rpm. The condensed pollen portion of the sample was dehydrated with acetic acid for 5min. at room temperature and then with an acetolysis treatment (a mixture of acetic anhydride and H_2SO_4 at a volume ratio of 9:1) at 95°C for 5 min. Acetic acid dehydration was repeated. Finally, the sample was rinsed 3 to 4 times with distilled water and mounted on a slide with glycerin jelly. The fossil pollen was identified and counted under a microscope (400x) until more than 250 arboreal pollen (AP) grains were tabulated. Non-arboreal pollen (NAP) and fern spores were also recorded. Each AP ratio is the percentage of all AP, and each NAP is the percentage of all pollen (AP+NAP).

For the TOC and TN analyses, the NJ88 core was cut into 1 to 2cm segments, while NJ95 was cut into 1cm segments. The analysis was performed on sediments from 0 to 1,021cm and 1,621 to 3,398cm in the NJ88 core and from 2,129 to 3,075cm in the NJ95 core (Fig. 2). The NJ95 core provides us with detailed information on late MIS3 conditions because of its high sedimentation rate and short sample segments compared to the NJ88 core. Both cores were correlated using marker tephra beds such as AT, Joichi-Pink and Akasuko. Each sediment sample was crushed in an agate mortar and carbonate minerals in the sediment were removed with 3% diluted hydrochloric acid (HCl). After drying, 20 to 30mg of the HCl-treated sediments were used to measure TOC and TN contents in a Yanako MT-5 CHN analyzer combusting at 900°C.

Results

The results of the pollen, TOC, and TN analyses for the upper one-third of the NJ88 core, which cover the last 28ka, have already been reported by Kumon *et al.* (2003). However, we report all results to elucidate continuous changes in the paleoclimate over the past 72ka. These results were reported recently as a preliminary proceeding (Kumon *et al.*, 2009) and reviewed by Kumon and Tawara (2009) to show the usefulness of TOC as a paleoclimate proxy.

1) Pollen analysis

Pollen assemblages of the sediment core (NJ88) showed drastic changes as illustrated in Fig. 4, and the local pollen zones NP-1 to NP-8 are identified from top to bottom on the basis of pollen composition. The characteristics of the pollen zone are described briefly from the base to the top. Sediments from horizons above a depth of 1.71m were not analyzed in this study because Tsukada (1966) already reported that the recent pollen composition is not suitable for climate reconstruction due to the strong influence of human activities in Japan.

NP-8 (33.98-31.53 m = 72.0-67.2 ka): *Picea* dominated. *Pinus* (*Haploxylon*), *Abies*, and *Tsuga* were also significant contributors. The sum of the conifers varied from 70% to more than 90% of all AP. *Cryptomeria* was present and boggy trees such as *Alnus* and *Myrica gale* were abundant in the lowest part of this zone.

NP-7 (31.30-27.91 m = 67.2-62.2 ka): *Picea* and *Pinus* (*Haploxylon*) were predominant (30%-50%), and *Abies* and *Tsuga* were also abundant. The sum of these conifers was more than 90% of all AP. *Quercus* (*Lepidobalanus*) and *Fagus* were almost absent.

NP-6 (27.71-24.78 m = 61.9-57.6 ka): *Picea* and *Pinus* *(Haploxylon)* were abundant, as were *Abies* and *Tsuga*. *Larix* was constinuously present although as a small percentage. In total, subarctic conifer trees represented about 80% of all AP. *Quercus* (*Lepidobalanus*) began to increase upward through this zone. *Betula* was subordinate, and NAP was abundant at the location of the Santen-setto tephra.

NP-5 (24.72-18.01 m = 57.4-44.7 ka): *Quercus* (*Lepidobalanus*) and *Fagus* were relatively abundant (10%-20%), together with *Juglans-Pterocarya* and *Ulmus-Zelkova*. The sum of these deciduous broad-leaved species increased to over 50% of AP at several locations. In contrast, *Picea* and *Tsuga* had low ratios.

NP-4 (17.98-10.63 m = 44.6-29.4 ka): *Picea, Abies, Pinus* (*Haploxylon*) and *Tsuga* were abundant, and the sum of these taxa was 70 to 90% of all AP. However, *Quercus* (*Lepidobalanus*), *Fagus, Juglans-Pterocarya, Ulmus-Zelkova,* and *Betula* occurred in variously small amounts. *Lepidobalanus* and *Haploxylon* were present at 14.0-13.5m just above the Joichi-Pink tephra. The Akasoko tephra existed at the lower horizon of this zone, and NAP such as Artemisia, Gramineae and Cyperaceae increased above this tephra.

NP-3 (10.59-7.35 m = 29.2-18.6 ka): *Picea* was dominant comprising more than 50% of total AP. *Abies* and *Tsuga* were also abundant. *Quercus* (*Lepidobalanus*), *Fagus* and *Ulmus-Zelkova* were low, at less than 5%, but they were constantly present. The AT tephra was located at the lowest part of this zone, and NAP, particularly Cyperaceae, increased just above the tephra.

NP-2 (7.34-6.04 m = 18.5-13.7 ka): This was a transitional zone. *Quercus* (*Lepidobalanus*) abruptly increased to approximately 50% of all AP at the base of this zone and decreased upwards. In contrast, *Picea* and *Abies* were rare at the base and increased upwards. *Alnus* was slightly abundant.

NP-1 (6.02-1.71 m = 13.6-3.5 ka): *Quercus* (*Lepidobalanus*) was predominant along with *Fagus, Juglans-Pterocarya, Carpinus* and *Ulmus-Zellkova*. *Fagus* reached its maximum at approximately 30% from 5.29 to 4.53m depth. Conifers such as *Pinus* (*Haploxylon*), *Picea*, *Abies* and *Tsuga* were at their lowest comprising less than 5% of AP. *Quercus* (*Lepidobalanus*) decreased slightly in the upper part, but still dominated. *Cryptomeria* decreased in the later stage of this zone. The K-Ah tephra was identified at 3.71m and was located slightly above the *Fagus* peak.

2) TOC and TN contents

The TOC, TN, and C/N ratios of 2,200 samples are shown in Fig. 5A. The data were set on a time scale to combine both the NJ88 and NJ95 cores. Some very low values for the TOC and TN were due to dilution by tephra and other volcanic materials lacking organic matter. On the other hand, the C/N ratio sometimes had high values of more than 15 in association with slightly high TOC. Most of these peaks were caused by intermittent intercalation of leaf or plant material in the sediment.

Generally speaking, TOC and TN vary sligntly, depending on the sampling location even in the same lake. The depositional center (usually the deepest part of the basin) has the highest concentration (Adhikari and Kumon, 2001). Then, there may be a slight difference between NJ88 and NJ95. The difference, however, seemed to be small compared to the fluctuation range of the values. The concentration of TOC and TN is used without calibration for the difference in locations.

As shown in Fig. 5A, three major modes of TOC, TN and C/N ratio variations can be recognized. From 0 to 12ka, TOC fluctuated significantly by around 5%, and TN also varied around 0.7%. These values were largest among the three modes. The C/N ratio was as high as 8 to 11, and fluctuated directly with TOC. Rapid changes in TOC, TN and the C/N ratio were recognized from 12 to 14ka including a slight retreat midway. From 14 to 29ka, TOC and TN were constantly low, ranging around 2% and 0.3%, respectively. The C/N ratio was also low at 6 to 8. There were some obscure peaks and depressions for the values. The AT tephra was situated at the lowest part of this interval.

TOC and TN for 29 to 50ka frequently fluctuated with a slightly high amplitude. The averages seemed to be a higher than those above them. The C/N ratio was relatively high

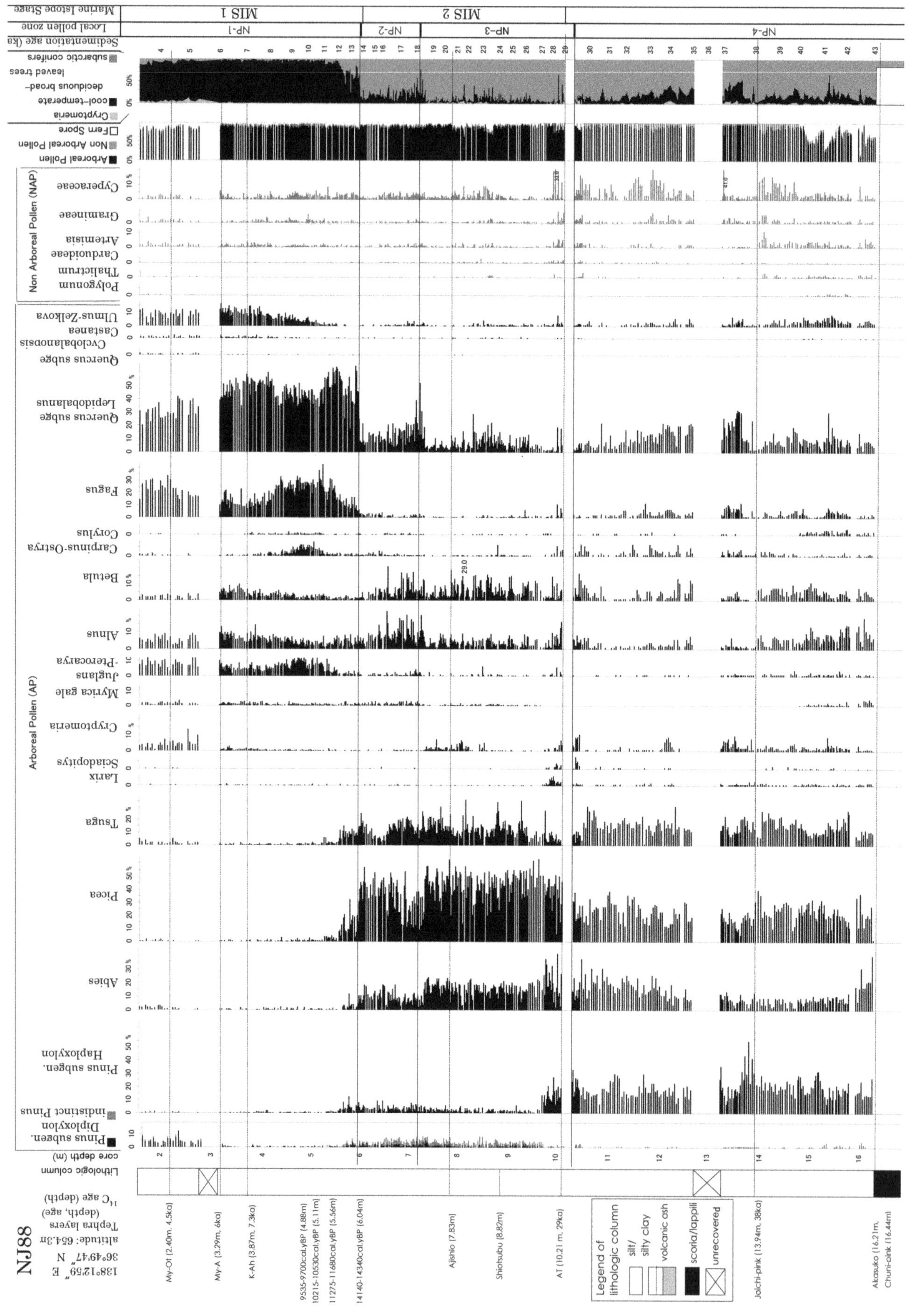

Figure 4 A pollen diagram of the major taxa from the NJ88 sediment core, Lake Nojiri, central Japan
The data are shown on a depth scale, and the ages on the right only show approximate estimates.

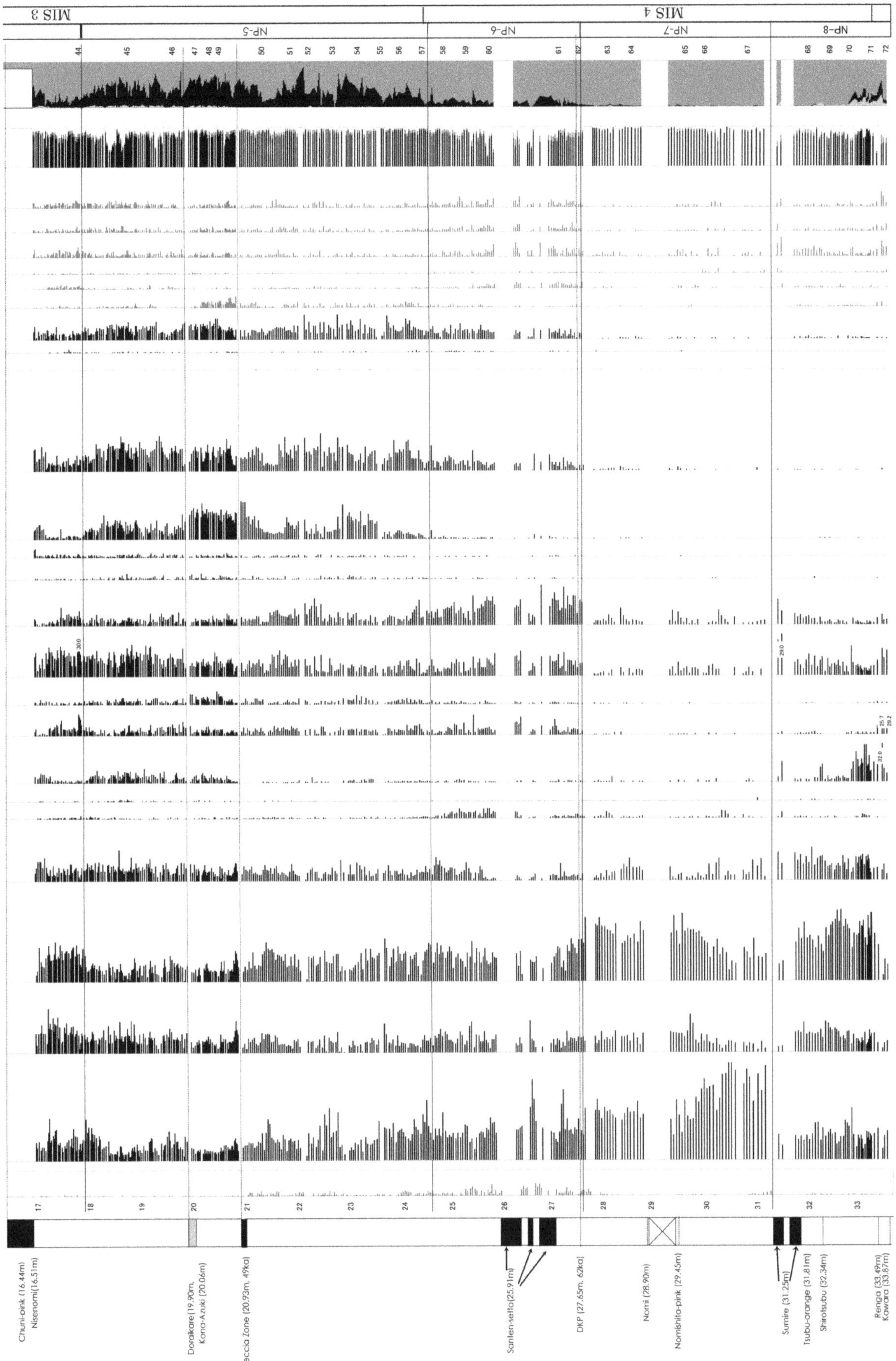
MIS 3
MIS 4
NP-5
NP-6
NP-7
NP-8
Chuni-pink (16.44m)
Nisenomi(16.51m)
Doraikare (19.90m, Kona-Azuki (20.06m)
Breccia Zone (20.93m, 49ka)
Santen-setto(25.91m)
DKP (27.65m, 62ka)
Nomi (28.90m)
Nomishita-pink (29.45m)
Sumire (31.25m)
Tsubu-orange (31.81m)
Shirotsubu (32.34m)
Renga (33.49m)
Kawara (33.87m)

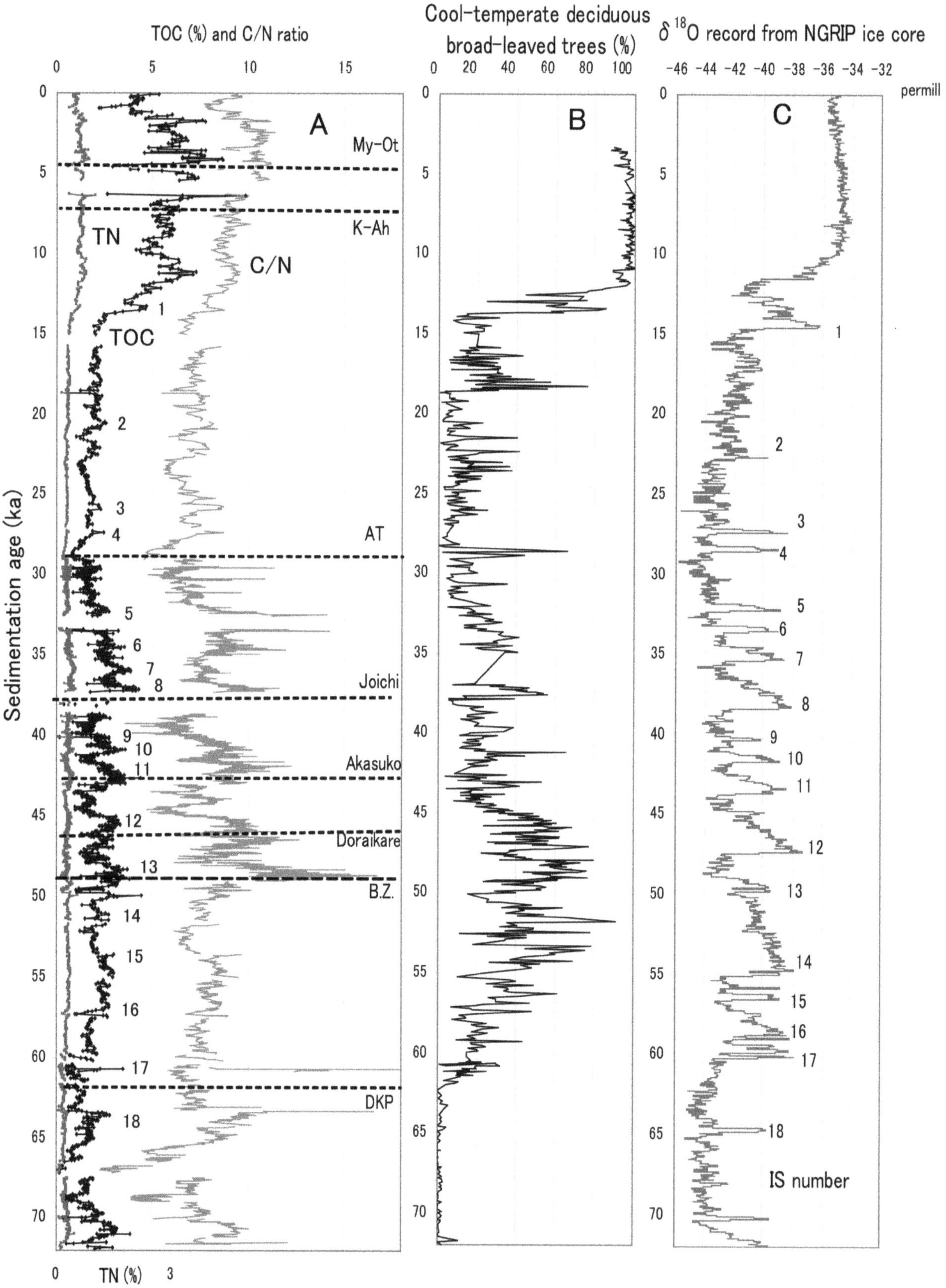

Figure 5 Temporal variations of the climate proxies from the cored sediments of Lake Nojiri compared with the $\delta^{18}O$ curve of NGRIP ice core (after Kumon and Tawara, 2009)

A. Profiles of TOC and TN contents with the C/N ratio from the composite data of the NJ88 and NJ95 sediment cores. Nos.1 to 18 are probable counterparts of the interstadials in Greenland.

B. Abundance of cool-temperate deciduous broad-leaved trees which include *Quercus (Lepidobalanus), Fagus, Juglans-Pterocarya, Ulmus-Zelkova* and *Carpinus*. The abundance means percentage of the sum of cool-temperate deciduous broad-leaved trees and subarctic conifers (i.e., *Pinus [Haploxylon], Abies, Picea, Tsuga*, and *Larix*).

C. $\delta^{18}O$ curve of NGRIP ice core (North Greenland Ice Core Project member, 2004) in association with the interstadial (IS) numbers of Dansgaard *et al.* (1993).

and showed fluctuations from 5 to 15. For 50 to 61ka. TOC was as high as 1.5% to 3.0%, with less variation. TN was relatively constant around 0.5%. From 61 to 72ka, TOC was low with a few small peaks, and TN was consistently low. The C/N ratio was highly variable from 3 to 12.

Discussion

1) Pollen assemblage comparisons

The stratigraphic changes of pollen assemblages in and around Lake Nojiri were elucidated by Tsukada (1966, 1967) for the late Last Glacial to the Holocene, and by the Palynological Research Group for Nojiri-ko Excavation (1990, 1993) for the latest Pleistocene. The results of this study are very similar to their results in general, but are much more detailed in temporal resolution. In addition, the boundary ages of the pollen zone were confirmed much more precisely in this study.

Several palynological studies covering the Last Glacial in Japan have been conducted (e.g., Yanahara Bog in northwest Japan [Kanauchi, 1988], Lake Suwa in central Japan [Oshima *et al.*, 1997], the Kashima-oki marine core off central Japan [Igarashi and Oba, 2006], and the Kurota Lowland [Takahara and Kitagawa, 2000], and the Kamiyoshi Basin [Takahara *et al.*, 2000] and Lake Biwa [Miyoshi *et al.*, 1999] in western Japan). Although compositions of the pollen assemblages differ from south to north, or from low to high altitudes, in general, the depicted vegetation changes are similar to results in this study. The vegetation from 70ka to 15ka was characterized by the dominance of arctic conifers, including an increase in deciduous broad-leaved trees in central and northwestern Japan for around 55 to 45ka. The pollen assemblages in western Japan from 60 to 30ka were also characterized by an abundance of *Cryptomeria japanica*, a temperate conifer peculiar to Japan (Takahara *et al.*, 2000; Takahara and Kitagawa, 2000).

Among the analyses, Igarashi and Oba (2006) used the Kashima-oki marine core MD01-2421 to show that vegetation changes occurred in the near-land areas such as the northern Kanto district. They identified two cold, three cool and two warm periods with several minor fluctuations over the last 144ka. In addition, the results were directly compared to the span from MIS5e to MIS1 using $\delta^{18}O$ of benthic foraminifera from the same core (Oba *et al.*, 2006). Their results are correlated with temporal changes of pollen assemblages in the sediment core from Lake Nojiri, namely, NP-8 to 6 with MIS4, NP-5 and 4 with MIS3, NP-3 and 2 with MIS2, and NP-1 with MIS1. A zone of abundant deciduous broad-leaved trees in early MIS3 is clearly identified in both studies.

The pollen zone for the central Japanese islands from the latest Last Glacial age to the Holocene is well established, and was classified into five zones labeled by Tsukada (1967b, 1981) LI, LII, RI, RII and RIII. The pollen assemblage of the NP-3 zone (7.35-10.59m) corresponds well with that of the LI zone and indicates the vegetation of the Last Glacial Maximum in central Japan as elucidated by Tsukada (1967b, 1981). The assemblage of NP-2 (6.04-7.34m) can be correlated with the LII zone, which is the transition between the Last Glacial Maximum and the Holocene. The pollen assemblage of NP-1 (2.63-6.02m) corresponding well with the RI and RII zones in central Japan. Although we have no data in this work, the pollen assemblage of the RIII zone is characterized by a slight decrease in deciduous broad-leaved trees and a small increase of conifer trees, which are adapted to a slightly cool climate; this is associated with artificial influences such as deforestation and agriculture (Tsukada, 1967b, 1981).

The transitional change of pollen composition between the Last Glacial and the Holocene was recently revealed in detail using the varved sediments of Lake Suigetsu, southwest Japan (Nakagawa *et al.*, 2003). On the basis of the Suigetsu varve age, post-glacial warming began there at 15.0ka and a Younger Dryas-like cooling event commenced at 12.3ka. The timing seems to have slight discordance with the Lake Nojiri core. The age of the *Quercus* (*Lepidobalanus*) increase in Lake Nojiri is around 18ka. If this is a sign of post-glacial warming, then it may have begun earlier at Lake Nojiri than at Lake Suigetsu. A slight cooling that may correspond to the Younger Dryas event is indicated by a depression in the abundance of cool-temperate deciduous trees from 13.5 to 11ka. These discordances may partly be due to errors of the age models of Lake Suigetsu or Lake Nojiri, and/or local differences in climate patterns for Japan. This problem needs to be resolved in the future.

As discussed above, the pollen zones identified in the sediment core from Lake Nojiri have features similar to other local pollen zones. Lake Nojiri is located in the central part of the Japanese islands, and at the foot of the Japanese Alps. As pointed out by Tsukada (1981), these geographical conditions are favorable for recording the sensitive vegetation changes controlled by climate change. Therefore, we can use the stratigraphic changes in the pollen assemblage elucidated here as a standard for pollen zoning, which may represent aspects of paleoclimate in the Japanese islands.

2) Implication of TOC and TN contents

Recently, Inouchi *et al.* (1996), Meyer (1997), Adhikari and Kumon (2001), Kumon *et al.* (2003), Kumon (2003), Tawara *et al.* (2006), Iwamoto and Inouchi (2007), and Kumon and Tawara (2009) demonstrated the usefulness of TOC and TN for paleoclimate reconstruction. The relationship between TOC in lake sediments and climate factors was partly explained by Kumon *et al.* (2005) at Lake Kizaki, central Japan. They examined relationships among climate factors, the biological productivity of lake water (the chlorophyll *a* amount), and TOC of lake sediment over the past 21 years. They found a good relationship between the average winter temperature, annual chlorophyll *a* amount, and TOC. They concluded that warm winter temperatures corresponding to a short, cold midwinter season produced high biomass in lake water during the early and/or late winter seasons. This winter biological productivity strongly influences the annual biological productivity, although its ecological mechanism is not yet fully understood.

In addition to biological productivity in lake water, the input of organic materials from the surrounding land area is another factor affecting TOC content. Organic matter from land area is mostly of vesicular plant origin, and has a high ratio of C/N. As high TOC often occurs with a high C/N ratio, land sources seem to be other contributors of organic matter. In the study area, the warm climate associated with abundant precipitation and the development of deciduous broad-leaved forests may increase the input of organic matter from land to lake. Other factors that affect TOC and TN are sedimentation rates, preservation potential, and dilution effect. These issues have been discussed in detail by Kumon (2003) and Kumon and Tawara (2009).

In conclusion, we can regard the TOC and TN contents of the Lake Nojiri sediments as proxies for the paleotemperature, particularly in the winter. This conclusion is concordant with the paleoclimate reconstructed from pollen data as discussed later.

3) Paleoclimate reconstruction for the past 72 ka

To indicate warmth, the abundance of cool-temperate deciduous broad-leaved trees is shown in Fig. 5B. The abundance was calculated as a percentage of the pollen grains of *Quercus* (*Lepidobalanus*), *Fagus*, *Juglans-Pterocarya*, *Ulmus-Zelkova* and *Carpinus* divided by the total of these genera plus the number of subarctic conifer taxa such as *Pinus* (*Haploxylon*), *Abies*, *Picea*, *Tusga*, and *Larix*. The $\delta^{18}O$ profile of the Greenland ice core (NGRIP; North Greenland Ice Core Project members, 2004) is also illustrated in Fig.5C for comparison.

As shown in Fig. 4, the pollen assemblages of the NJ88 core from 1.71 to 6.02m (3.5 to 13.6ka) indicate cool-temperate deciduous broad-leaved forests, which is a natural potential vegetation around Lake Nojiri. The assemblages for 7.34-10.39m (18.6 to 29.2ka) and 27.91-33.88m (62 to 72ka) represent subarctic coniferous forests of very cold climate. Those from 27.71 to 10.46m (29 to 62ka) are characterized by an abundance of *Quercus* (*Lepidobalanus*) and *Fagus*; this indicates mixed flora transitional between boreal coniferous and temperate broad-leaved forests with considerable temporal fluctuations. These temporal changes of vegetation were common in central and northwestern Japan as discussed before, and are concordant also with the temporal changes of ocean surface temperature for the northwest Pacific off the Japanese islands (MD01-2421 core: Yamamoto *et al.*, 2004; Oba *et al.*, 2006).

When the abundance of deciduous broad-leaved trees is used as a temperature proxy (Fig. 5B), its fluctuation curve is concordant with the $\delta^{18}O$ curve of the Greenland ice core (Fig. 5C). Guiot *et al.* (1993) compiled pollen data in Western Europe for the last Glacial/Interglacial cycle and reconstructed the land climate using transfer functions. The reconstructed paleo-temperature is similar to the results from Lake Nojiri. The TOC and TN curves are also correlated with the abundance of deciduous broad-leaved trees and $\delta^{18}O$ as shown in Fig. 5. These temporal changes of vegetation correspond well to global climate changes recorded in marine sediments; that is, MIS1 to MIS4 (Martinson *et al.*, 1987). Short-term fluctuations of pollen composition can imply short-term climate variations such as the Dansgaard-Oeschgar cycle (Dansgaard *et al.*, 1993). The TOC peaks are also well correlated with the interstadial stages (ISs) from 1 to 18 as shown in Fig. 5A, however, some discordances in age exist for a few hundreds to one thousand years maximum. These discordances may be partly due to estimation errors of the age model, and partly due to local variations of climate in the Japanese islands.

Scientific excavations were carried out at and around the western coast of Lake Nojiri by the Nojiri-ko Excavation Research Group since 1962. Their results have been published in many reports (e.g., Utashiro, 1980; Nojiri-ko Excavation Research Group, 1984, 1987, 2006). They reported abundant mammal remains such as *Palaeoloxodon naumanni*, *Sinomegaceros yabei*, and *Cervus* sp., and Palaeolithic artifacts. Although controversy has arisen over whether the bone materials are natural or modified by humans, Ono (2001) recognized several bone artifacts, and the Anthropology and Archaeology Research Group for Nojiri-ko Excavation (2006) confirmed 12 lithic artifacts after careful reexamination. These artifacts and fossil remains were at the horizons between the Breccia Zone and Joichi-Pink tephra beds in the middle Nojiri-ko Formation where temperate deciduous broad-leaved trees were relatively dominant and the climate was warm. This implies that animals and humans visited the site under optimum climate condition around 48ka and left just before the coldest stage around 35ka.

Conclusions

We elucidated vegetation changes around Lake Nojiri, central Japan for 80 to 130year on the basis of pollen composition. TOC and TN analyses support the results of our pollen analysis more detailed time resolution. Both the abundance of cool-temperate deciduous broad-leaved trees andTOC, as temperature proxies vary concordantly showing semi-periodic fluctuations. Both profile patterns are similar to that of $\delta^{18}O$ for marine sediment. MIS1 to MIS4 are recognized in the samples with good age assignment, and even IS1 to IS18 can be correlated with the small peaks of TOC.

This reconstructed paleoclimate from the early Last Glacial to post glacial times is much detailed than previous studies for Japan. The differences in timing and duration of cool and warm phases in central Japan compared with Europe and polar climate data seem to be partly due to local variability the Japanese islands to erroneous age determination. These variations found in this study may be a key to elucidate the fluctuations in global climate in association with many local climate variations. The age model of the NJ88 core should be improved in future because of a scarcity of radiometric dates of marker tephras. Our study, however, has shown a stable outline of detailed climate changes through the Last Glacial and Holocene. The results are expected to be very useful for archaeology in Japan because climate change can be correlated using the wide-spread marker tephra beds in and around the Japanese islands.

References

Acoustic Research Subgroup for Nojiri-ko Excavation [Nojiri-ko Chishitu Group Onpa Tansa Subgroup] 1987 Acoustic stratigraphy in Lake Nojiri. *Chidanken Senpo*, no.32, pp.23-36. (in Japanese with English abstract)

Adhikari, D.P. and Kumon, F. 2001 Climatic changes during the past 1300 years as deduced from the sediments of Lake Nakatsuna, central Japan. *Limnology*, 2:157-168.

Adhikari, D.P., Kumon, F., and Kawajiri, K. 2002, Holocene climate variability as deduced from the organic carbon and diatom records in the sediments of Lake Aoki, central Japan. *Chishitsugaku Zasshi*, 108:249-265.

Anthropology and Archaeology Research Group for Nojiri-ko Excavation [Nojiri-ko Jinrui Koko Group] 2006 Results of the archaeological investigation of the 15th Nojiri-ko Excavation and a reexamination of the stone artifacts from the Tategahana site. *Bulletin of the Nojiri-ko Museum*, no.14, pp.31-53. (in Japanese with English abstract)

Dansgaard, W., Johnsen S.J., Clausen, H.B., Dahl-Jensen, D., Gundestrup, N., Hammer, C.U., Hvidberg,,C.S., Steffensen, J.P., Sveinbjornsdottir, A.E., Jouzel, J., and Bond, G. 1993 Evidence for general instability of past climate from a 250-kyr ice-core record. *Nature*, 364:128-220.

Fairbanks, R.G., Mortlock, R.A., Chiu, T.-C., Cao, L., Kaplan, A., Guilderson, T.P., Fairbanks, T.W., Bloom, A.L., Grootes, P.M. and Nadeau, M.-J. 2005 Radiocarbon calibration curve spanning 0 to 50,000 years BP based on paired $^{230}Th/^{234}U/^{238}U$ and ^{14}C dates on pristine corals. *Quaternary Science Review*, 24:1781-1796.

Guiot, J., de Beaulieu, J.L., Cheddadi, R., David, F., Ponel, P. and Reille, M. 1993 The climate in Western Europe during the last Glacial/Interglacial cycle derived from pollen and insect remains. *Palaeogeography, Palaeoclimatology, Palaeoecology*, 103:73-93.

Inouchi, Y., Yokota, S. and Terashima, S. 1996 Climatic change around Lake Biwa during the past 300,000 years and 2,000 years. *In* Mikami, T., Matsumoto, E., Ohta, S. and Sweda, T. (eds.), *Proceedings of the 1995 Nagoya IGBP-PAGES/PEP-II Symposium*. pp.109-114.

Iwamoto, N, and Inouchi, Y. 2007 Reconstruction of the millennial-scale variations in the East Asian summer monsoon over the past 300 ka based n the total carbon content of sediment from Lake Biwa, Japan. *Environmental Geology*, 52:1607–1616.

Kanauchi, A. 1988 The vegetational history during the Last Glacial period from Yanohara Bog, Fukushima Prefecture, northeastern part of Japan. *Daiyonki Kenkyu*, 27:177-186. (in Japanese with English abstract)

Kumon, F. 2003 Total organic carbon and total nitrogen contents in lake sediment as a proxies of paleoclimate. *Daiyonki kenkyu*, 42:195-204. (in Japanese with English abstract)

Kumon, F. and Iniouchi, Y. 1990 All-core boring and the correlation with acoustic reflectors in Lake Nojiri, central Japan. *Chishitsugaku Ronshu*, no.36, pp.167-178. (in Japanese with English abstract)

Kumon, F., Kawai, S. and Inouchi, Y. 2003 Climate changes between 25,000 and 6,000 years BP deduced from TOC, TN, and fossil pollen analyses of a sediment core from Lake Nojiri, central Japan. *Daiyonki Kenkyu*, 42:13-26. (in Japanese with English abstract)

Kumon, F., Kanamaru, K., Tawara, T., Kakuta, N., Yamamoto, M. and Hayashi, H. 2005 Relationships among weather factors, biological productivity and TOC content of sediments in Lake Kizaki, central Japan. *Chishitsugaku Zasshi*, no.111, pp.599-609. (in Japanese with English abstract)

Kumon, F., Kawai, S., and Inouchi, Y. 2009 High-resolution reconstruction of paleoclimae during the last 72ka on the basis of the drilled sediments from Lake Nojiri, central Japan. *Kyusekki Kenkyu*, 5:3-10. (in Japanese with English abstract)

Kumon, F., Tawara, T. 2009 Deatailed reconstruction of paleoclimate based on total organic carbon proxy of lake sediment during the past 160 ka in central Japan. *Chishitsugaku Zasshi*, 116:344-356. (in Japanese with English abstract)

Machida, H. and Arai, F. 2003 *Atlas of Tephra in and around Japan (revised edition)*. University of Tokyo Press, Tokyo, 336p. (in Japanese)

Martinson, D.G., Pisias, N.G., Hays, J.D., Imbrie, J., Moore, T.C. and Shackleton, N.J. 1987 Age dating and orbital theory of the ice ages: development of a high resolution o to 300,000-year chronostratigraphy. *Quaternary Research*, 27:1-29.

Meyers, P.A. 1997 Organic geochemical proxies of paleoceanographic, paleolimnologic and paleoclimatic processes. *Organic Geochemistry*, 27:213-250.

Miyoshi, N., Fujiki, T. and Morita, Y. 1999 Palynology of a 250m core from Lake Biwa: a 430,000 years record of glacial-interglacial vegetation changes in Japan. *Review of Palaeobotany and Palynology*, 104:267-283.

Nagahashi, T., Sato, T., Takeshita, Y., Tawara, T. and Kumon, F. 2007 Stratigraphy and chronology of widespread tephra beds intercalated in the TKN-2004 core sediment obtained from the Takano Formation, central Japan. *Daiyonki Kenkyu*, 46:305-325. (in Japanese with English abstract)

Nakagawa, T., Kitagawa, H., Yasuda, Y., Tarasov, P.E., Nishida, K., Gotanda, K. Sawai, Y. and Yangtze River Civilization Program members 2003 Asynchronous climate changes in the North Atlantic and Japan during the last termination. *Science*, 299:688-691.

Nojiri-ko Excavation Research Group [Nojiri-ko Hakkutsu Chosa Dan] (ed.) 1984 The Lake Nojiri Excavation 3 (1978-1983). *Chidanken Senpo*, no.27, 267p. (in Japanese with English abstract)

Nojiri-ko Excavation Research Group [Nojiri-ko Hakkutsu Chosa Dan] (ed.) 1984 The Lake Nojiri Excavation 4 (1984-1986). *Chidanken Senpo*, no.32, 213p. (in Japanese with English abstract)

Nojiri-ko Excavation Research Group [Nojiri-ko Hakkutsu Chosa Dan] (ed.) 2006 The Lake Nojiri Excavation 10. *Bulletin of the Nojiri-ko Musem*, no.14, 123p. (in Japanese with English abstract)

North Greenland Ice Core Project Members 2004 High-resolution record of Northern Hemisphere climate extending into the last interglacial period. *Nature*, 431:147-151.

Oba, T., Murayama, M., Matsumoto, E. and Nakamura, T. 1995 AMS-^{14}C ages of the Japan Sea cores from the Oki Ridge. *Daiyonki Kenkyu*, 34:289-296. (in Japanese with English abstract)

Ono, A. 2001 *Flaked Bone Tools: An alternative perspective on the Palaeolithic.* University of Tokyo Press, Tokyo, 290 p. (in Japanese)

Oshima, H., Tokunaga, S., Shimokawa, K., Mizuno, K. and Yamazaki, H. 1997 Fossil pollen assemblages of core samples from Lake Suwa, Nagano Prefecture, and their correlation to other pollen assemblages in central Japan. *Daiyonki Kenkyu*, 36:165-182. (in Japanese with English abstract)

Palynological Research Group for Nojiri-ko Excavation [Nojiri-ko Kafun Group] 1990 Fossil pollen assemblages of the Kannoki Formation and the lower Nojiri-ko Member and the change of paleoenvironment during the period after the deposition of the Ajishio volcanic ash layer. *Chidanken Senpo*, no.37, pp.61-76. (in Japanese with English abstract)

Palynological Research Group for Nojiri-ko Excavation [Nojiri-ko Kafun Group] 1993 Fossil pollen assemblages of a drilling core from the bottom of Lake Nojiri with reference to Late Pleistocene environments. *Chidanken Senpo*, no.41, pp.39-52. (in Japanese with English abstract)

Sawada, K., Arita, Y., Nakamura, T., Akiyama, M., Kamei, T., and Nakai, N. 1992 ^{14}C dating of the Nojiriko Formation using accelerator mass spectrometry. *Chikyu Kagaku*, 46:133-142. (in Japanese with English abstract)

Seltzer, G.O., Rodbell, D.T., Baker, P.A., Fritz, S.C., Tapia, P.M., Rowe, H.D., and Dunbar, R.B. 2002 Early warming of tropical South America at the Last Glacial-Interglacial transition. *Science*, 296:1685-1686.

Stuiver, M., Reimer, P.J., Bard, E., Beck, J.W., Burr, G.S., Hughen, K.A., Kromer, B., McCormac, G., Van Der Plicht, J. and Spurk, M. 1998 INTCAL98 radiocarbon age calibration, 24,000-0calBP. *Radiocarbon*, 40:1041-1083.

Takahara, H. and Kitagawa, H. 2000 Vegetation and climate history since the last interglacial in Kurota Lowland, western Japan. *Palaeogeography, Palaeoclimatology, Palaeoecology*, 155:123-134.

Takahara, H., Uemura, Y. and Danhara, T. 2000 The vegetation and climate history during the early and mid Last Glacial period in Kamiyoshi basin, Kyoto. *Japanese Journal of Palynology*, 46:133-146.

Tawara, T., Kumon, F., Nagahashi, Y., Kakuta, N. and Nozue, Y. 2006 Reconstruction of Late Pleistocene climate based on total organic carbon (TOC) contents in TNK-2004 core drilled from Takano Formation, central Japan. *Chishitsugaku Zasshi*, 112:568-579. (in Japanese with English abstract)

Tsukada, M. 1966 Late postglacial absolute pollen diagram in Lake Nojiri. *Shokubutsugaku Zasshi*, 79, pp.179-184.

Tsukada, M. 1967a Vegetation and climate around 10,000 B.P. in central Japan. *American Journal of Sciences*, 265:562-585.

Tsukada, M. 1967b The last 12,000 years: A vegetation history of Japan I. *Shokubutsugaku Zasshi*, 80:323-336. (in Japanese with English abstract)

Tsukada, M. 1981 The last 12,000 years—the vegetation history of Japan II. new pollen zones. *Nihon Seitai Gakkaishi*, 31:201-215. (in Japanese with English abstract)

Tsukada, M. 1983 Vegetation and climate during the last glacial maximum in Japan. *Quaternary Research*, 19:212-235.

Utashiro, T. (ed.) 1980 The Paleolithic site and paleoenvironment in and around the Lake Nojiri. *Chishitsugaku Ronshu*, no.19, 268p. (in Japanese with English abstract)

Yamamoto, M., Oba, T., Shimamune, J. and Ueshima, T. 2004 Orbital-scale anti-phase variation of sea surface temperature in mid-latitude North Pacific margins during the last 145,000 years. *Geophysical Research Letters*, 31, L16311.

Absolute Chronology of Archaeological and Paleoenvironmental Records from the Japanese Islands, 40–15 ka BP

Yuichiro Kudo

National Museum of Japanese History,
117 Jonai-cho, Sakura, Chiba, 285-8502 Japan.
E-mail: kudo@rekihaku.ac.jp

Abstract: *This paper clarifies the temporal correspondences between archaeological and paleoenvironmental records during the late Pleistocene, that is, ca. 40–15 ka cal BP, in eastern Honshu, Japanese Islands. Three primary issues are discussed: (1) the archaeological chronology during the Upper Palaeolithic in eastern Honshu, (2) technical problems regarding the use of radiocarbon dates by archaeological studies and the use of calibration curves for the period before 12.4 ka cal BP, and (3) the correspondence between geological and archaeological chronologies based on calibrated radiocarbon dates. Available data place the earliest lithic assemblages of Homo sapiens roughly between 38 and 34 ka cal BP. These dates correspond to GIS-8 of NGRIP. The second half of the backed blade industry, the point industry, and the earlier phase of the microblade industry appear to have been associated with cool and dry climatic conditions. The use of pottery precedes the abrupt warming of the late glacial. The emergence of pottery and the expansion of cool-temperate deciduous broadleaved forest of late glacial show no obvious causal connection.*

Keywords: *Japanese Upper Palaeolithic, radiocarbon dating, glacial calibration, absolute chronology*

Introduction

Relationships between human activities and environmental changes have become an important research topic in Japanese prehistoric archaeology. Establishing reliable correlations between archaeological and geological chronologies is critical, especially where the archaeological chronology depends mainly on relative dating based on stratigraphy and typology. In recent years, the use of ^{14}C dating for Palaeolithic sites in the Japanese Islands has progressed (Ono *et al.*, 2002; Kudo, 2004, 2005, 2006; Izuho and Akai, 2005; Izuho and Takahashi, 2005).

In this paper, I examine correlations between the geological and archaeological chronologies from the second half of Oxygen Isotope Stage (OIS)-3 to the end of OIS-2, that is, ca. 40–15 ka cal BP, in eastern Honshu, Japanese Islands. This period corresponds to the latter part of the last glacial, and to the Japanese Upper Palaeolithic and Incipient Jomon periods.

Landscape and vegetation during the last glacial

The Japanese Islands lie approximately between 24ºN and 46ºN latitude and between 122ºE and 148ºE longitude, and experience a very wide range of climates. Generally speaking, four climate zones can be distinguished in the Japanese Islands today: subtropical, warm-temperate, cool-temperate, and subarctic. Southwestern Hokkaido and eastern Honshu are covered by cool-temperate deciduous broadleaved forests. Coastal areas of central Honshu and most of western Honshu, Shikoku, and Kyushu, except for mountainous areas, are covered by warm-temperate evergreen forests. In addition, climate and vegetation exhibit contrasts between the Sea of Japan and Pacific Ocean sides of the islands, largely because of heavier snowfall on the Sea of Japan side. Winter precipitation is high on the Sea of Japan side owing to moisture from the Tsushima Warm Current and occurs mainly as snow.

The coastline and vegetation of the Japanese Islands during the last glacial maximum (LGM) are shown in general outline in Fig. 1 (after Nasu, 1985). Hokkaido Island was part of the Eurasian continent, to which it was connected by a land bridge between Sakhalin and Hokkaido, during the LGM. Hokkaido and Honshu, however, were not connected even during the LGM. The Tsushima Strait between Kyushu and the Korean Peninsula was also open during the LGM, although the climate of these lands was more continental than at present because of the weakened flow of the Tsushima Warm Current. Central to western Honshu, Shikoku, and Kyushu were covered by a mixed forest of subarctic conifers and cool-temperate tree species around the time of the LGM, whereas central to northern Honshu and western Hokkaido were covered by a subarctic coniferous forest and eastern Hokkaido by an open, subarctic forest.

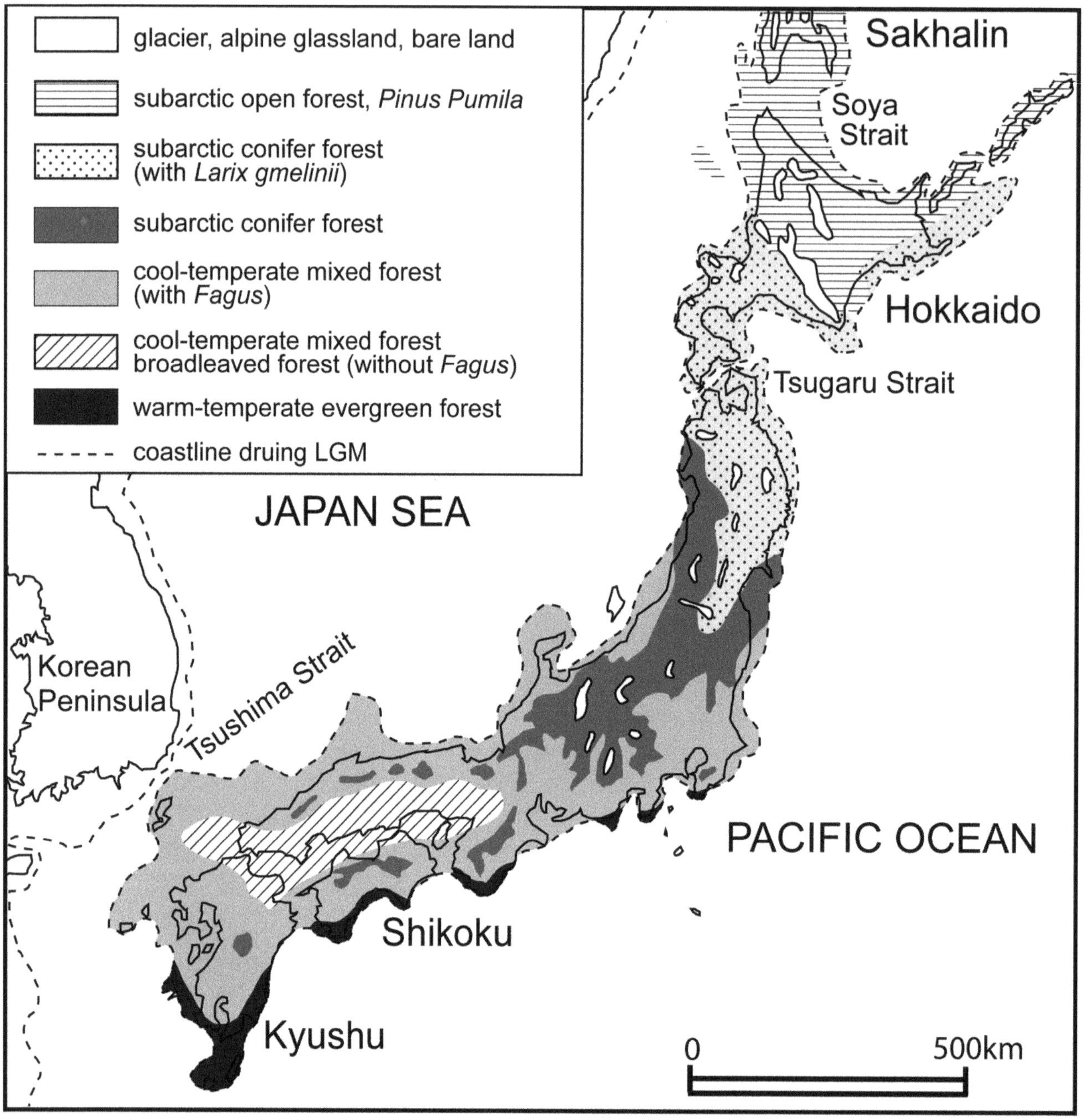

Fig. 1 Coastline and vegetation during the last glacial maximum (LGM) (modified after Nasu, 1985).

Archaeological chronology and radiocarbon dates from the Upper Palaeolithic and Incipient Jomon in the Japanese Islands

The Upper Palaeolithic chronology of the eastern Japanese Islands is based mainly on the lithic typology and stratigraphy of sites on the Musashino and Sagamino plateaus in the southern Kanto plain, where relatively thick accumulations of loam are clearly stratified (Fig. 3). In addition, ash layers from contemporaneous volcanic eruptions are traceable in many archaeological strata (Machida, 2005). Unfortunately, the acid soils derived from volcanic ash preserve no organic artifacts or food remains except for charcoal. Thus, little progress has been made regarding subsistence during the Upper Palaeolithic in the past five decades.

In the area surrounding the Kanto plain, detailed local stratigraphy and local sequences of the Palaeolithic assemblages are lacking. Instead, the lithic assemblages have been correlated by morphological and typological comparison with assemblages from the Musashino and Sagamino plateaus (Fig. 2).

Ono *et al.* (2002) have compiled more than 400 radiocarbon dates from Japanese Upper Palaeolithic sites, and Izuho and Akai (2005) have discussed the geochronology in relation to archaeological sites in Hokkaido. Here, I focus on the relationship between the archaeological chronology and environmental history in central and eastern Honshu on the basis of ^{14}C dates compiled from 40 archaeological sites (Tab. 1, 2) and a widely distributed key tephra layer, Aira-Tn (AT) (Fig. 2, Tab. 3).

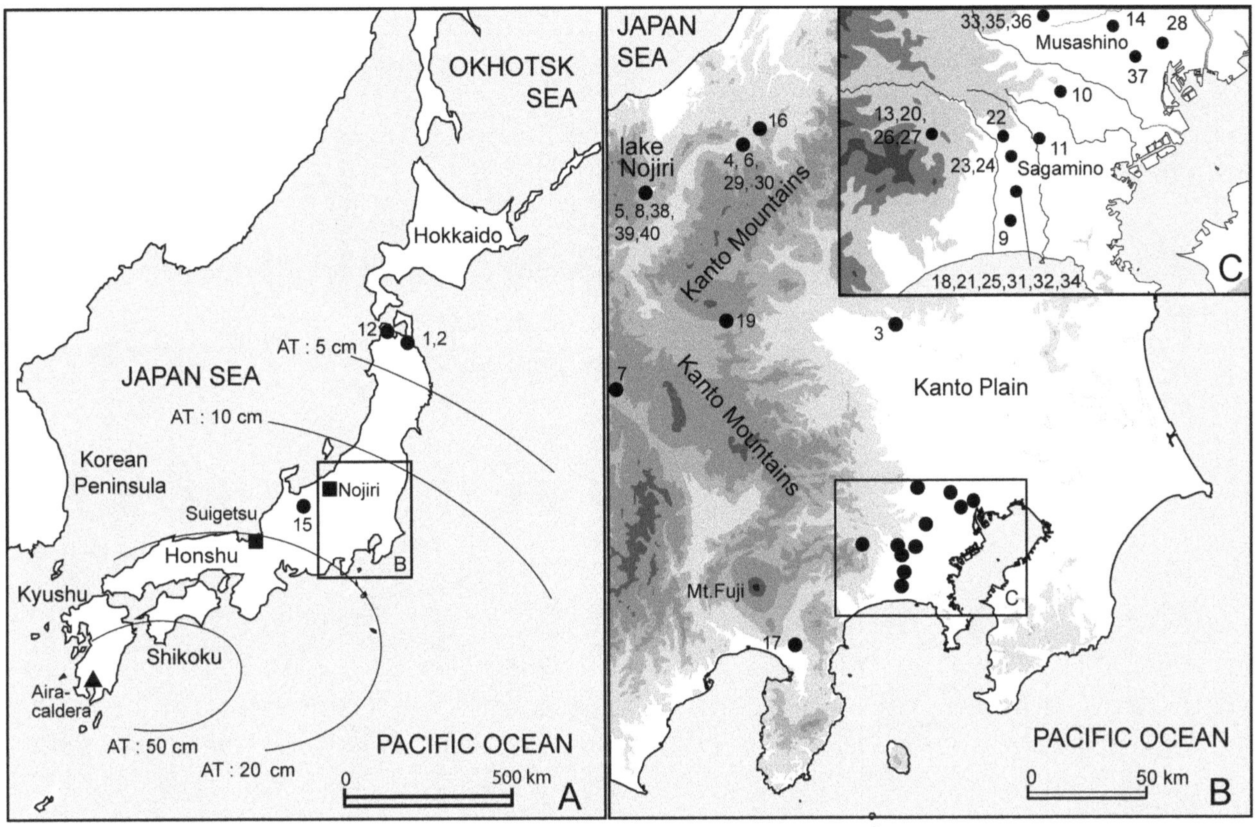

Fig. 2 Location map of the archeological sites, Lake Nojiri, Lake Suigetsu, and the thickness distribution of the AT tephra (contours). 1. Kushibiki, 2. Takihashi, 3. Saishikada-Nakajima, 4. Unoki-Minami, 5. Seiko-sanso B, 6. Kubodera Minami, 7. Nakajima B, 8. Kannoki, 9. SFC, 10. Manpukuji, 11. Tsukimino-kamino 2, 12. Odai-Yamamoto I, 13. Miyagase-Kitahara, 14. Gotenyama, 15. Miyanomae, 16. Araya, 17. Yasumiba, 18.Yoshioka B, 19. Shimomouchi, 20. Miyagase-Sazaranke, 21. Yoda-Minamihara, 22. Tana-Mukaihara, 23. Fukudahei-Ninoku 1, 24. Fukudahei-Ninoku 2, 25. Yoda-Toriimae, 26. Miyagase-Nakappara, 27 .Miyagase-Ueppara, 28. Kitashinjyuku-2, 29. Mukoubara A, 30. Mukoubara B, 31. Yoshioka B (B2), 32. Yoda-Ogouchi, 33. Musashidai-W CL4 (Vb), 34.Yoshioka B (B3), 35. Musashidai-W CL2 (XIIa-IXa), 36. Musashidai-W CL1 (IX-X), 37.Komaba, 38. Hinatabayashi B(Va), 39. Hinatabayashi B(Vb), 40. Kannoki

The earliest reliable evidence of Upper Palaeolithic human activity in Honshu is in Tachikawa Loam layer X on the Musashino plateau and layer B5 on the Sagamino plateau (Fig. 3). ^{14}C dates of charcoal associated with the lithic assemblage are ca. 33–30 ka ^{14}C BP. In general, in the Japanese Islands, the Upper Palaeolithic is distinguished from the Jomon period by the absence of pottery (Kobayashi, 1982). The oldest pottery has been dated at ca. 13.5–13.0 ka ^{14}C BP (Odai Yamamoto I Site Excavation Team, 1999; Taniguchi and Kawaguchi, 2001; Kudo, 2004, 2005) and assigned to the Incipient Jomon. Thus, the recognized time range of the Japanese Upper Palaeolithic is ca. 33–13 ka ^{14}C BP, which corresponds geochronologically to late OIS-3 and all of OIS-2.

1) Early Upper Palaeolithic

The Upper Palaeolithic on the Musashino and Sagamino plateaus has been divided into several archaeological phases on the basis of stratum number and lithic typology (Fig. 2). The Upper Palaeolithic stratigraphic sequence on the Musashino plateau comprises Tachikawa Loam layer III to X, and that on the Sagamino plateau layers L1S to L6. Suwama (1988, 2004) divided lithic assemblages on the Sagamino plateau into Stage I to XII. Thus, archaeological phases are often referred to by the Tachikawa Loam layers of Musashino Plateau and layer of the Sagamino plateau (Fig. 3).

The earliest phase of the Upper Palaeolithic: Trapezoid industry: The earliest phase of the Upper Palaeolithic, found in Tachikawa Loam layer X and IX-lower and in layer B5 to B4 (Stage I and II) on the Sagamino plateau is known as the trapezoid industry, which is characterized by the use of trapezoid tools and edge-ground stone tools (Sato, 1992). Radiocarbon dating has been conducted at various sites using charcoal from hearth features or charcoal concentrations associated with trapezoid industries. At the Musashidai site (No. 36 in Fig. 3 and Tab. 1), charcoal samples from two hearths have been dated to ca. 30 ka ^{14}C BP (Tokyo Archaeological Research Center, 2004), and a charcoal sample from a hearth at the Tokyo University Komaba campus site has been dated to ca. 31 ka ^{14}C BP (No. 37 in Tab. 1) (Hara *et al.*, 2004) from Tachikawa Loam layer X to IX-lower. Similar ^{14}C dates have been obtained from other sites in the vicinity. Charcoal samples from the Kannoki and Hinata-bayashi B sites (No. 39 and 40 in Fig. 3 and Tab. 1) near Lake Nojiri (Fig. 2) have been dated to

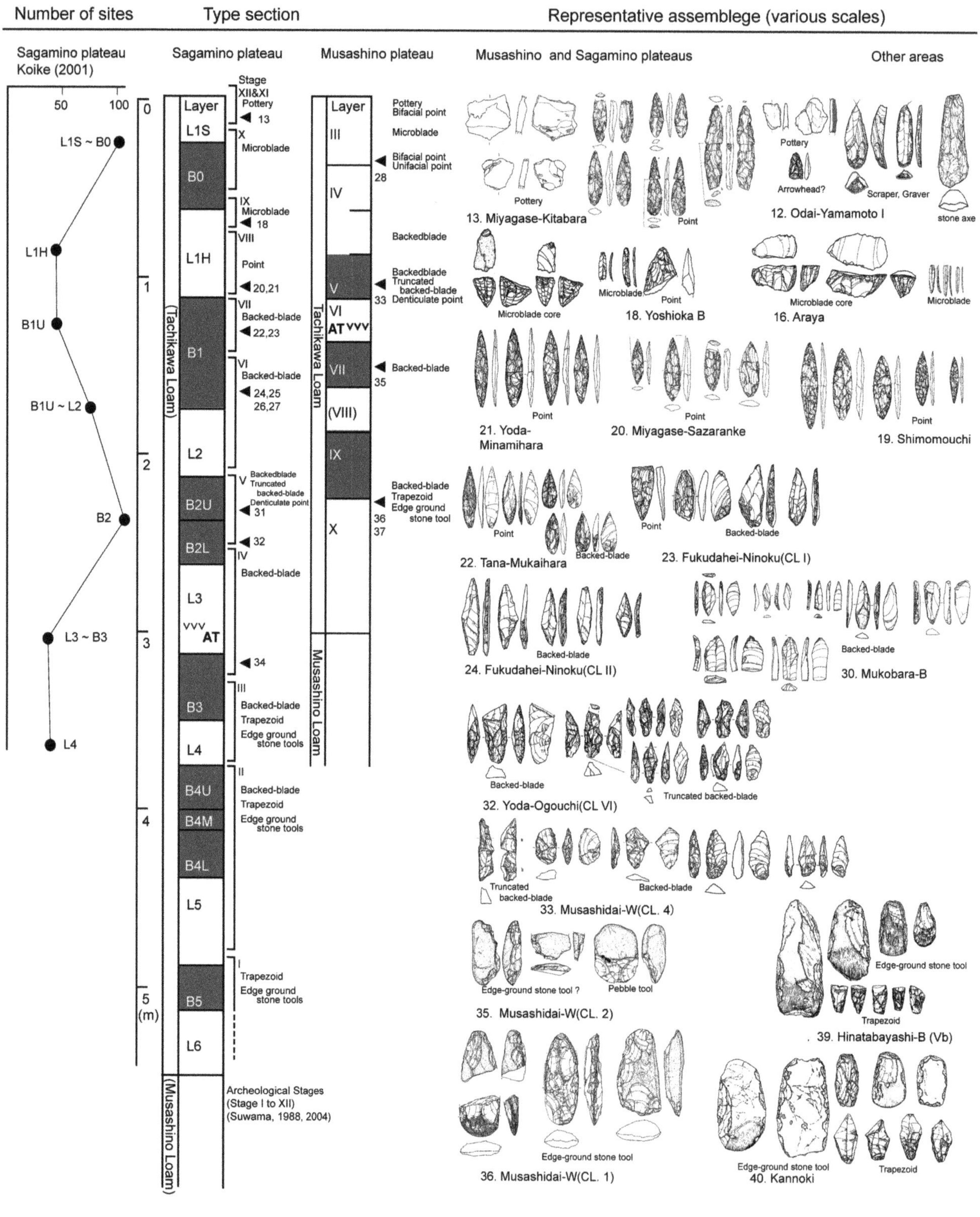

Fig. 3 Type sections of the Musashino and Sagamino plateaus and typical artifacts from archaeological sites.

Table 1 Radiocarbon dates from early Upper Palaeolithic to Incipient Jomon period archaeological sites in eastern Honshu.

No.	*site name*	*Context*	^{14}C BP (1σ)	$\delta^{13}C$ (‰)	*Lab Number*	*Material*	*Method*	*Cultural features*	*Musashino and Sagamino stratum*	*Reference*
1	Kushibiki	Pit No.4	10030±50	—	Beta-113349	charcoal	AMS	Cord-marked pottery	—	Aomori Prefecture, 1999
2	Takihashi	Pebble concentration	10260±40	—	Beta-138898	charcoal	AMS	Nail-impressed pottery	—	Hashikami town, 2000
3	Saishikada-Nakajima	Dwelling structure No.11	10070±70	—	Beta-128025	charcoal	AMS	Cord-marked pottery	—	Kasakakeno town, 2003
		Pit No.71	11180±40	—	Beta-130326	charcoal	β	Nail-impressed pottery	—	Kasakakeno town, 2003
4	Unoki-Minami	potsherd	11000±50	-24.8	Beta-136739	pottery adhesion	AMS	Nail-impressed and Cord marked pottery	—	Imamura (ed.), 2004
		potsherd	11040±50	-24.5	Beta-136740	pottery adhesion	AMS	Nail-impressed and Cord marked pottery	—	Imamura (ed.), 2004
		potsherd	11130±50	-25.2	Beta-136741	pottery adhesion	AMS	Nail-impressed and Cord marked pottery	—	Imamura (ed.), 2004
		potsherd	11630±50	-21.2	Beta-136742	pottery adhesion	AMS	Nail-impressed and Cord marked pottery	—	Imamura (ed.), 2004
5	Seiko-sanso B	potsherd	12000±40	—	Beta-133848	pottery adhesion	AMS	Linear-relief pottery	—	Nagano prefecture, 2000b
		potsherd	12160±40	—	Beta-133849	pottery adhesion	AMS	Linear-relief pottery	—	Nagano prefecture, 2000b
		potsherd	12340±50	—	Beta-133847	pottery adhesion	AMS	Linear-relief pottery	—	Nagano prefecture, 2000b
6	Kubodera-Minami	potsherd	12280±50	-24.9	Beta-136743	pottery adhesion	AMS	Linear-relief pottery	—	Imamura (ed.), 2004
		potsherd	12420±50	-24.8	Beta-136744	pottery adhesion	AMS	Linear-relief pottery	—	Imamura (ed.), 2004
		potsherd	12490±60	-23.8	Beta-136745	pottery adhesion	AMS	Linear-relief pottery	—	Imamura (ed.), 2004
		potsherd	12620±50	-23.6	Beta-136746	pottery adhesion	AMS	Linear-relief pottery	—	Imamura (ed.), 2004
		potsherd	12510±40	-23.9	Beta-136747	pottery adhesion	AMS	Linear-relief pottery	—	Imamura (ed.), 2004
		potsherd	12520±50	-25.2	Beta-140494	pottery adhesion	AMS	Linear-relief pottery	—	Imamura (ed.), 2004

No.	site name	Context	^{14}C BP (1σ)	$\delta^{13}C$ (‰)	Lab Number	Material	Method	Cultural features	Musashino and Sagamino stratum	Reference
		potsherd	12630±50	-26.5	Beta-140495	pottery adhesion	AMS	Linear-relief pottery	—	Imamura (ed.), 2004
7	Nakajima B	pottery concentration	12460±310	—	I-13767	charcoal	β	Linear-relief pottery	—	Nagano prefecture, 1987
8	Kannnoki	potsherd	12350±50	-25.8	NUTA2-6885	pottery adhesion	AMS	Linear-relief pottery	—	Nagano prefecture, 2004
		potsherd	12360±50	-22.8	NUTA2-6883	pottery adhesion	AMS	Linear-relief pottery	—	Nagano prefecture, 2004
		potsherd	12490±50	-24.6	NUTA2-6883	pottery adhesion	AMS	Linear-relief pottery	—	Nagano prefecture, 2004
		potsherd	12870±110	-24.8	PLD-1845	pottery adhesion	AMS	Linear-relief pottery	—	Nagano prefecture, 2004
		potsherd	13010±110	-25.0	PLD-1844	pottery adhesion	AMS	Linear-relief pottery	—	Nagano prefecture, 2004
9	Keio-SFC	Cultural layer 2	11350±160	—	GaK-15904	charcoal	β	Linear-relief pottery	—	Okamoto (ed.),1993
10	Manpukuji-No.1	potsherd	12330±40	-26.0	Beta-191840	pottery adhesion	AMS	Linear-relief pottery	—	Kobayashi *et al.*, 2005a
11	Tsukimino-Kamino 2	potsherd	12480±50	-26.6	Beta-158196	pottery adhesion	AMS	Linear-relief pottery	—	Kobayashi *et al.*, 2005b
12	Odai-yamamoto 1	potsherd	13210±160	—	NUTA-6515	pottery adhesion	AMS	Plain Pottery	—	Odai Yamamoto, 1999
		potsherd	13030±170	-30.5	NUTA-6507	pottery adhesion	AMS	Plain Pottery	—	Odai Yamamoto, 1999
		potsherd	12720±160	—	NUTA-6509	pottery adhesion	AMS	Plain Pottery	—	Odai Yamamoto, 1999
		potsherd	12680±140	-29.6	NUTA-6506	pottery adhesion	AMS	Plain Pottery	—	Odai Yamamoto, 1999
		potsherd	13780±170	—	NUTA-6510	pottery adhesion	AMS	Plain Pottery	—	Odai Yamamoto, 1999
		Layer III	13480±70	-26.1	Beta-125550	charcoal	AMS	Plain Pottery	—	Odai Yamamoto, 1999
13	Miyagase-Kitappara	CL I ,Pebble concentration no.2	13060±80	—	Beta-105398	charcoal	AMS	Plain Pottery	Sagamino L1S	Kanagawa, 1998
		CL I ,Pebble concentration no.2	13050±80	—	Beta-105400	charcoal	AMS	Plain Pottery	Sagamino L1S	Kanagawa, 1998
		CL I ,Pebble concentration no.5&6	13060±100	—	Beta-105401	charcoal	AMS	Plain Pottery	Sagamino L1S	Kanagawa, 1998
		CL I ,Pebble concentration no.5&6	13020±80	—	Beta-105402	charcoal	AMS	Plain Pottery	Sagamino L1S	Kanagawa, 1998

No.	site name	Context	^{14}C BP (1σ)	δ^{13}C (‰)	Lab Number	Material	Method	Cultural features	Musashino and Sagamino stratum	Reference
		CL I ,Pebble concentration no.5&6	13050±80	—	Beta-105403	charcoal	AMS	Plain Pottery	Sagamino L1S	Kanagawa, 1998
14	Gotenyama	potsherd	13560±40	—	—	pottery adhesion	AMS	Plain Pottery	—	Kato Kensetsu, 2004
		Layer IIc	13200±70	—	—	charcoal	AMS	Plain Potttry	—	Kato Kensetsu, 2004
15	Miyanomae	Layer 17	12860±160	—	NUTA-3644	Bark	AMS	Microblade	—	Miyagawa town,1998
		Layer 17	14550±160	—	NUTA-3637	Strobile	AMS	Microblade	—	Miyagawa town,1998
16	Araya	—	13200±350	—	GaK-948	charcoal	β	Microblade	—	Tohoku Univ. 1990
		Pit no.01	14050±110	—	GrA-5701	charcoal	AMS	Microblade	—	Serizawa and Sudo, 2003
		Pit no.01	14150±110	—	GrA-5702	charcoal	AMS	Microblade	—	Serizawa and Sudo, 2003
		Pit no.06	14100±100	—	GrA-5703	charcoal	AMS	Microblade	—	Serizawa and Sudo, 2003
		Pit no.06	14200±120	—	GrA-5704	charcoal	AMS	Microblade	—	Serizawa and Sudo, 2003
		Pit no.06	14150±110	—	GrA-5705	charcoal	AMS	Microblade	—	Serizawa and Sudo, 2003
		Pit no.06	14100±110	—	GrA-5706	charcoal	AMS	Microblade	—	Serizawa and Sudo, 2003
		Pit no.06	14100±110	—	GrA-5707	charcoal	AMS	Microblade	—	Serizawa and Sudo, 2003
		Pit no.14	14150±110	—	GrA-5708	charcoal	AMS	Microblade	—	Serizawa and Sudo, 2003
		Pit no.14	14100±110	—	GrA-5709	charcoal	AMS	Microblade	—	Serizawa and Sudo, 2003
		Pit no.14	14150±120	—	GrA-5710	charcoal	AMS	Microblade	—	Serizawa and Sudo, 2003
		Pit no.14	14200±110	—	GrA-5711	charcoal	AMS	Microblade	—	Serizawa and Sudo, 2003
		Pit no.14	14200±110	—	GrA-5712	charcoal	AMS	Microblade	—	Serizawa and Sudo, 2003
		Dwelling structure?	14250±110	—	GrA-5713	charcoal	AMS	Microblade	—	Serizawa and Sudo, 2003
		Dwelling structure?	13690±80	—	GrA-5715	charcoal	AMS	Microblade	—	Serizawa and Sudo, 2003
		Dwelling structure?	13700±290	—	GrA-5716	charcoal	AMS	Microblade	—	Serizawa and Sudo, 2003
17	Yasumiba	Hearth 1	14300±700	—	Gak-604	charcoal	β	Microblade	—	Sugihara and Ono, 1967
18	Yoshioka B (L1H)	L1H upper	16490±250	—	TKa-11613	charcoal	AMS	Microblade	Sagamino stratum L1H upper	Kanagawa, 1999a
		L1H upper	16860±160	—	TKa-11599	charcoal	AMS	Microblade	Sagamino stratum L1H upper	Kanagawa, 1999a
19	Shimomouchi	Charcoal concentration no.1	16250±180	—	NUTA-1515	charcoal	AMS	Point	—	Nagano Prefetcure, 1992
20	Miyagase-Sazaranke	CL.3 hearth P1	17460±330	—	Gak-18281	charcoal	β	Point	—	Kanagawa, 1996
		CL.3	15470±290	—	Gak-18282	charcoal	β	Point	—	Kanagawa, 1996

No.	*site name*	*Context*	^{14}C BP (1σ)	δ^{13}C (‰)	*Lab Number*	*Material*	*Method*	*Cultural features*	*Musashino and Sagamino stratum*	*Reference*
21	Yoda-Minamihara	CL.2 charcoal concentration no.4	17130±90	-27.7	Beta-116828	charcoal	AMS	Point	Sagamino stratum L1H	Kanagawa, 2004a
		CL.2 charcoal concentration no.4	17150±90	-27.4	Beta-116829	charcoal	AMS	Point	Sagamino stratum L1H-B1	Kanagawa, 2004a
		CL.2 charcoal concentration no.4	16880±90	-28.0	Beta-116830	charcoal	AMS	Point	Sagamino stratum B1	Kanagawa, 2004a
		CL.2 charcoal concentration no.2	16960±80	-24.4	Beta-183093	charcoal	AMS	Point	Sagamino stratum L1H	Kanagawa, 2004a
		CL.2 charcoal concentration no.2	16970±80	-24.6	Beta-183094	charcoal	AMS	Point	Sagamino stratum L1H	Kanagawa, 2004a
22	Tana-Mukaihara	Dwelling structure	17650±60	-27.0	Beta-127792	charcoal	AMS	Backed blade	Sagamino stratum B1 upper	Sagamihara city, 2004
		Dwelling structure	17630±50	-26.6	Beta-127793	charcoal	AMS	Backed blade	Sagamino stratum B1 upper	Sagamihara city, 2004
23	Fukudahei-Ninoku	CL.1 pebble concentration no.1	17920±320	—	TKa-11612	charcoal	AMS	Backed blade	Sagamino stratum B1 upper	Kanagawa, 1999b
		CL.1 pebble concentration no.1	18100±210	—	TKa-11607	charcoal	AMS	Backed blade	Sagamino stratum B1 upper	Kanagawa, 1999b
		CL.1 pebble concentration no.1	17880±220	—	TKa-11666	charcoal	AMS	Backed blade	Sagamino stratum B1 upper	Kanagawa, 1999b
		CL.1	18960±480	—	TKa-11525	charcoal	AMS	Backed blade	Sagamino stratum B1 upper	Kanagawa, 1999b
		CL.1	19410±250	—	TKa-11594	charcoal	AMS	Backed blade	Sagamino stratum B1 upper	Kanagawa, 1999b
		CL.1	19480±490	—	TKa-11608	charcoal	AMS	Backed blade	Sagamino stratum B1 upper	Kanagawa, 1999b
		CL.1	18380±470	—	TKa-11609	charcoal	AMS	Backed blade	Sagamino stratum B1 upper	Kanagawa, 1999b
		CL.1	18820±290	—	TKa-11597	charcoal	AMS	Backed blade	Sagamino stratum B1 upper	Kanagawa, 1999b
24	Fukudahei-Ninoku	CL.2 pebble concentration no.2	19440±430	—	TKa-11591	charcoal	AMS	Backed blade	Sagamino stratum B1 lower	Kanagawa, 1999b
		CL.2 pebble concentration no.2	18970±440	—	TKa-11598	charcoal	AMS	Backed blade	Sagamino stratum B1 lower	Kanagawa, 1999b
		CL.2 pebble concentration no.4	19240±700	—	TKa-11601	charcoal	AMS	Backed blade	Sagamino stratum B1 lower	Kanagawa, 1999b
		CL.2 pebble concentration no.5	18770±330	—	TKa-11602	charcoal	AMS	Backed blade	Sagamino stratum B1 lower	Kanagawa, 1999b
		CL. 2	19220±330	—	TKa-11603	charcoal	AMS	Backed blade	Sagamino stratum B1 lower	Kanagawa, 1999b

No.	site name	Context	^{14}C BP (1σ)	$δ^{13}C$ (‰)	Lab Number	Material	Method	Cultural features	Musashino and Sagamino stratum	Reference
		CL. 2	19460±350	—	TKa-11605	charcoal	AMS	Backed blade	Sagamino stratum B1 lower	Kanagawa, 1999b
		CL.2 pebble concentration no.1	19300±270	—	TKa-11611	charcoal	AMS	Backed blade	Sagamino stratum B1 lower	Kanagawa, 1999b
		CL.2 pebble concentration no.1	19660±440	—	TKa-11537	charcoal	AMS	Backed blade	Sagamino stratum B1 lower	Kanagawa, 1999b
25	Yoda-Toriimae	CL.4	19740±190	-25.9	NUTA-5451	charcoal	AMS	Backed blade	Sagamino stratum B1 lower	Kanagawa, 2002
		CL.4	18980±160	-24.9	NUTA-5490	charcoal	AMS	Backed blade	Sagamino stratum B1 lower	Kanagawa, 2002
		CL.4	19370±150	-25.9	NUTA-5296	charcoal	AMS	Backed blade	Sagamino stratum B1 lower	Kanagawa, 2002
		CL.4	19570±150	-26.1	NUTA-5297	charcoal	AMS	Backed blade	Sagamino stratum B1 lower	Kanagawa, 2002
		CL.4	17910±130	-25.9	NUTA-5300	charcoal	AMS	Backed blade	Sagamino stratum B1 lower	Kanagawa, 2002
26	Miyagase-Nakappara	CL.5 hearth P1	18920±100	-26.7	Beta-97116	charcoal	AMS	Backed blade	Sagamino stratum B1 lower	Kanagawa, 1997a
27	Miyagase-Ueppara	CL.5 hearth P1	19240±100	—	Beta-97117	charcoal	AMS	Backed blade	Sagamino stratum B1 lower	Kanagawa, 1997b
		CL.5 pebble concentration	19470±100	—	Beta-97118	charcoal	AMS	Backed blade	Sagamino stratum B1 lower	Kanagawa, 1997b
28	Kitashinjyuku-2chome	Pebble concentration no.1	19360±180	-27.2	Beta-132645	charcoal	AMS	Backed blade	Musashino stratum III lower	Shinjyuku-ku №107, 2000
29	Mukoubara A	Trench no.1	19120±140	-24.2	Beta-141743	charcoal	AMS	Backed blade	—	Tsunan town, 2005
		Trench no.1	19050±150	-23.1	Beta-141744	charcoal	AMS	Backed blade	—	Tsunan town, 2005
		Trench no.1	19100±140	-23.0	Beta-141745	charcoal	AMS	Backed blade	—	Tsunan town, 2005
		Trench no.1	19240±140	-23.2	Beta-141746	charcoal	AMS	Backed blade	—	Tsunan town, 2005
30	Mukoubara B	Trench no.5	19090±140	-23.2	Beta-141747	charcoal	AMS	Backed blade	—	Tsunan town, 2005
31	Yoshioka B	Lithic assemblage V	21970±80	—	Beta-130330	charcoal	AMS	Backed blade	Sagamino stratum B2 upper	Kanagawa, 2003
32	Yoda-Ogochi	CL.6 hearth	22880±80	-27.4	Beta-125304	charcoal	AMS	Backed blade	Sagamino stratum B2L middle	Kanagawa, 2004b
		CL.6 hearth	21840±120	-27.7	Beta-125305	charcoal	AMS	Backed blade	Sagamino stratum B2L middle	Kanagawa, 2004b

No.	site name	Context	^{14}C BP (1σ)	$\delta^{13}C$ (‰)	Lab Number	Material	Method	Cultural features	Musashino and Sagamino stratum	Reference
		CL.6 hearth	21330±50	-26.5	Beta-125306	charcoal	AMS	Backed blade	Sagamino stratum B2L middle	Kanagawa, 2004b
		CL.6 hearth	22850±180	-28.2	Beta-128829	charcoal	AMS	Backed blade	Sagamino stratum B2L middle	Kanagawa, 2004b
		CL.6 hearth	22790±200	-25.0	Beta-130869	charcoal	AMS	Backed blade	Sagamino stratum B2L middle	Kanagawa, 2004b
33	Musashidai-west area	CL.4 hearth SX-48	24100±200	-28.0	Beta-182635	charcoal	AMS	Backed blade	Musashino stratum V b	Tokyo Metropolitan, 2004
		CL.4 hearth SX-72	23930±150	-25.6	Beta-156136	charcoal	AMS	Backed blade	Musashino stratum V b	Tokyo Metropolitan, 2004
		CL.4 hearth SX-98	24530±300	-25.0	Beta-182636	charcoal	AMS	Backed blade	Musashino stratum V b	Tokyo Metropolitan, 2004
34	Yoshioka B (B3)	Lithic assemblage VI	25520±110	—	Beta-130331	charcoal	AMS	Backed blade	Sagamino stratum B3 upper	Kanagawa, 2003
		Lithic assemblage VI	23280±100	—	IAAA-10703	charcoal	AMS	Backed blade	Sagamino stratum B3 upper	Kanagawa, 2003
		Lithic assemblage VI	25780±110	—	IAAA-10704	charcoal	AMS	Backed blade	Sagamino stratum B3 upper	Kanagawa, 2003
35	Musashidai-west area (CL2)	CL.2 charcoal concentration	27390±250	-27.4	Beta-182637	charcoal	AMS	—	Musashino stratum VII a-IX a	Tokyo Metropolitan, 2004
36	Musashidai-west area (CL1)	CL.1 hearth SX-47	30400±400	-23.8	Beta-156135	charcoal	AMS	Edge-ground stone tool	Musashino stratum IX b-Xc	Tokyo Metropolitan, 2004
		CL.1 hearth SX-104	29860±150	-24.7	Beta-182638	charcoal	AMS	Edge-ground stone tool	Musashino stratum IX b-Xc	Tokyo Metropolitan, 2004
37	Tokyo Univ. Komaba	Hearth SK142	30800±200	—	MTC-03710	charcoal	AMS	Trapezoid	Musashino stratum IX	Hara *et al.*, 2004
38	Hinata-bayashiB(Va)	Layer Va upper	27950±210	—	Beta-120862	charcoal	AMS	Edge-ground stone tool	—	Nagano prefecture, 2000a
		Layer Va	28400±210	—	Beta-120865	charcoal	AMS	Edge-ground stone tool	—	Nagano prefecture, 2000a
		Layer Va	27940±210	—	Beta-120867	charcoal	AMS	Edge-ground stone tool	—	Nagano prefecture, 2000a
		Layer Va	28540±220	—	Beta-120868	charcoal	AMS	Edge-ground stone tool	—	Nagano prefecture, 2000a
		Layer Va upper	27940±200	—	Beta-120869	charcoal	AMS	Edge-ground stone tool	—	Nagano prefecture, 2000a
39	Hinata-bayashiB(Vb)	Layer Vb upper	29870±250	—	Beta-120858	charcoal	AMS	Edge-ground stone tool	—	Nagano prefecture, 2000a

No.	site name	Context	^{14}C BP (1σ)	$\delta^{13}C$ (‰)	Lab Number	Material	Method	Cultural features	Musashino and Sagamino stratum	Reference
		Layer Vb lower	31420±280	—	Beta-120861	charcoal	AMS	Edge-ground stone tool	—	Nagano prefecture, 2000a
		Layer Vb upper	28230±210	—	Beta-120863	charcoal	AMS	Edge-ground stone tool	—	Nagano prefecture, 2000a
		Layer Vb upper	29820±250	—	Beta-120864	charcoal	AMS	Edge-ground stone tool	—	Nagano prefecture, 2000a
		Layer Vb upper	29640±240	—	Beta-120866	charcoal	AMS	Edge-ground stone tool	—	Nagano prefecture, 2000a
40	Kannnoki	Layer Vb	33070±540	-25.5	Beta-82577	charcoal	AMS	Edge-ground stone tool	—	Nagano prefecture, 2000c
		Layer Vc	33040±530	-26.1	Beta-82580	charcoal	AMS	Edge-ground stone tool	—	Nagano prefecture, 2000c
		Layer V?	32260±590	—	Beta-109414	charcoal	AMS	Edge-ground stone tool	—	Nagano prefecture, 2000c
		Layer V?	32110±610	—	Beta-109416	charcoal	AMS	Edge-ground stone tool	—	Nagano prefecture, 2000c
		Layer Vb	30510±510	—	Beta-109417	charcoal	AMS	Edge-ground stone tool	—	Nagano prefecture, 2000c
		Layer Vb	32410±340	—	Beta-109419	charcoal	AMS	Edge-ground stone tool	—	Nagano prefecture, 2000c

ca. 33–30 ka ^{14}C BP and ca. 29–28 ka ^{14}C BP, respectively (Nagano Prefecture Archaeological Research Center, 2000a). Lithic assemblages from these sites are contrasted with those of Tachikawa Loam layer X to IX-lower in the Musashino plateau.

The early backed blade industry: A later phase of the early Upper Palaeolithic is found in the upper part of Tachikawa Loam layer IX, VII, and VI on the Musashino plateau, and in layer L4 to B2-lower (Stage III and IV) on the Sagamino plateau. This phase is known as the early backed blade industry and is characterized by the blade flaking technique and a greater variety of tool types, especially standardized backed blades of both side retouched. Trapezoids are still present in the lithic assemblage in Tachikawa Loam layer VII, but the blade flaking technique and backed blades become dominant from Tachikawa Loam layer VI (Yamaoka, 2006).

At the Musashidai site (Tachikawa Loam layer VII), a charcoal concentration was dated to ca. 27.4 ka ^{14}C BP (No. 35 in Fig. 3 and Tab. 1) (Tokyo Archaeological Research Center, 2004). Edge-ground stone tools and pebble tools were excavated from this site (Fig. 3), but the association between these lithics and the charcoal is ambiguous. Because of the scarcity of radiocarbon dates, the detailed chronology of the lithic industries of Tachikawa Loam layer IX-upper and VII to VI cannot be ascertained.

The widely distributed tephra Aira-Tn (AT) is an important chronological marker in the Japanese Upper Palaeolithic. This tephra occurs in Tachikawa Loam layer VI and Sagamino layer L3 (Stage IV) (Fig. 3). The tephra is from the eruption of Aira caldera in southern Kyushu, which spread volcanic ash from Kyushu to northern Honshu (Machida and Arai, 2003) (Fig. 2). Charred natural wood fragments from directly below the AT tephra have been dated to ca. 24.5–24.0 ka ^{14}C BP (Tab. 2) (Ikeda *et al.*, 1995). These dates almost coincide with the transition from OIS-3 to OIS-2 (Machida, 2005). This key tephra is the marker used to separate the early Upper Palaeolithic from the late Upper Palaeolithic in the Japanese Islands (Sato, 1992).

2) Late Upper Palaeolithic

The late Upper Palaeolithic has also been divided into several phases: Tachikawa Loam layer V to III-upper on

Table 2 Radiocarbon dates for AT tephra from charred wood in the Osumi pumice fall and the Ito pyroclastic flow (Ikeda et al., 1995).

Tephra	*Sampling Unit*	^{14}C BP(1σ)	δ^{13}C (‰)	*Lab Number*	*Material*	*Method*	*Reference*
Aira-Tn (AT)	Osumi pumice fall	24790 ± 350	-24.3	NUTA-2564	Solid carbon	AMS	Ikeda *et al.*, 1995
	Osumi pumice fall	24050 ± 350	-24.3	NUTA-2580	Humic acid	AMS	Ikeda *et al.*, 1995
	Osumi pumice fall	24750 ± 360	-24.4	NUTA-2581	Humic acid	AMS	Ikeda *et al.*, 1995
	Ito pyroclastic flow	24240 ± 250	-24.4	NUTA-2563	Solid carbon	AMS	Ikeda *et al.*, 1995
	Ito pyroclastic flow	23730 ± 390	-24.5	NUTA-2565	Humic acid	AMS	Ikeda *et al.*, 1995
	Ito pyroclastic flow	23630 ± 250	-24.4	NUTA-2572	Humic acid	AMS	Ikeda *et al.*, 1995
	Ito pyroclastic flow	23870 ± 300	-24.4	NUTA-2573	Humic acid	AMS	Ikeda *et al.*, 1995
	Ito pyroclastic flow	24420 ± 320	-24.3	NUTA-2634	Humic acid	AMS	Ikeda *et al.*, 1995
	Ito pyroclastic flow	24840 ± 300	-24.3	NUTA-2635	Humic acid	AMS	Ikeda *et al.*, 1995
	Ito pyroclastic flow	24650 ± 390	-24.3	NUTA-2636	Humic acid	AMS	Ikeda *et al.*, 1995
	Ito pyroclastic flow	25270 ± 290	-24.3	NUTA-2637	Humic acid	AMS	Ikeda *et al.*, 1995
	Ito pyroclastic flow	23590 ± 390	-24.4	NUTA-2638	Humic acid	AMS	Ikeda *et al.*, 1995

the Musasino plateau (Suzuki and Yajima, 1978), and Sagamino Stage V to XI on the Sagamino plateau (Fig. 3) (Suwama, 1988, 2004). These phases have been roughly classified on the basis of lithic typologies as follows: (1) Late backed blade industry (Tachikawa Loam layers V to III-lower, Sagamino layer B2 to B1-upper [Stage V to VII]), (2) Point industry (Tachikawa Loam layers III-lower, Sagamino layer L1H-middle [Stage VIII]), as well as (3) Microblade industry (Tachikawa Loam layer III-middle, Sagamino layers L1H to L1S-lower [Stage IX to X]).

Late backed blade industry: The lithic assemblage of Tachikawa Loam layer V to IV-lower and Sagamino layer B2 (Stage V) consists of truncated backed blades (*Kiridashigata-sekki*), denticulate points (*Kakusuijo-sekki*), and round scrapers (ex. No.32 and 33 in Fig. 3), and has been dated to ca. 23–21 ka ^{14}C BP at the Yoda-Ogouchi site (No. 32 in Fig. 3 and Tab. 1) (Kanagawa Archaeology Foundation, 2004b), and to ca. 23 ka ^{14}C BP at the Yoshioka B site (No.34 in Tab. 1) (Kanagawa Archaeology Foundation, 2003). This phase is also characterized by a *Ko*-type backed blade, which is strongly related to the lithic industries of western Honshu.

The backed blade industry of Sagamino layer L2 and B1-lower (Stage VI) has been dated to ca. 19–18 ka ^{14}C BP at the Fukudahei-Ninoku site (No.24 in Fig. 3 and Tab. 1) (Kanagawa Archaeology Foundation, 1999b), the Yoda-Toriimae site (No.25 in Tab. 1) (Kanagawa Archaeology Foundation, 2002), the Miyagase-Nakappara site (No.26 in Tab. 1) (Kanagawa Archaeology Foundation, 1997a), and the Miyagase-Ueppara site (No.27 in Tab. 1) (Kanagawa Archaeology Foundation, 1997b). The backed blade industry from Tachikawa Loam layer IV-middle to IV-upper has been dated to ca. 19.3 ka ^{14}C BP at the Kitashinjhuku-Nichome site (No.28 in Tab. 1) (Shinjhuku-ku No. 107 Site Research Group, 2000). This phase is characterized by the *Sunagawa* type blade technique and the standardized backed blade of both side retouched (Figure3) made on blades (ex. No.24 in Fig. 3).

The final phase of the backed blade industry, when the characteristic assemblage also includes small geometric backed blades and bifacial and unifacial points, was excavated from Tachikawa Loam layers IV-upper to III-lower and Sagamino layer B1-upper (Stage VII) and is dated to ca. 19.5–18.4 ka ^{14}C BP at the Fukudahei-Ninoku site (No.23 in Fig. 3 and Tab. 1) (Kanagawa Archaeology Foundation, 1999b), and to ca. 17.7 and 17.6 ka ^{14}C BP at the Tana-Mukaihara site (No.22 in Fig. 3 and Tab. 1) (Sagamihara City Board of Education, 2004). The dated charcoal samples at the Tana-Mukaihara site were excavated from a hearth within a dwelling feature.

Point Industry: A lithic assemblage consisting of bifacial, unifacial, and edge-retouched points has been found in Tachikawa Loam layer III-lower and Sagamino layer L1H (Stage VIII). Associated dates are ca. 17.2–16.9 ka ^{14}C BP from Yoda-Minamihara site (No.21 in Fig. 3 and Tab. 1), two discrepant dates of 17.5 and 15.5 ^{14}C BP from the Miyagase-sazaranke site (No.20 in Fig. 3 and Tab. 1) (Kanagawa Archaeology Foundation, 1996), and ca. 16.2 ka ^{14}C BP from the Shimomouchi site (No.19 in Fig. 3 and Tab. 1) on the northern Kanto plain (Nagano Prefecture Archaeological Research Center, 1992). Overall, this point industry is roughly dated to 17–16 ka ^{14}C BP.

Microblade Industry: The microblade industry assemblage, which is characterized by use of the microblade technique, occurs in Tachikawa Loam layer III-middle and Sagamino layer L1H-upper to L1S-lower (Stage IX to X). Two main types of microblade core have been identified in Honshu: subconical microblade cores of the *Nodake-Yasumiba* type and wedge-shaped microblade cores made by the *Yubetsu* technique. *Nodake-Yasumiba* type cores are distributed mainly in western and eastern Honshu, and those made by the *Yubetsu* technique are found in eastern and northern Honshu.

The microblade industry has been dated to ca. 17–12 ka ^{14}C BP in Honshu (Sano, 2007). Charcoal samples associated

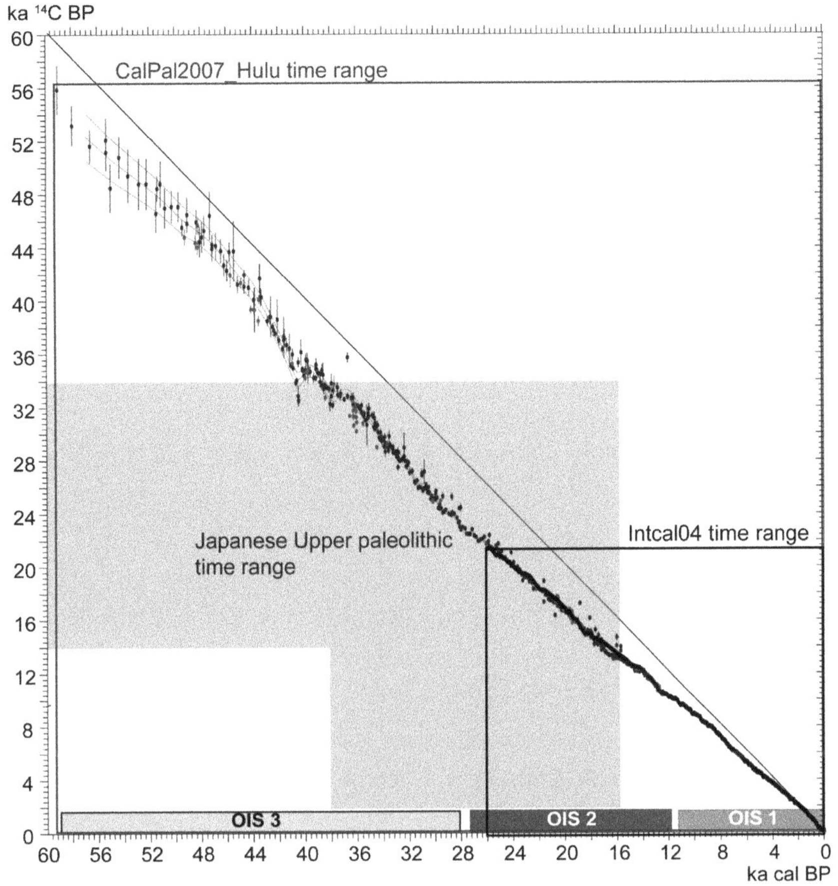

Fig. 4 Time ranges of the IntCal04 (Reimer et al, 2004) and CalPal-2007$_{Hulu}$ (Weninger and Jöris, 2008) data sets.

with subconical microblade cores have been dated to ca. 16.9 and 16.5 ka ^{14}C BP at the Yoshioka B site (No.18 in Fig. 3 and Tab. 1) (Kanagawa Archaeological Foundation, 1999a) and to ca. 14 ka ^{14}C BP at the Yasumiba site (No.17 in Tab. 1) (Sugihara and Ono, 1965). Two discrepant dates of 14.6 and 12.9 ^{14}C BP were obtained at the Miyanomae site (No.15 in Tab. 1)(Miyagawa Town Board of Education, 1998). In the area neighboring the Kanto plain, an assemblage of *Yubetsu* technique microblades has been dated to ca. 14.2–13.7 ka ^{14}C BP at the Araya site (No.16 in Tab. 1), using charcoal samples excavated from a dwelling structure and storage pits (Tohoku University Archaeology Laboratory, 1990; Serizawa and Sudo, 2003). It is still uncertain whether the oldest microblade assemblage in Honshu is as old as ca. 16.9 ka ^{14}C BP because association of the lithic assemblage and charcoal samples at the Yoshioka B site was unclear. Overall, the microblade industry of Honshu seems to range in age from ca. 15(16?) to 13.5 ka ^{14}C BP (Kudo, 2005).

The microblade industry first appeared about 5000 years earlier in Hokkaido, where the earliest microblade industry has been dated to ca. 20 ka ^{14}C BP at the Kashiwadai site (Hokkaido Archaeological Research Center, 1999).

3) Incipient Jomon

Bifacial point industry with plain pottery: This phase is characterized by bifacial points, edge-ground stone axes, and a high proportion of blade tools. It is found in Tachikawa Loam layer III-upper and Sagamino layer L1S-upper (Stage XI). Sometimes plain pottery is present, but the amount at any given site is very small, perhaps just one or two vessels.

The oldest pottery has been dated to ca. 13.5–13.0 ka ^{14}C BP at the Odai-Yamamoto I site (No.12 in Fig. 3 and Tab. 1) on the northern coast of Honshu (Odai Yamamoto I Site Excavation Team, 1999). This date is based on the average of five samples of organic material adhering to pottery (pottery adhesion samples) and on a charcoal sample. The mean age of the pottery adhesion samples is ca. 13.1 ka ^{14}C BP, and the charcoal sample was dated to 13.5 ka ^{14}C BP (See Tab. 1). Charcoal from a fireplace was dated to ca. 13.1 ka ^{14}C BP at the Miyagase-Kitappara site (No.13 in Fig. 3 and Tab. 1) on the Kanto plain (Kanagawa Archaeological Foundation, 1998b). Thus, the first use of pottery on Honshu seems to have occurred at around 13.5–13.0 ka ^{14}C BP.

Linear-relief pottery: Linear-relief pottery, which is found from Kyushu to northern Honshu, is dated to ca. 13.0–11.4 ^{14}C BP. Reliable dates are concentrated around 12.5–12.0 ^{14}C BP. Linear-relief pottery is present at more sites than plain pottery. The lithic assemblage of this phase is characterized by tanged points. In addition, the tanged points become smaller and arrowheads become an element of the lithic assemblage during the latter half of this phase. These arrowheads mark the first use of the bow and arrow in this region.

Glacial calibration curves before 12.4 ka

Three calibration curves, IntCal04 (Reimer *et al.*, 2004), Fairbanks0805 (Fairbanks *et al.*, 2005), and CalPal-2007$_{Hulu}$ (Weninger and Jöris, 2008), are commonly used to calibrate radiocarbon dates from Japanese Upper Palaeolithic sites. IntCal04 covers 0–26 ka cal BP; its upper part, from 0 to12.4 ka cal BP, is based on dendrochronology, and its lower part from 12.4 to 26.0 ka cal BP is based on a U/Th-^{14}C coral data set and from 12.4 to 14.7 ka cal BP on the marine foraminiferal ^{14}C-calibration data set of the Cariaco basin grayscale. Thus, IntCal04 does not cover the first half of the Japanese Upper Palaeolithic (Fig. 4). This period is covered by Fairbanks0805, a U/Th-^{14}C-coral-based calibration curve covering the period from 0 to ca. 50 ka cal BP. Unfortunately, this calibration curve has low sample density during ca. 26–36 ka cal BP, a time range that is critical for dating the emergence of the Japanese Upper Palaeolithic assemblages.

The newest CalPal-2007$_{Hulu}$ calibration curve (Weninger and Jöris, 2008) uses the "Hulu age model," employing U/Th dates synchronized with the ^{18}O profile of stalagmites from Hulu cave, China (Wang *et al.*, 2001) (Fig. 4). The CalPal-2007$_{Hulu}$ glacial data pairs measurements of U/Th and ^{14}C from pristine corals (Fairbanks *et al.*, 2005) and also uses ^{14}C measurements on planktonic foraminifera from the Cariaco Basin calendric ages using the downcore grayscale (Hughen *et al.*, 2006) and correlated with the U/Th-dated Hulu oxygen isotope profile (Wang *et al.*, 2001). Further ^{14}C age calibration data were derived by "Hulu-tuning" the north Atlantic marine cores PS2644 (Voelker *et al.*, 2000) and MD952042 (Bard *et al.*, 2004).

Discrepancies among the available records used for calibration means that calibration of the ^{14}C time scale before 26 ka cal BP is still problematic (cf. Notcal04: van der Plicht *et al.*, 2004). However, if the Hulu age model is sufficiently accurate, then the resulting calibration curve should be very useful for correlating the archaeological and paleoenvironmental records from the Japanese Upper Palaeolithic.

Discussion

Fig. 5 shows ^{14}C dates from archaeological sites alongside environmental history trends during the second half of OIS-3 and OIS-2. The radiocarbon dates from the archeological sites were calibrated by using CalPal_2007$_{Hulu}$ and the AT tephra. Four kinds of data are used as proxies for environmental change (Fig. 5): Greenland NGRIP ice core $\delta^{18}O$ data (North Greenland Ice Core Project members, 2004), U/Th-dated Hulu cave stalagmite $\delta^{18}O$ data (Wang *et al.*, 2001), and total organic carbon (TOC) and pollen composition (relative abundance of deciduous broadleaved tree pollen) profiles during 50–10 ka cal BP at Lake Nojiri, central Honshu (Kumon *et al.*, this volume).

Kumon *et al.* (this volume) documented the paleoenvironmental changes at Lake Nojiri since 72 ka cal BP on the basis of pollen and fluctuations in TOC and nitrogen (TN) contents and the C/N ratio. The proportion of cool-temperate deciduous broadleaved tree pollen (total of *Betula*, *Fagus*, *Quercus*, and *Ulmus-Zelkova*) relative to total pollen (sum of broadleaved tree and subarctic conifer [*Abies*, *Picea*, and *Tsuga*, but excluding *Cryptomeria*] pollen) gives a good general picture of the vegetation changes around Lake Nojiri. In addition, the TOC profile is compared to Greenland Interstadial (GIS) numbers. Kumon *et al.* (this volume) established the "Lake Nojiri sedimentation age model" by using the calibrated radiocarbon ages of eight key tephra layers before 49 ka (four tephras during 38–49 ka were calibrated after Fairbanks *et al.*, 2005), and the estimated age of the DKP tephra (62 ka).

1) Second half of OIS-3

The expansion of *Homo sapiens* into Asia is not well understood, though the oldest fossil evidence suggests that *Homo sapiens* entered East Asia at ca. 45–40 ka cal BP (Kaifu, 2009). The earliest occupation of the Japanese Islands occurred during the dispersal of *Homo sapiens* into East Asia.

In the Japanese Islands, the earliest human occupation is found in Tachikawa Loam layer X, at ca. 33–30 ka ^{14}C BP (ca. 38–34 ka cal BP). In the NGRIP time scale, these dates correspond to GIS-8 or earlier, before the coldest phase of the last glacial (Fig. 5). The temporal placement of the lithic industry from Tachikawa Loam layer IX, VII, and VI, or Sagamino layer B5 to L3 (Stage II to IV) is still uncertain.

In this phase, edge-ground stone tools are found at many archaeological sites. Opinion is divided as to whether these stone tools were used for felling wood or for butchering big game. Function and application studies of edge-ground stone tools, along with reliable dating of extinct game species (*Palaeoloxodon naumanni, Sinomegaceros yabei*, etc.) and reliable information on the vegetation of this period, are needed.

Lake Nojiri is one of the most important archaeological areas in central Honshu. At the Kannoki and Hinata-bayashi sites near Lake Nojiri, many edge-ground stone tools were excavated. Faunal remains of *Paleaoloxodon naumanni* and *Sinomegaceros yabei* found in the lake sediments have been dated to ca. 49–33 ka ^{14}C BP (ca. 53–37 ka cal BP) (Sawada *et al.*, 1992). The temporal correspondence between humans of the earliest Upper Palaeolithic and hunting of such big game is still uncertain. Intensive study of the timing of faunal extinction is in progress (Takahashi *et al.*, 2006; Takahashi, 2007).

Paleoenvironmental studies of Lake Nojiri sediments have indicated that cool-temperate deciduous broadleaved trees dominated during ca. 50–45 ka cal BP, and decreased ca. 45–38 ka cal BP. Cool-temperate deciduous broadleaved

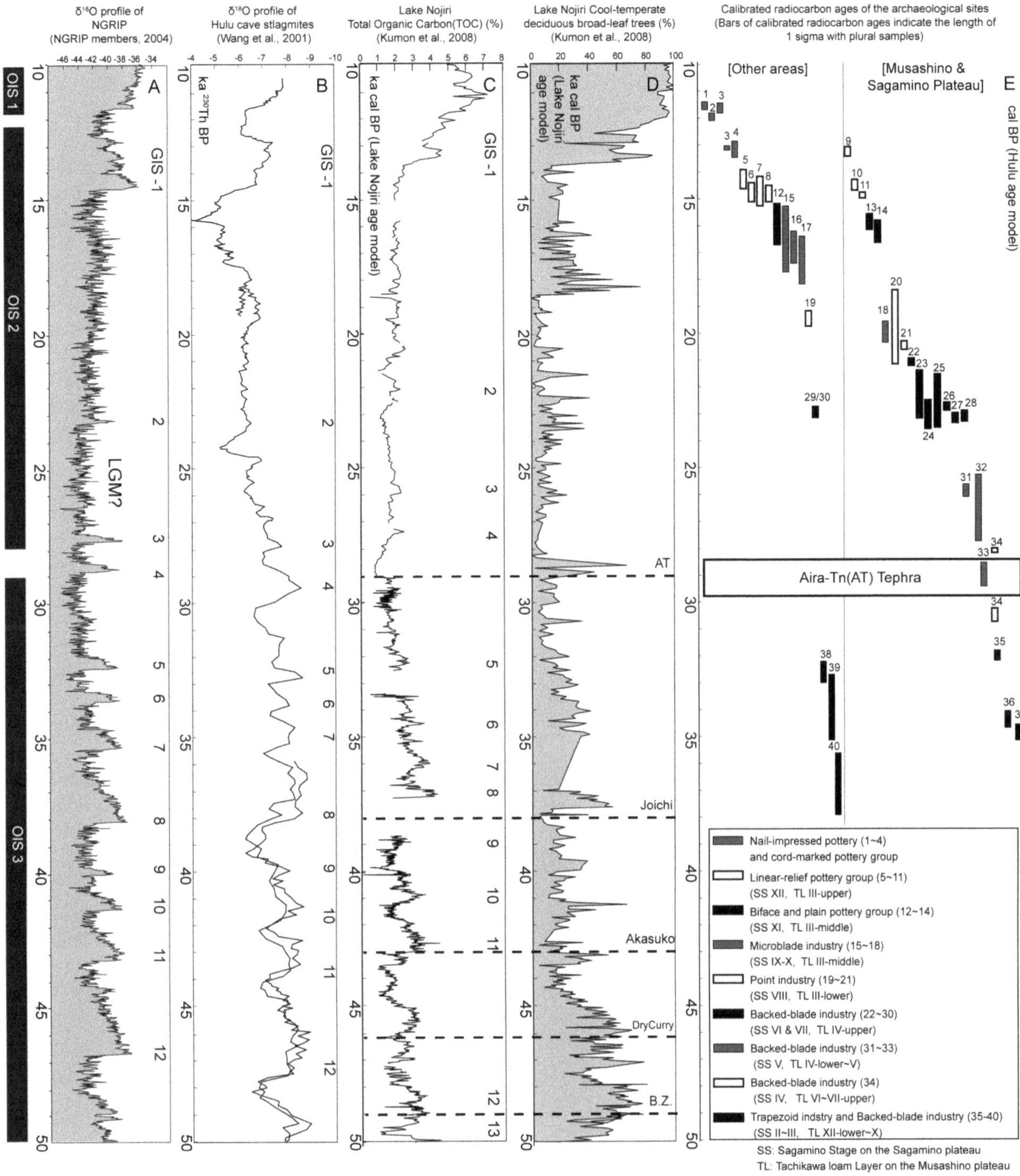

1. Kusibiki, 2. Takihashi, 3. Saishikada-Nakajima, 4. Unoki-Minami, 5. Seiko-sanso B, 6. Kubodera Minami, 7. Nakajima B, 8. Kannoki, 9. SFC, 10. Manpukuji, 11. Tsukimino-kamino 2, 12. Odai-Yamamoto I, 13. Miyagase-Kitabara, 14. Gotenyama, 15. Miyanomae, 16. Araya, 17. Yasumiba, 18.Yoshioka B, 19. Shimomouchi, 20. Miyagase-Sazaranke, 21. Yoda-Minamihara, 22. Tana-Mukaihara, 23. Fukudahei-Ninoku 1, 24. Fukudahei-Ninoku 2, 25. Yoda-Toriimae, 26. Miyagase-Nakappara, 27 .Miyagase-Ueppara, 28. Kitashinjyuku-2, 29. Mukoubara A, 30. Mukoubara B, 31. Yoshioka B (B2), 32. Yoda-Ogouchi, 33. Musashidai-W CL4 (Vb), 34.Yoshioka B (B3), 35. Musashidai-W CL2 (XIIa-IXa), 36. Musashidai-W CL1 (IX-X), 37.Komaba, 38. Hinatabayashi B(Va), 39. Hinatabayashi B(Vb), 40. Kannoki

Fig. 5 Archaeological chronology and paleoenvironmental sequences during the late Pleistocene in eastern Honshu.
A: $\delta^{18}O$ profile of Hulu-tuned NGRIP (North Greenland Ice Core Project members, 2004)
B: $\delta^{18}O$ profile of Hulu cave stalagmites (Wang *et al.*, 2001)
C: Lake Nojiri total organic carbon (TOC) (Kumon *et al.*, 2009)
D: Lake Nojiri cool-temperate deciduous broadleaved tree pollen (Kumon *et al.*, 2009)
E: Calibrated radiocarbon ages of the archaeological phases (bars indicate 1 sigma when more than one date was available).

trees again increased temporarily around GIS-8 (Fig. 5) (Kumon *et al.*, this volume). However, the climate gradually became colder toward the LGM, when subarctic coniferous trees became dominant around Lake Nojiri.

2) Last Glacial Maximum

The late backed blade industry of Tachikawa Loam layer V to IV-lower and Sagamino layer B2 (Stage V) dates to ca. 28–25 ka cal BP, geochronologically the LGM in the narrow sense. The backed blade industry of Tachikawa Loam layer IV to III-lower and Sagamino layer L2 and B1 (Stage VI and VII) dates to ca. 24–21 ka cal BP. The succeeding point industry dates to ca. 22–18 ka cal BP, and the microblade industry to ca. 18 (20?)–15 ka cal BP. The microblade industry appeared in eastern Honshu several thousand years after its appearance in Hokkaido (ca. 25–24 ka cal BP), the northernmost Japanese island (Izuho and Akai 2005; Izuho and Takahashi 2005; Kudo, 2005).

Tsuji and Kosugi (1991) showed that the change from a cool-temperate deciduous broadleaved forest to a subarctic conifer forest began in western and eastern Honshu before the AT eruption. TOC and pollen records from Lake Nojiri indicate that very cold and dry climatic conditions prevailed from ca. 32 to 18 ka cal BP (Kumon *et al.*, 2009). The number of archaeological sites on the Kanto plain increased during the LGM (Fig. 3).

The late backed blade industry, the point industry, and probably the earlier phase of the microblade industry also appear to correspond to cool and dry conditions. Cool-temperate deciduous broadleaved trees increased from ca. 18 to 17 ka cal BP, and then declined from ca. 15 or 14 ka cal BP. This period corresponds to the later microblade industry and the bifacial point industry with plain pottery.

Information about game fauna during the LGM and the late glacial period is scarce. Takahashi (2007) estimated that elephants and giant deer were widely extinct by 23 ka ^{14}C BP (ca. 27 ka cal BP) on Honshu Island.

3) Late Glacial

Exactly when pottery began to be used has long been a major question in Japanese archaeology. The development of radiocarbon dating techniques for charred carbonaceous residues on potsherds has added much more certainty to the chronology of pottery production. The earliest pottery has been dated to ca. 16–15 ka cal BP, preceding the abrupt warming and vegetation changes of the late glacial (Taniguchi and Kawaguchi, 2001; Kudo, 2004, 2005). On northern Honshu Island, where the oldest pottery was found (Odai-Yamamoto I site), the late glacial forest consisted of subarctic coniferous trees dominated by *Picea*, *Abies*, and *Larix* (Noshiro *et al.*, 1997). Thus, the environment of northern Tohoku apparently remained cold, keeping its glacial vegetation, after the LGM.

Abrupt vegetation changes to taxa associated with warmer climatic conditions occurred at 15 ka cal BP at Lake Suigetsu in central Honshu (Nakagawa *et al.*, 2003, 2005). This warm interval coincided with the linear-relief pottery group (ca. 15–13 ka cal BP), which succeeded the earlier plain pottery. In the Lake Nojiri pollen record, however, the increase in cool-temperate deciduous broadleaved tree pollen does not coincide with the calibrated ages of the linear-relief pottery phase (Fig. 5). There may be a problem of the "Lake Nojiri age model" between K-Ah tephra (ca. 7.3 ka) and AT tephra (ca. 29 ka). A date of 12,310 ± 100 ^{14}C BP was obtained from 604 cm depth, when cool-temperate deciduous broadleaved tree pollen abruptly increased (Kumon *et al.*, 2003). The linear-relief pottery phase at Seiko-sanso B site near Lake Nojiri dates to 12.3–12.0 ka ^{14}C BP, which accords well with the pollen core dates. Thus, linear-relief pottery phase corresponds to the warmer part of the late glacial.

The use of pottery is generally considered to have enabled prehistoric peoples to utilize many plant foods of the cool-temperate deciduous broadleaved forest, such as nuts and acorns, which require leaching to remove tannins before eating. Contrary to this generally accepted premise of Japanese archaeologists, the emergence of pottery and the expansion of cool-temperate deciduous broadleaved forest show no obvious causal connection (Kudo, 2005), though we must draw attention to the relationship between the spread of linear-relief pottery and the warming and vegetation changes of the late glacial. Detailed investigation of the correlation between pottery use and environmental conditions is required.

Conclusion

This paper presents a general outline of the temporal relationships between archaeological chronology and environmental history during ca. 40–15 ka cal BP with respect to three primary issues: early evidence of *Homo sapiens*, the general course of cultural change during the Upper Palaeolithic, and the emergence of pottery in the Japanese Islands.

Available data place the earliest lithic assemblages of *Homo sapiens* roughly between 38 and 34 ka cal BP. These dates correspond to GIS-8 of NGRIP.

The second half of the backed blade industry, the point industry, and the earlier phase of the microblade industry appear to have been associated with cool and dry climatic conditions. The second half of the backed blade industry dates to ca. 28–21 ka cal BP, the point industry to ca. 22–18 ka cal BP, and the microblade industry to ca. 18(20?)–15 ka cal BP.

Important radiocarbon dates of ca. 16–15 ka cal BP are associated with the earliest pottery at the Odai-yamamoto I site. These dates precede the abrupt warming and vegetation changes of the late glacial, although radiocarbon dates of pottery adhesions from the Odai-Yamamoto I site have large uncertainty.

Better dating of archaeological and environmental sequences is needed to facilitate further comparison between changes in human activities and the environment. Interdisciplinary studies should be promoted more proactively. Overall, the existing archaeological framework can be related to the various records of environmental change, but the correlations must be improved to clarify the association between human subsistence and their responses to changes in climate, fauna, and flora.

Acknowledgements

This is an expanded and revised version of the paper "Environmental Changes and Human Occupation in East Asia during OIS-3 and OIS-2" presented at the International Symposium on Quaternary Environmental Changes and Humans in Asia and the Western Pacific, held during 19–22 November 2007. I thank professors A. Ono, S. Tsuji, T. Nakamura, Dr. O. Jöris, and K. Sano for many valuable comments, and Prof. P. Matthew for improving the English draft of this paper. Special thanks are due to Prof. F. Kumon for profitable discussion and for providing TOC and pollen data from Lake Nojiri. However, I am entirely responsible for any remaining errors.

References

Aomori Prefecture Archaeological Research Center (ed.) 1999 *Kushibiki iseki* [Kushibiki site]. Aomori Prefecture Archaeological Research Reports 263. (in Japanese)

Bard, E., Menot-Combes, G., Rostek, F. 2004 Present status of radiocarbon calibration and comparison records based on Polynesian corals and Iberian Margin sediments. *Radiocarbon* 46, pp. 1189-1202.

Fairbanks, R. G., Mortlock, R. A., Chiu, T., Cao, L., Kaplan, A., Guilderson, T. P., Fairbanks, T. W. and Bloom, A. L. 2005 Marine radiocarbon calibration curve spanning 0 to 50,000 years B.P. based on paired $^{230}Th/^{234}U/^{238}U$ and ^{14}C dates on pristine corals. *Quaternary Science Reviews*, 24, 1781-1796.

Hara, Y., Kawaguchi, T., Ibi, H., Matsuzaki, H. and Haruhara, Y. 2004 *Tokyo daigaku Komaba kounai iseki Kokusai gakujyutsu kouryu-to chiten* [Tokyo University Komaba Campus Site: Komaba faculty house] *Dai 10 kai Sekki-bunka kenkyu-kai happyo yoshi*,1-6. (in Japanese)

Hokkaido Archaeological Research Center (ed.) 1999 *Kashiwadai 1 iseki* [Kashiwadai 1 site] Hokkaido Archaeological Research Center, Sapporo.

Hughen, K., Southon, J., Lehman, S., Bertrand, C., Turnbull, J., 2006 Marine-derived ^{14}C calibration and activity record for the past 50,000 years updated from the Cariaco Basin. *Quaternary Science Reviews* 25, pp. 3216-3227.

Ikeda, A. Okuno, M. Nakamura, T. Tsutsui, M and Kobayashi, T. 1995 Accelerator mass spectrometric ^{14}C Dating of charred wood in the Osumi Pumice Fall and Ito Ignimbrite from Aira Caldera, Southern Kyushu, Japan. *Daiyonki Kenkyu* [The Quaternary Research] 34, pp. 377-379. (in Japanese with English abstract)

Imamura, M. (ed.) 2004 *Jomon jidai Yayoi jidai no Kouseido Hennen-taikei no Kouchiku* [Establishment of the highly accurate chronology of the Jomon and Yayoi period]. Report of the Grants in Aid for Scientific Research (A) 2001-2003. (in Japanese)

Izuho, M., and Akai, F. 2005 Geochronology of Palaeolithic sites in Hokkaido, Japan. *Kyusseki Kenkyu* [Palaeolithic Research] 1, pp. 39-55 (in Japanese with English abstract).

Izuho, M. and Takahashi, K. 2005 Correlation of Palaeolithic Industries and Paleoenvironmental Changes in Hokkaido (Japan). Current Research in the Pleistocene 22, pp. 19-21.

Kanagawa Archaeology Foundation (ed.) 1997a *Miyagase Iseki-gun X, Nakappara (No.13c,iseki)* [Nakappara (No. 13c site), in the Miyagase site group]. Kanagawa Archaeology Foundation, Yokohama. (in Japanese).

Kanagawa Archaeology Foundation (ed.) 1996 *Miyagase Iseki-gun I* [Miyagase site group I]. Kanagawa Archaeology Foundation, Yokohama (in Japanese)

Kanagawa Archaeology Foundation (ed.) 1997b *Miyagase Iseki-gun XII, Ueppara (No.13,iseki)* [Ueppara (No. 13site), in the Miyagase site group]. Kanagawa Archaeology Foundation, Yokohama. (in Japanese).

Kanagawa Archaeology Foundation (ed.) 1998a *Yoshioka iseki-gun V* [Yoshioka site group V]. Kanagawa Archaeology Foundation, Yokohama. (in Japanese).

Kanagawa Archaeology Foundation (ed.) 1998b *Miyagase Iseki-gun XV, Kitappara (No.10, 11 kita) iseki* [Kitahara (No. 10, 11 North) site, in the Miyagase site group]. Kanagawa Archaeology Foundation, Yokohama. (in Japanese)

Kanagawa Archaeology Foundation (ed.) 1999a *Yoshioka iseki-gun IX* [Yoshioka site group IX] Kanagawa Archaeology Foundation, Yokohama. (in Japanese).

Kanagawa Archaeology Foundation (ed.) 1999b *Fukudahei-Ninoku iseki* [Fukudahei-Ninoku site] Kanagawa Archaeology Foundation, Yokohama. (in Japanese).

Kanagawa Archaeology Foundation (ed.) 2002 *Yoda-Toriimae iseki* [Yoda-Toriimae site] Kanagawa Archaeology Foundation, Yokohama. (in Japanese).

Kanagawa Archaeology Foundation (ed.) 2003 *Yoshioka iseki-gun X* [Yoshioka site group X] Kanagawa Archaeology Foundation, Yokohama. (in Japanese).

Kanagawa Archaeology Foundation (ed.) 2004a *Yoda-Minamihara iseki* [Yoda-Minamihara site] Kanagawa Archaeology Foundation, Yokohama. (in Japanese).

Kanagawa Archaeology Foundation (ed.) 2004b *Yoda-Ogouchi iseki* [Yoda-Ogouchi site] Kanagawa Archaeology Foundation, Yokohama. (in Japanese).

Kasakakeno Town Board of Education (ed.) 2003 *Saishikada-Nakajima Iseki Hakkutsu Chosa Houkokusyo(1)* [Archaeological research report of Saishikada-Nakajima site (1)], Kasakakeno Town Board of Education. (in Japanese)

Kobayashi, T. 1982 *Soron* [General Description]. In Kato,

S., Kobayashi, T. and Fujimoto, T. (eds.) *Jomon bunka no kenkyu 3. Jomon doki I* [The studies of Jomon culture 3. Jomon pottery I], pp.3-15, Yuzankaku Press, Tokyo (in Japanese).

Kobayashi, K., Sakamoto, M., Ozaki, T., Shinmen, T. and Muramoto, S. 2005a *Kanagawa-ken Manpukuji No.1iseki syutsu-do Sousouki Doki Futyaku-butu no ^{14}C Nendai Sokutei* [^{14}C dating of the pottery adhesion on the Incipient Jomon pottery of the Manpukuji No.1 site, Kanagawa. Manpukuji Site Research Group (ed.) *Manpukuji iseki-gun*, pp. 517-520, Manpukuji Site Research Group.

Kobayashi, K. Imamura, M. and Harunari, H. 2005b *Yamato-shi Kamino iseki syutsudo Jomon sousouki doki fucyaku-butu no ^{14}C nendai.* [^{14}C dates of adhesion on the incipient Jomon pottery excavated at the Kamino site, Yamato city]. *Yamato shi-shi Kenkyu* 31, pp. 1-11.

Koike, S. 2001 *Sagamino daiti no ritti to bunkaso* [Location and cultural layers on the Sagamino Plateau] Suwama. J (ed.) *Sagamino Kyusekki Hennen no Totatu-ten.* pp. 21-34. Kanagawa Koukogaku kai, Yokohama.

Kudo, Y. 2004 Reconsidering the geochronological and archaeological framework of the late Pleistocene - early Holocene transition on the Japanese Islands. *Internationale Archäologie Arbeitsgemeinschaft, Tagung, Symposium, Kongress* 5, 253-268.

Kudo, Y. 2005 The temporal correspondence between archaeological chronology and environmental changes in the final Pleistocene in eastern Honshu Island. *Daiyonki Kenkyu* [The Quaternary Research] 44, 51-64 (in Japanese with English abstract)

Kudo, Y. 2006 The temporal correspondence between archaeological chronology and environmental changes 28,000-11,000 CALYBP in eastern Honshu. *Current Research in the Pleistocene* 23, 12-15.

Kumon, F., Kawai, S. and Inouchi, Y. 2003 Climate changes between 25,000 and 6,000 years BP deduced from TOC, TN, and fossil pollen analyses of a sediment core from Lake Nojiri, central Japan. *Daiyonki-Kenkyu* (The Quaternary Research), 42, 13-26. (in Japanese with English abstract)

Kumon, F., Kawai, S. and Inouchi, Y. 2009 High-resolution climate reconstruction during the last 72 ka from total organic carbon (TOC), total nitrogen (TN) and pollen analyses of the drilled sediments in Lake Nojiri, central Japan. BAR international series, this volume.

Machida, H. 2005 A tephrochronological review on the earlier Palaeolithic archaeological remains in South Kanto. *Kyusekki Kenkyu* [Palaeolithic Research] 1, 7-16. (in Japanese with English abstract)

Machida, H. and Arai, F. 2003 Atlas of tephra in and around Japan (revised edition). University of Tokyo Press, Tokyo.

Minakami Town Board of Education (ed.) 2000 *Takihashi iseki hakkutu chosa hokoku sho* [Archaeological research report of Takihashi site]. Minakami town board of education. (in Japanese)

Miyagawa Town Board of Education (ed.) 1998 Miyanomae iseki hakkutsu cho-sa Hokoku-syo [Archaeological research report of Miyanomae site]. 422p, Miyagawa Town Board of Education, Miyagawa Town. (in Japanese)

Nagano Prefecture Archaeological Research Center (ed.) 1987 *Chuo Jidosha-do Nagano-sen Maizou Bunkazai Hkkutsu chosa Houkoku-sho 1* [Chuo Highway Nagano Line Archaeological Research report 1] . Nagano Prefecture Archaeological Research Center, Nagano.

Nagano Prefecture Archaeological Research Center (ed.) 1992 *Shimomouchi iseki* [Shimomouchi site], Nagano Prefecture Archaeological Research Center, Nagano.

Nagano Prefecture Archaeological Research Center (ed.) 2000a *Hinata-bayashi B iseki, Hinata-bayashiA iseki, Nanatsuguri iseki, Oohira iseki* [Hinata-bayashi B site, Hinata-bayashi A site, Nanatsuguri site, Oohira site]. Nagano Prefecture Archaeological Research Center, Nagano. (in Japanese)

Nagano Prefecture Archaeological Research Center (ed.) 2000b *Jyo-shin-etsu Jidosha-do Maizou-BunkazaiHakkutu ChosaHoukokusyo 16-Shinano machi nai sono 2* [Jyo-shin-etsu Highway Archaeological research report 16: in Shinano-machi no. 2]. Nagano Prefecture Archaeological Research Center Nagano. (in Japanese)

Nagano Prefecture Archaeological Research Center (ed.) 2000c *Kannoki iseki, Nishioka A iseki* [Kannoki site, Nishioka A site]. Nagano Prefecture Archaeological Research Center, Nagano. (in Japanese)

Nagano Prefecture Archaeological Research Center (ed.) 2004 *Ippan Kokudo 18 gou (Nojiri bypass) Maizou-BunkazaiHakkutu ChosaHoukokusyo -Shinano machi nai sono 2* [Jyo-shin-etsu Highway Archaeological research report: in Shinano-machi no.2]. Nagano Prefecture Archaeological Research Center Nagano. (in Japanese)

Nakagawa, T., Kitagawa, H., Yasuda, Y., Tarasov, P. E., Nishida, K., Gotanda, K., Sawai, Y., Yangtze River Civilization Program Members. 2003 Asynchronous climate changes in the north Atlantic and Japan during the last termination. *Science* 299, pp. 688-691.

Nakagawa, T., Kitagawa, H., Yasuda, Y., Tarasove, P. E., Gotanda, K., Sawai, Y. 2005. Pollen/event stratigraphy of the varved sediment of Lake Suigetsu, central Japan from 15,701 to 10,217 SG vyr BP (Suigetsu varve years before present): description, interpretation, and correlation with other regions. *Quaternary Science Reviews,* 24, pp. 1691–1701.

Nasu, T. 1985 Sendoki jidai no kankyo [Environment of the pre-ceramic period]. *Nihon Kokogaku. Dai X-ji kanko* [Japanese archaeology. Series X. Iwanami koza] vol. 2, pp. 51-109. Iwanami shoten, Tokyo. (in Japanese)

North Greenland Ice Core Project members. 2004 High-resolution record of Northern Hemisphere climate extending into the last interglacial. *Nature* 431, 147-151.

Noshio, S., Terada, K., Tsuji, S., and Suzuki, M. 1997 *Larix-Picea* forests of the Last Glacial Age on the eastern slope of Towada Volcano in northern Japan. *Review of Palaeobotany and Palinology*, 98, pp. 207-222.

Odai Yamamoto I site Excavation Team (ed.) 1999 *Archaeological research at Odai Yamamoto I site.*

Inquiry into the question of the End of Palaeolithic culture and the beginning of the Jomon culture, Odai Yamamoto I site Excavation Team, Tokyo. (in Japanese with English abstract) Okamoto, T. (ed.) 1993 *Archaeological research at the Keio Shonan Fujisawa Campus 1*. Keio University. (in Japanese)

Ono, A., Sato, H., Tsutsumi, T and Kudo, Y. 2002 Radiocarbon dates and archaeology of the Late Pleistocene in the Japanese Islands. *Radiocarbon* 44, pp. 477-494.

Reimer, P.J., Baillie, M.G.L., Bard, E., Bayliss, A., Beck, J.W., Blackwell, P.G., Buck, C.E., Burr, G.S., Cutler, K.B., Damon, P.E., Edwards, R.L., Fairbanks, R.G., Friedrich, M., Guilderson, T.P., Herring, C., Hughen, K.A., Kromer, B., McCormac, F.G., Manning, S.W., Bronk Ramsey, C., Reimer, R.W., Remmele, S., Southon, J.R., Stuiver, M., Talamo, S., Taylor, F.W., van der Plicht, J., Weyhenmeyer, C.E., 2004. IntCal04. Terrestrial radiocarbon age calibration, 0-26 cal kyr BP. *Radiocarbon* 46, pp. 1029-1058.

Sagamihara City Board of Education (ed.) 2004 *Tana-Mukaihara iseki* [Tana-Mukaihara site II] Sagamihara City Board of Education, Sagamihara. (in Japanese)

Sano, K. 2007 Emergence and mobility of microblade industries in the Japanese Islands. In Kuzumin, Y. Keates S. G.. and Shen, C. (eds) *Origin and Spread of Microblade Technology in Northern Asia and Northern America*, pp. 79-90, Archaeology Press Simon Fraser University. Burnaby, B.C.

Sato, H. 1992 *Structure and evolution of Japanese Palaeolithic culture*. Kashiwa-Shobo, Tokyo. (in Japanese)

Sawada, K. Arita, Y. Nakamura, T. Akiyama, M. Kamei, T. and Nakai, N. ^{14}C dating of the Nojiri-ko Formation using accelerator mass spectrometry. *Earth Science* [Chikyu Kagaku] 46, pp. 133-142.

Serizawa, T. and Sudo, T. 2003 *The Araya site: Report of the second and third term excavations, 1988-1989*. Department of Archaeology Tohoku University, Kawaguchi Board of Education, Sendai. (in Japanese with English abstract)

Shinjyuku-ku No. 107 Site Research Group 2000 *Tokyo-to Shinjyuku-ku Kita-Shinjyuku 2 chome iseki 2* [Kita-Shinjyuku 2 chome site 2, Shinkjuku borough, Tokyo] Shinjyuku-ku no.107 Site Research Group, Tokyo. (in Japanese)

Sugihara, S. and Ono, S. 1965 *Shizuoka ken Yasumiba iseki ni okeru Saisekijin bunka*[Microblade culture at the Yasumiba site, Shizuoka prefecture]. Kokogaku-Shukan 3-2, pp. 1-33. (in Japanese)

Suwama, J. 1988 *Sagamino daiti ni okeru Sekkigun no Hensen ni tsuite* [Transition of the lithic industries on the Sagamino plateau]. *Kanagawa Kouko* 24, pp. 1-30. (in Japanese)

Suwama, J. 2004 Historical changing of Obsidian use in Sagamino uplands: an overview. Ambiru, M., Yajima, K., Sasaki, K., Shimada, K. and Yamashina, A. (eds.) Obsidian Summit International Workshop Meiji University Session, Obsidian and its use in stone age of east Asia, Proceedings, pp.15-23.

Suzuki, J. and Yajima, K. 1978 *Sendoki jidai no Sekkigun to sono Hennen* [Lithic industries and its chrlonology of the pre-ceramic period]. Otsuka, H. Tozawa, M. and Aahara, M. (eds.) *Nihon Koukogaku wo Manabu(1)* [Study of the Japanese Archaeology (1)], pp.154-182, Yuhikaku press, Tokyo. (in Japanese)

Takahashi, K. 2007 The formative history of the terrestrial mammalian fauna of the Japanese Islands during the Plio-Pleistocene. *Palaeolithic Research* [Kyusekki Kenkyu] 3, pp. 5-13.

Takahashi, K., Soeda, Y., Izuho, M. Aoki, K., Yamada, G. Akamatsu, G. and Chang, C. H. 2006 The chronological record of the woolly mammoth (*Mammuthus primigenius*) in Japan, and its temporary replacement by *Palaeoloxodon naumanni* during MIS 3 in Hokkaido (northern Japan). *Palaeogeography, Palaeoclimatology, Palaeoecology* 233, pp. 1-10.

Taniguchi, Y. and Kawaguchi, J. 2001 ^{14}C ages and calibrated dates of the oldest pottery culture in the Chojakubo-Mikoshiba period. *Daiyonki Kenkyu* [The Quaternary Research] 40, pp. 485-498. (Japanese with English abstract)

Tohoku University Archaeology Laboratory (ed). 1990. *The Araya site: the results of the second and the third term excavations*. Kawaguchi Town Board of Education. (in Japanese with English abstract), Kawaguchi. (in Japanese)

Tokyo Archaeological Research Center (ed.) 2004 *Musashi Kokubunji-ato Kanren iseki (Musashidai-Nishi-tiku)* [Associated sites with the remains of Musashi provincial monastery (Musashidai-west area)], Tokyo Metropolitan Archaeological Research Center, Tokyo. (in Japanese)

Tsuji, S. and Kosugi, M. 1991 Influence of Aira-Tn ash (AT) eruption on ecosystem. *Daiyonki Kenkyu* [The Quaternary Research] 30, pp. 419-426. (in Japanese with English abstract)

Tsunan Town Board of Education (ed.) 2005 *Chonai iseki Shikutsu Kakunin Chosa Hokokusyo (5)* [Archaeological research report of the sites in the town (5)], Tsunan Town Board of Education, Tsunan. (in Japanese)

van der Plicht, J., Beck, J.W., Bard, E., Baillie, M.G.L., Blackwell, P.G., Buck, C.E., Friedrich, M., Guilderson, T.P., Hughen, K.A., Kromer, B., McCormac, F.G., Bronk Ramsey, C., Reimer, P.J., Reimer, R.W., Remmele, S., Richards, D.A., Southon, J.R., Stuiver, M., Weyhenmeyer, C.E., 2004. NotCal04. Comparison/ Calibration ^{14}C records 26–50 cal kyr BP. *Radiocarbon* 46, pp. 1225-1238.

Voelker, A.H.L., Grootes, P.M., Nadeau, M.-J., Sarnthein, M., 2000 ^{14}C levels in the Iceland sea from 25–53 kyr and their link to the Earth's magnetic field intensity. *Radiocarbon* 42, pp. 437–452.

Wang, Y.J., Cheng, H., Edwards, R.L., An, Z.S., Wu, J.Y., Shen, C., Dorale, J.A., 2001 A high-resolution absolute-dated late Pleistocene monsoon record from Hulu cave, China. *Science* 294, pp. 2345-2348.

Weninger, B. and Jöris, O., 2008. A ^{14}C age calibration curve for the last 60 ka: the Greenland-Hulu U/Th timescale and its impact on understanding the Middle to Upper Palaeolithic. *Journal of Human Evolution.*55, pp.772-781.

Weninger, B., O. Jöris, and U. Danzeglocke 2007 Cologne Radiocarbon Calibration & Paleoclimate Research Package. http://www.calpal.de/, Universität zu Köln Institut für Ur- und Frühgeschichte Radiocarbon Laboratory.

Yamaoka, T. 2006 An innovative process of lithic raw material utilization during the early Upper Palaeolithic on the Musashino upland. *Kodai Bunka* [Cultura Antiqua] 58, pp.107-125

Editor's note: This article was submitted and accepted before the issue of IntCal09.

Terrestrial Mammal Faunas in the Japanese Islands during OIS 3 and OIS 2

Yoshinari Kawamura

Aichi University of Education
1 Hirosawa, Igaya-cho, Kariya, Aichi Prefecture, 448-8542, Japan
E-mail: yskawamr@auecc.aichi-edu.ac.jp

Ryohei Nakagawa

Mie Prefectural Museum
Komei-cho 147-2, Tsu, Mie Prefecture, 514-0006, Japan
E-mail: nakagr00@pref.mie.jp

Abstract: *The terrestrial mammal faunas of OIS 3 and OIS 2 ages are described separately for the three biogeographic regions in Japan (Hokkaido, Honshu-Shikoku-Kyushu and the Ryukyu Islands), on the basis of data from remains reliably dated to these ages. In Hokkaido, the known members of the OIS 3 fauna are restricted to only three extinct species of large herbivores including Mammuthus primigenius which probably immigrated from the northern part of the Eurasian Continent. These three seem to have coexisted during OIS 3 in Hokkaido. The OIS 2 fauna of this region is unknown except for M. primigenius.*

In Honshu-Shikoku-Kyushu, the OIS 3 fauna was dominated by extant species now inhabiting this region, and was of the temperate-forest type. It also included several extinct species, as well as a few species now extinct in the region but extant elsewhere. The OIS 3 fauna of Honshu-Shikoku-Kyushu differed greatly from that of the adjacent Eurasian Continent, and thus it is inferred that this region was isolated from the continent during OIS 3. The OIS 2 fauna of Honshu-Shikoku-Kyushu was identical to the OIS 3 fauna of the same region except for the presence of a few visiting elements in the former exemplified by Alces alces and Bison priscus. The OIS 2 fauna also differed greatly from the contemporaneous faunas of the adjacent continent, but the visiting elements were mostly present in the faunas of northeast China and Primorye. It is therefore inferred that Honshu-Shikoku-Kyushu was isolated from the adjacent continent during OIS 2, but ice bridges formed across the Tsugaru Strait in winter that enabled the immigration of the visiting elements from Hokkaido, which was connected to the continent.

The OIS 3 and OIS 2 faunas of the Ryukyu Islands were of insular type, and the components of the faunas were considerably different from those of the adjacent continent. Thus, it is considered that the islands were separated from the continent by open sea during these periods. The faunas also showed remarkable regional difference between the central and southern parts of the islands. This indicates that both regions were also separated from each other during OIS 3 and OIS 2.

Introduction

In the Japanese Islands, terrestrial mammal remains of Late Pleistocene age are more abundant and better dated than those of earlier periods of the Pleistocene. Thus, these later remains have enabled more detailed and more precise reconstruction of the terrestrial mammal fauna of this period, which was summarily described in Kawamura *et al.* (1989) and Kawamura (1991, 1994, 2004). Updating these works, Kawamura (2007a) presented a brief review on the fauna of the last glacial period including oxygen isotope stages 3 and 2 (OIS 3 and OIS 2), but detailed descriptions were not provided in the publication.

In this paper, therefore, we present more detailed descriptions of terrestrial mammal faunas of OIS 3 and OIS 2, and compare them with the faunas of the same periods in the adjacent Eurasian Continent and on Sakhalin Island. Descriptions are given separately for three biogeographic regions in the Japanese Islands, namely Hokkaido, Honshu-Shikoku-Kyushu and the Ryukyu Islands (Fig. 1).

Hokkaido

Pleistocene terrestrial mammal remains from Hokkaido are much fewer than those from the other regions, and occur exclusively in fluvial or lacustrine sediments and from the bottom of the adjacent sea. Cave and fissure sediments generally yield abundant and diverse remains, but these are unknown in Hokkaido. The localities yielding remains from OIS 3 and OIS 2 are shown in Fig. 2 and Appendix 1. Most of the localities yield only one specimen of one species, and in total only three species are known, namely *Mammuthus primigenius* (woolly mammoth), *Palaeoloxodon naumanni* (Naumann's elephant) and *Sinomegaceros yabei* (Yabe's giant deer). All are extinct species of large herbivores. Our knowledge of the faunas of OIS 3 and OIS 2 is therefore quite limited. Among the

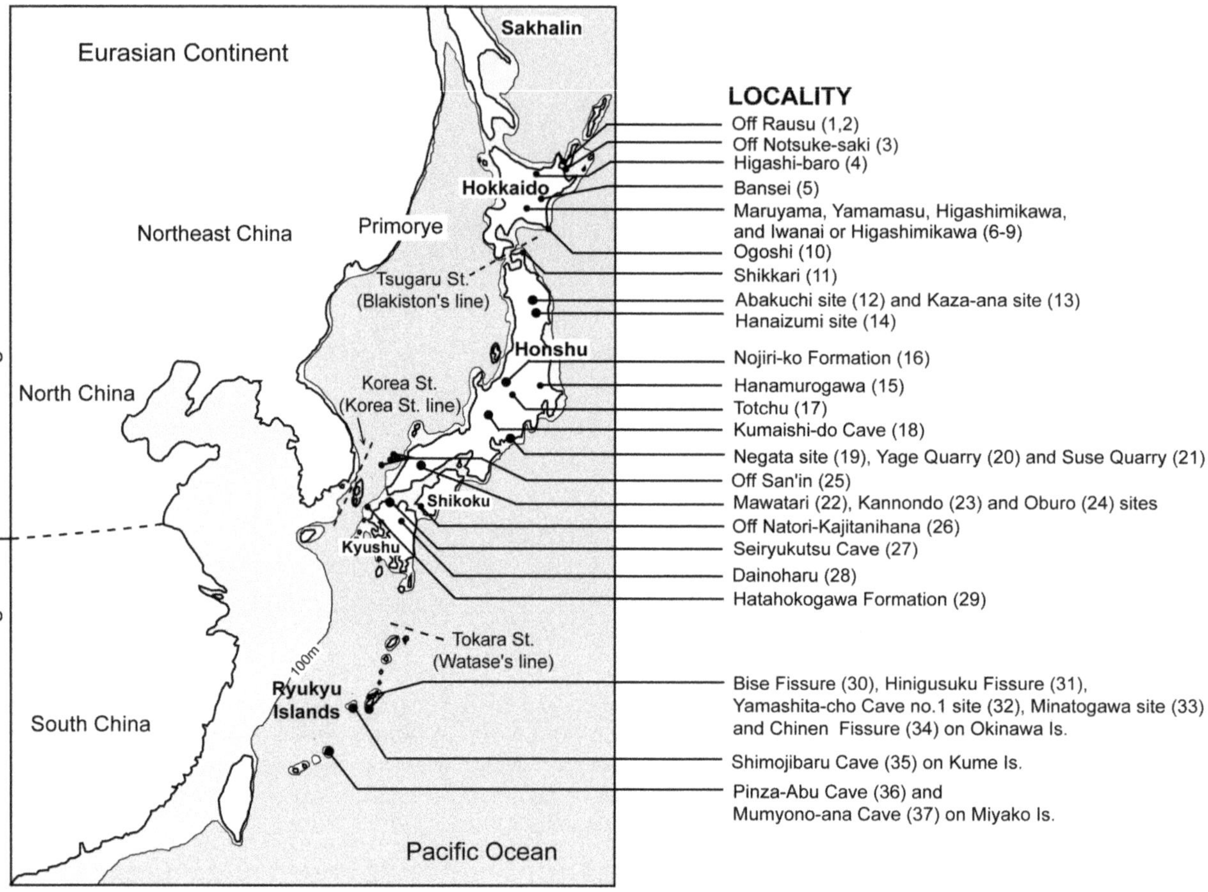

Fig. 1 Geographic distribution of the localities yielding mammal remains of OIS 3 and OIS 2 in the Japanese Islands. The present bathymetric contour of 100m and biogeographic boundary lines are also drawn in and around the islands. From a biogeographic point of view, Honshu, Shikoku and Kyushu Islands formed a single land mass sometimes called the Honshu-Shikoku-Kyushu complex. The numbers in parenthesis indicate the locality numbers given in Fig. 2 and Appendices 1 to 3. Large solid circle: localities yielding more than two forms of mammals, small solid circle: localities yielding less than three forms.

localities listed in Appendix 1, eight are assigned to OIS 3, although two of the eight are uncertain in chronology. These eight localities have yielded the above-mentioned three species, which comprise the OIS 3 fauna of Hokkaido as far as the available data indicate (Fig. 3). Among the species, *M. primigenius* is a representative of the Mammoth Fauna which flourished in the northern part of the Eurasian Continent during the Late Pleistocene (Vereshchagin, 1979; Kahlke, 1994 and others), and its most preferable habitat is generally considered to have been the Mammoth Steppe which is now gone (Guthrie, 1990; Lister and Bahn, 1994; and others). In the Late Pleistocene, *M. primigenius* could migrate southward into Hokkaido from eastern Siberia, because Hokkaido was connected by land with the adjacent continent through Sakhalin. On the other hand, *P. naumanni* and *S. yabei* are the typical elements of the Late Pleistocene fauna of Honshu-Shikoku-Kyushu (see Figs. 4 and 5). These are considered to have inhabited temperate forests as well as subfrigid coniferous forests (Nasu, 1991 and others). We prefer to consider that the three species could coexist in Hokkaido during OIS 3, because it is likely that the climatic and vegetation conditions in Hokkaido were much milder and more wooded than farther north where the main distribution of the Mammoth Fauna existed. This inference is, of course, not decisive until the remains of the three species are found in localities synchronous with each other, or in the same horizon of a single locality.

In northeast China, many localities yield abundant terrestrial mammal remains of OIS 3 age (see the data given in Jin and Kawamura, 1996). The remains are characterized by the frequent occurrence of *M. primigenius* as well as those of larger herbivores such as *Equus przewalskii* (Przewalski's horse), *Coelodonta antiquitatis* (woolly rhinoceros) and *Gazella przewalskii* (Przewalski's gazelle), which are indicative of grassland environments. These herbivores have never been found in Hokkaido. This then seems to characterize the differences in faunal distribution between Hokkaido and the adjacent continent during OIS 3.

On the other hand, the remaining two localities in Appendix 1 are assigned to OIS 2, but their ages approach the boundary between OIS 3 and OIS 2 (Fig. 2). These localities yield *M. primigenius* only. Thus the faunal composition of OIS 2 is unknown except the presence of *M. primigenius*. It is also unknown whether *P. naumanni* and *S. yabei* survived during OIS 2. As pointed out by Kawamura (1991, 1994, 2004, 2007a), these species as well as *M. primigenius*

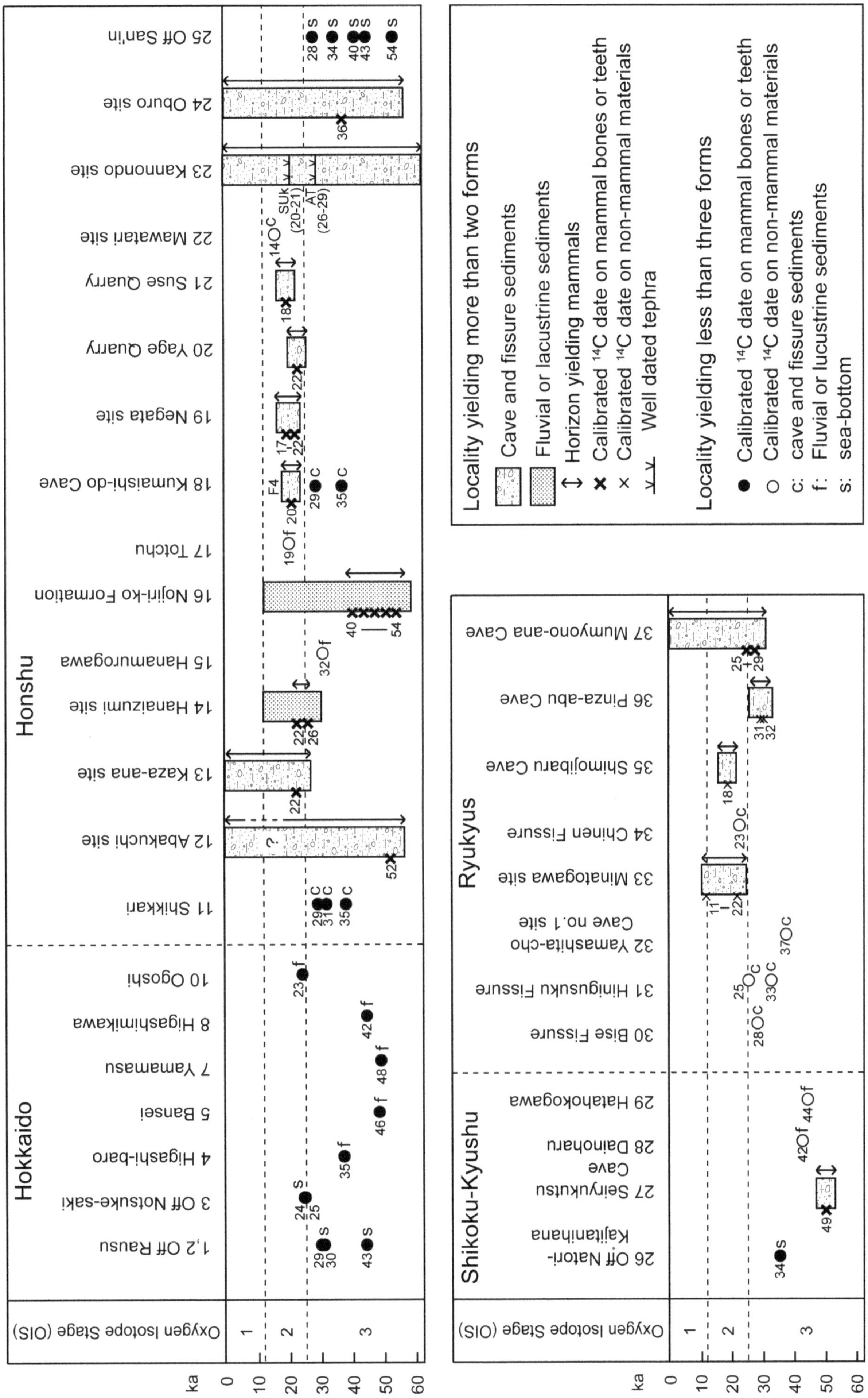

Fig. 2 Chronological distribution of the localities yielding mammal remains of OIS 3 and OIS 2. The number of each locality corresponds to those in Fig. 1 and Appendices 1 to 3. The numbers on the left sides of the geologic columns and the circles (open and solid) indicate approximate calendar dates in ka.

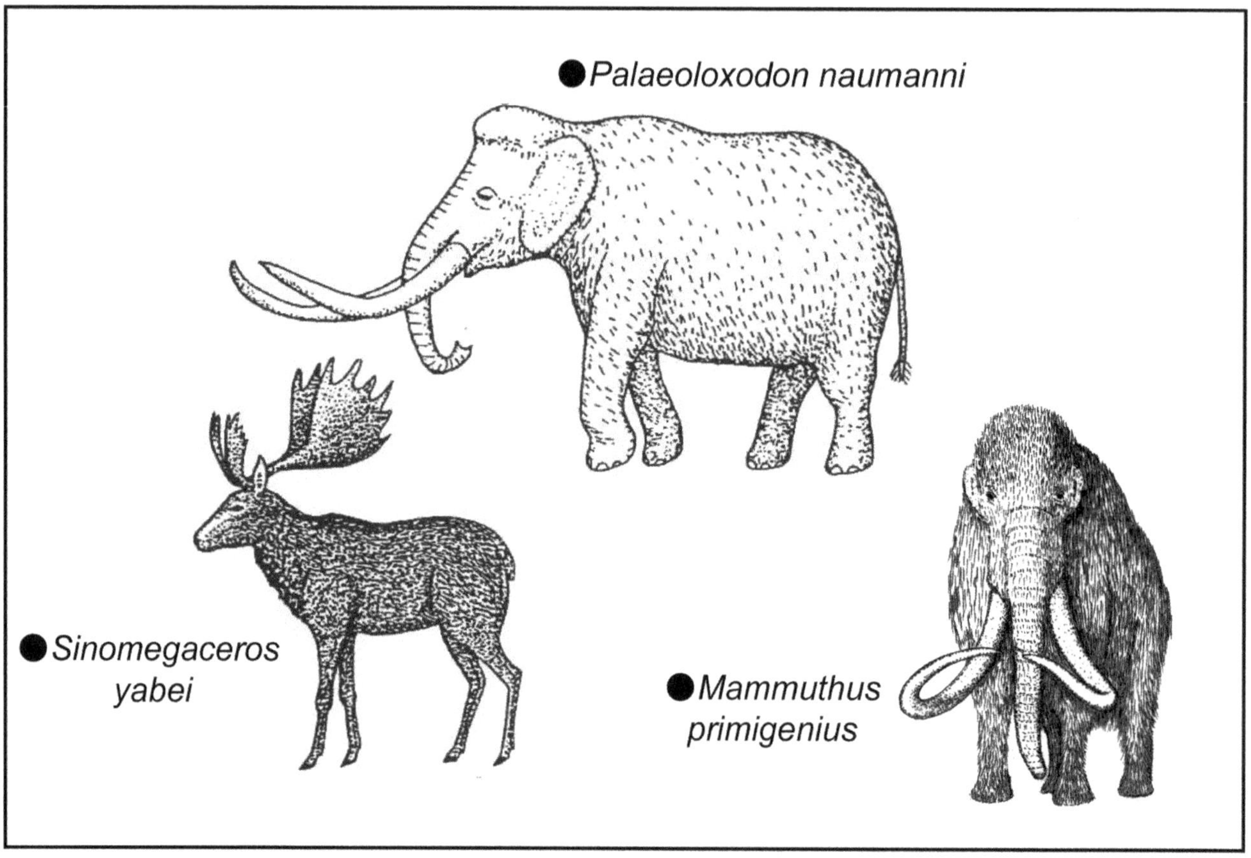

Fig. 3 Mammals of OIS 3 and OIS 2 known in Hokkaido. ●extinct species.

probably became extinct by the onset of the Holocene, because they are absent in the Holocene fauna of Hokkaido.

Takahashi *et al.* (2004, 2005) inferred that the replacement between *M. primigenius* and *P. naumanni* occurred in relation to climatic and vegetation changes during OIS 3 on the basis of the mammal data listed in Appendix 1, as well as composite and originally discontinuous pollen data obtained from several sites not so near the mammal localities. They considered that *M. primigenius* migrated southward with the expansion of open taiga forest during the cold phase of OIS 3, while *P. naumanni* possibly shifted northward from Honshu with the expansion of cool-temperate forests during the warmer phase (ca. 34 to 26 ka). This inference is, however, quite doubtful, because the mammal and pollen data are too scarce and uncertain to link the migration to the climatic and vegetation changes. Considering the pollen data (Igarashi, 1975, 1993; Igarashi and Kumano, 1981; Igarashi *et al.*, 1990, Ooi *et al.*, 1997), we believe that the vegetation changes were not so severe during the Late Pleistocene except OIS 2, as to induce the migration. The chronological positions of some pollen-bearing beds also need to be revised, taking the recent tephrochronology data into consideration (Machida and Arai, 2003 and others).

Honshu-Shikoku-Kyushu

Terrestrial mammal remains of OIS 3 and OIS 2 from Honshu-Shikoku-Kyushu are much more abundant and diversified than those from Hokkaido, and have been obtained from many localities scattered across a wide area from the northern end of Honshu to northern Kyushu (Fig. 1, Appendix 2). The remains occur in fluvial and lacustrine sediments and from the adjacent sea bed, as well as in cave and fissure sediments (Fig. 2). Some of these sediments have successive layers spanning OIS 3 and OIS 2. Thus, our knowledge of the mammal faunas is much more abundant and extensive in Honshu-Shikoku-Kyushu than in Hokkaido.

1) OIS 3 fauna

The mammal-bearing sediments of the following localities are assigned to OIS 3, as shown in Fig. 2 and Appendix 2: Shikkari, the Abakuchi site (lower part of the Late Pleistocene horizons), Hanamurogawa, the Noriji-ko Formation (mammal-bearing layers), Kumaishi-do Cave (two sites other than F_4 and F_3), the Kannondo site (Horizons N to P), the Oburo site (Layer 6), Seiryukutsu Cave (Naumann Branch), Dainoharu, and the Hatahokogawa Formation. Additionally, the remains from the sea-bottom off San'in and Natori-Kajitanihana are also assigned to OIS 3. On the basis of data from these localities, the mammal fauna of OIS 3 is explained below (Fig. 4).

The fauna of OIS 3 was dominated by extant species now inhabiting Honshu-Shikoku-Kyushu, which included many endemic species of Japan. They are exemplified by *Macaca*

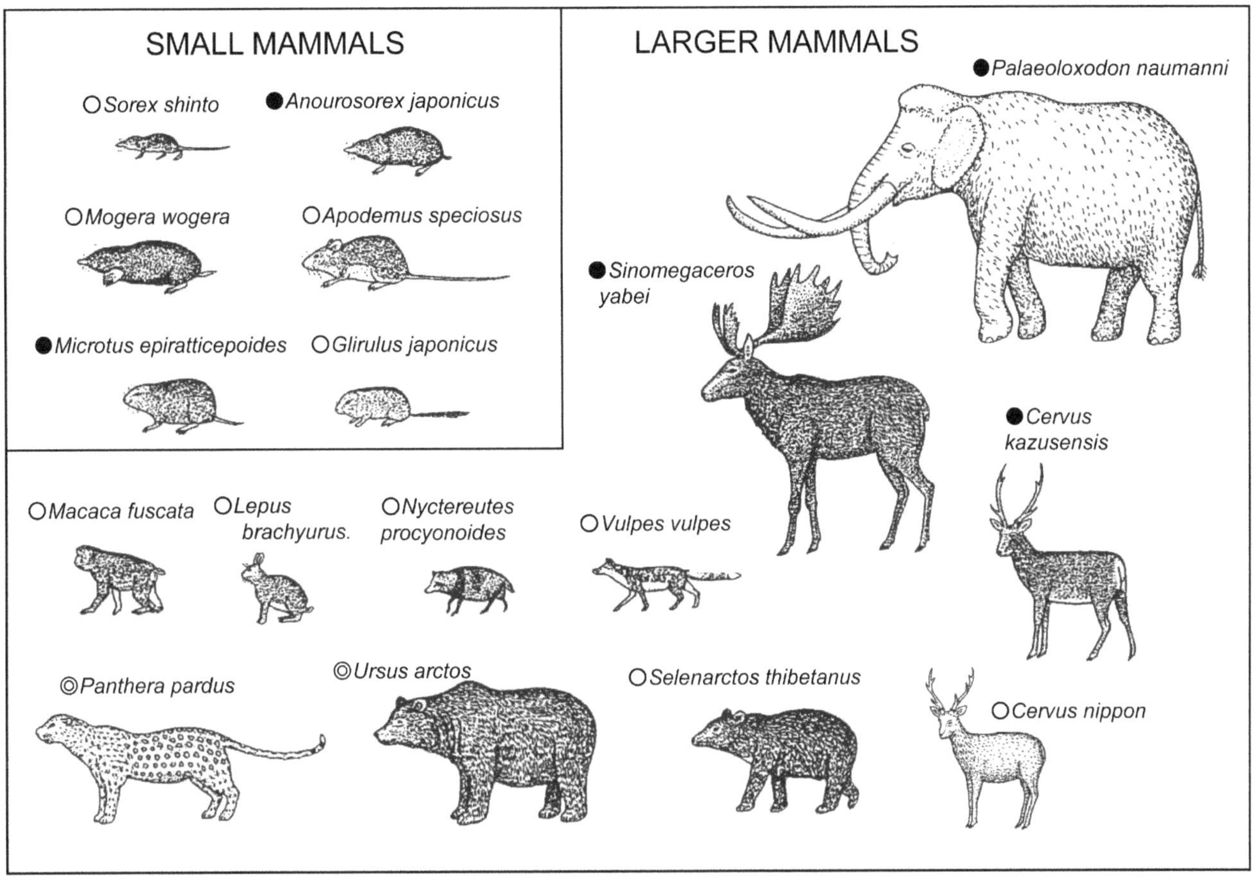

Fig. 4 Representative mammals of OIS 3 in Honshu-Shikoku-Kyushu. ○extant species now inhabiting Honshu-Shikoku-Kyushu, ●extinct species, ◎ extant species now extinct in Honshu-Shikoku-Kyushu.

fuscata (Japanese macaque), *Lepus brachyurus* (Japanese hare), *Apodemus speciosus* (large Japanese field mouse), *Glirulus japonicus* (Japanese dormouse), *Vulpes vulpes* (red fox), *Nyctereutes procyonoides* (raccoon-dog), *Selenarctos thibetanus* (Asiatic black bear) and *Cervus nippon* (sika deer). The predominance of these species indicates that the fauna of OIS 3 can be regarded as the temperate-forest type characteristic to Honshu-Shikoku-Kyushu. Among the extant taxa, *Sorex shinto* (Shinto shrew) and *Dymecodon pilirostris* (lesser Japanese shrew-mole) occur commonly in low-altitude localities. This indicates that the climate of OIS 3 was cooler than present, because these species now inhabit the high mountains of Honshu-Shikoku-Kyushu.

On the other hand, the fauna of OIS 3 included several extinct species, which were exemplified by *Anourosorex japonicus* (mole-shrew), *Microtus epiratticepoides* (grass vole), *Palaeoloxodon naumanni*, *Sinomegaceros yabei* and *Cervus kazusensis* (deer species including *C. praenipponicus* as mentioned by Kawamura, 2009a). Most were endemic to Japan. These species can be regarded as survivors from the Middle and early Late Pleistocene of Honshu-Shikoku-Kyushu. Besides the extinct species, the fauna also included species which are now extinct in Honshu-Shikoku-Kyushu, but survive in other regions. They are represented by *Ursus arctos* (brown bear) and *Panthera pardus* (leopard), which are eurytopic species in the present Eurasian Continent. They are also survivors from the earlier periods of Honshu-Shikoku-Kyushu as the extinct species mentioned above.

We now compare the OIS 3 fauna of Honshu-Shikoku-Kyushu with those of the adjacent areas of the Eurasian Continent, such as Primorye, northeast China, north China and south China (Fig. 1). In Primorye, Geographical Society Cave is the most representative locality, yielding mammal remains listed in Kuzmin (1992) and others. The mammal assemblage of this cave was radiocarbon dated and includes typical elements of the Mammoth Fauna exemplified by *Mammuthus primigenius*, *Equus caballus* (horse), *Coelodonta antiquitatis* and *Bison priscus* (steppe bison). They are not found in the OIS 3 fauna of Honshu-Shikoku-Kyushu (Fig. 4). Besides these, most of the major elements of the assemblage are also absent from the fauna of Honshu-Shikoku-Kyushu. They are *Hyaena* sp. (a hyaena), *Capreolus capreolus* (roe deer), *Cervus elaphus* (red deer), *Alces alces* (elk) and *Nemorhaedus goral* (goral). The remainders of the major elements are *Canis lupus* (wolf), *Ursus arctos* and *Cervus nippon*, which are found in the OIS 3 fauna of Honshu-Shikoku-Kyushu. They are extant species which currently have wide distribution on the Eurasian Continent.

In northeast China, many localities yielding mammal

remains were dated to OIS 3 by the radiocarbon method, and are exemplified by Mingyuegou, Zhoujiayoufang, Huangshan and Guxiangtun (see Jin and Kawamura, 1996). As for small mammals, all the species from these localities are not found in the OIS 3 record of Honshu-Shikoku-Kyushu. As for larger mammals, the fauna of northeast China is characterized by the frequent occurrence of typical elements of the Mammoth Fauna such as *Mammuthus primigenius*, *Equus przewalskii*, *Coelodonta antiquitatis*, *Alces alces* and *Bos primigenius* (aurochs). As mentioned above, they are absent from the OIS 3 record of Honshu-Shikoku-Kyushu. In addition to these, most of the large mammal species are absent from the fauna of Honshu-Shikoku-Kyushu, although two eurytopic extant species of carnivores (*Canis lupus* and *Ursus arctos*) are common to both of the faunas.

In north China, mammal remains of OIS 3 age have been recovered from several localities including Upper Cave of Zhoukoudian (Pei, 1940; Chen *et al.*, 1992), Tianyuan Cave (Tong *et al.*, 2006), and Sjara-osso-gol (Boule *et al.*, 1928; Wu *et al.*, 1989). They were dated by the radiocarbon and/or U-series methods. Almost all small mammal species from these localities are not found in the OIS 3 fauna of Honshu-Shikoku-Kyushu. Larger mammals from the localities usually lack the typical elements of the Mammoth Fauna, and contain a number of species which are absent from the fauna of Honshu-Shikoku-Kyushu such as *Pugma larvata* (masked palm-civet), *Crocuta ultima* (an extinct hyena), *Equus hemionus* (kulan), *Cervus elaphus* and *Capreolus capreolus*. On the other hand, several larger mammals from north China are common to Honshu-Shikoku-Kyushu. They are *Canis lupus*, *Nyctereutes procyonoides*, *Vulpes vulpes*, *Panthera pardus* and *Cervus nippon*, which are extant species now distributed widely on the Eurasian Continent.

In south China, sediments of several caves yielding abundant mammal remains are assigned to OIS 3 on the basis of radiocarbon and/or U-series dates (for example Liujiang, Xiaohuidong and Longtanshan, according to Xue and Zhang, 1991). Some of them, however, are still controversial in age (Shen and Fang, 2001; and others). The mammal assemblages are characterized by the typical elements of the *Stegodon-Ailuropoda* Fauna, which flourished in south China during the Middle and Late Pleistocene (Kahlke, 1961; Xue and Zhang, 1991; and others). The elements are exemplified by *Ailuropoda melanoleuca* (giant panda), *Stegodon orientalis* (an extinct relative of elephants), *Megatapirus augustus* (an extinct tapir) and *Rhinoceros sinensis* (an extinct rhinoceros). These are completely absent from the OIS 3 fauna of Honshu-Shikoku-Kyushu. Moreover, almost all the small mammal forms from south China are not found in Honshu-Shikoku-Kyushu. Only a few eurytopic species alive today are common in south China and Honshu-Shikoku-Kyushu (for example, *Canis lupus* and *Nyctereutes procyonoides*).

Summarizing, we can conclude that the OIS 3 fauna of Honshu-Shikoku-Kyushu was greatly different from those of the adjacent continent. These differences are also observed in the early Late Pleistocene and late Middle Pleistocene. Thus, it is inferred that Honshu-Shikoku-Kyushu was isolated from the continent during OIS 3 and in earlier periods, as already stated in the same context by Kawamura *et al.* (1989), Kawamura (1990, 1992b, 1998) and others.

During OIS 3 and OIS 2, Hokkaido was connected by land with the adjacent continent via Sakhalin, because the depths of the straits between them are much shallower than that of the Tsugaru Strait between Hokkaido and Honshu (Fig. 1). On the other hand, it is generally believed that the Tsugaru Strait was submerged during OIS 3 and OIS 2. Thus it is probable that open water blocked the faunal influx into Honshu-Shikoku-Kyushu from the continent through Hokkaido and Sakhalin, which resulted in the peculiar and conservative characters of the OIS 3 fauna on Honshu-Shikoku-Kyushu.

2) OIS 2 fauna

The mammal-bearing sediments of the following localities are assigned to OIS 2, as shown in Fig. 2 and Appendix 2: The Abakuchi site (upper part of the Late Pleistocene horizons), the Kaza-ana site (Layers 4 and 5), the Hanaizumi site (mammal-bearing layer), Totchu, F_4 of Kumaishi-do Cave, the Negata site, Yage Quarry (Locality 5), Suse Quarry (East Fissure), the Mawatari site (Layer 4), the Kannondo site (Horizons N and M), and the Oburo site (Layer 5). Based on data from these localities, the mammal fauna of OIS 2 is explained below (Fig. 5).

The fauna of OIS 2 was identical with that of OIS 3 except for the presence of four forms mentioned below. The fauna was dominated by temperate-forest species now inhabiting Honshu-Shikoku-Kyushu, similar to OIS3. It also included the same extinct species, and the same extant species now extinct in Honshu-Shikoku-Kyushu, as those in the OIS 3 fauna. The only difference between the OIS 3 and OIS 2 faunas is the presence of three large herbivores (*Alces alces*, *Bos primigenius* and *Bison priscus*) and a lemming (*Myopus* or *Lemmus* sp.) as shown in Fig. 5, all absent from the OIS 3 fauna. Among them, the herbivore species are part of the Mammoth Fauna, although most of its elements, including *Mammuthus primigenius* and *Coelodonta antiquitatis*, are absent in the OIS 2 fauna. The lemming is an inhabitant of cold areas of the Eurasian Continent, but is now absent from Japan including Hokkaido. These four forms occur from limited localities, and are considered to have been minor compornents of the OIS 2 fauna. Kawamura (2007b) regarded them as temporary visitors or visiting elements that appeared shortly in Honshu-Shikoku-Kyushu in association with cold climate.

We now compare the OIS 2 fauna of Honshu-Shikoku-Kyushu with those of Sakhalin and China. On Sakhalin Island, Ostantsevaya Cave yielded mammal remains radiocarbon dated to OIS 2 (Kirillova and Tesakov, 2008).

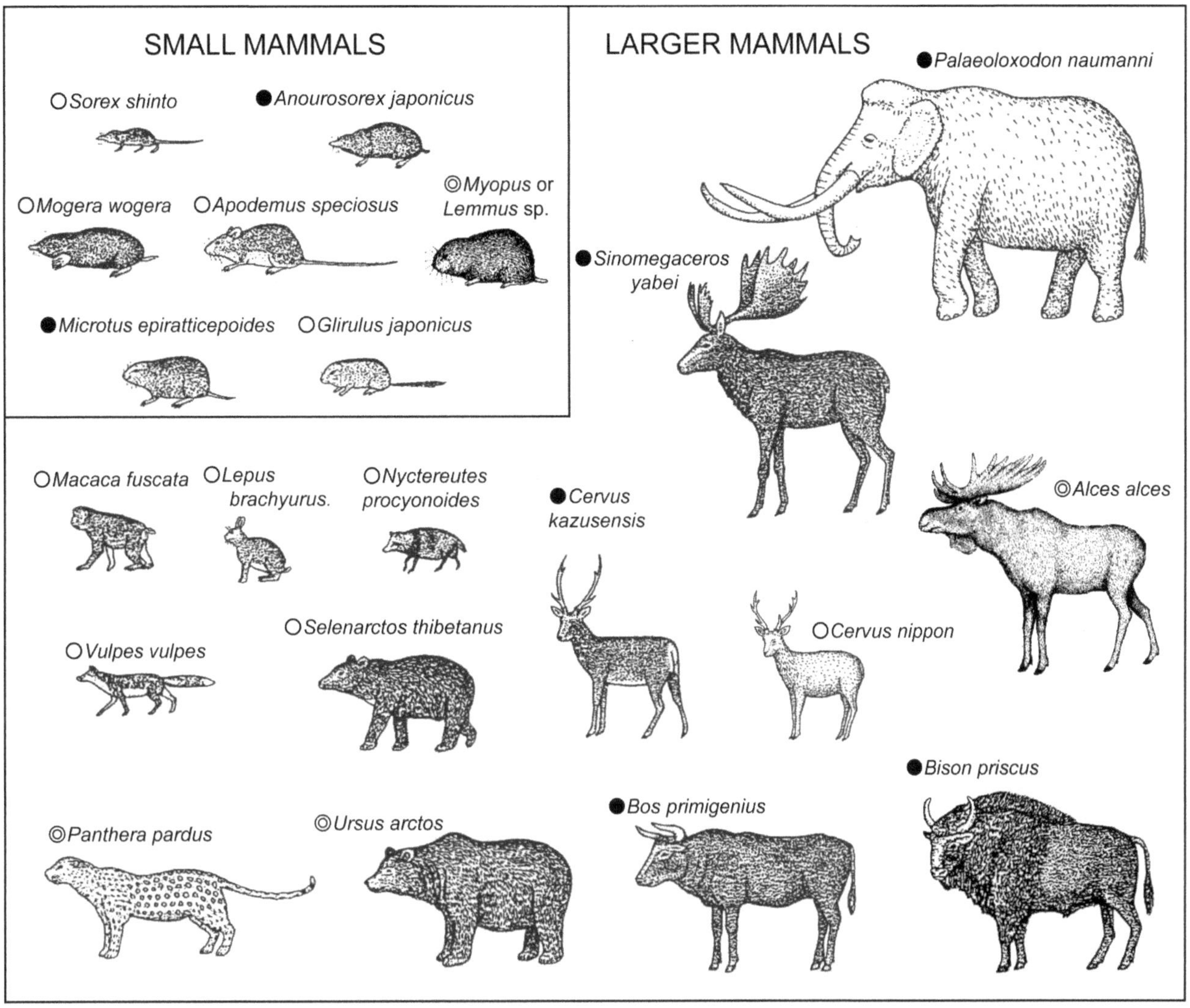

Fig. 5 Representative mammals of OIS 2 in Honshu-Shikoku-Kyushu. ○●◎ same as in Fig. 4.

The specific composition of the remains is greatly different from the OIS 2 fauna of Honshu-Shikoku-Kyushu. In northeast China, mammal remains from several localities were radiocarbon dated to OIS2 (see Jin and Kawamura 1996). These localities are exemplified by Gulongshan, Qingshantou and Yanjiagang. The mammal assemblages from the localities generally contain the typical elements of the Mammoth Fauna (for example, *Mammuthus primigenius* and *Coelodonta antiquitatis*), and are dominated by forms absent from the OIS 2 fauna of Honshu-Shikoku-Kyushu.

In north China, mammal assemblages radiocarbon and/or U-series dated to OIS 2 are known from limited localities, including Hutouliang (Gai and Wei, 1977; Wu *et al.*, 1989), Xiaonanhai (Chou, 1965; Wu *et al.*, 1989) and Xueguan (Wang *et al.*, 1983; Wu *et al.*, 1989). They resemble the OIS 3 fauna of the same area, and are dominated by forms absent from the OIS 2 fauna of Honshu-Shikoku-Kyushu. In south China, remains from several localities were radiocarbon and/or U-series dated to OIS 2 including Baiyanjiao Cave (Li and Cai, 1986), Ma'anshan (Zhang, 1988; Wu *et al.*, 1989) and Sanjiacun (Qiu *et al.*, 1984). The specific composition of the mammal remains from these localities strongly resembles that of the OIS 3 fauna of the same area. These remains are characterized by elements of the *Stegodon-Ailuropoda* Fauna, and most are absent from the OIS 2 fauna of Honshu-Shikoku-Kyushu, as in the case of the OIS 3 fauna of south China.

These comparisons indicate that the OIS 2 fauna of Honshu-Shikoku-Kyushu differs greatly from that of the adjacent continent, as well as that of Sakhalin. Thus, the immigration of mammals into Honshu-Shikoku-Kyushu from the continent could not have been during OIS 2, except for the visiting, temporary elements. These are cold-loving forms, which possibly have high migratory ability. Instead of a stable land bridge, seasonal ice bridges across the Tsugaru Strait (Fig. 1) were used to explain the immigration from the north (Kawamura, 1985, 1989, 1998, and others). It is generally believed that the Tsugaru Strait was not dried up even in OIS2, but it became much narrower than at the present. The ice bridges are thought to have formed in the winter during OIS 2, and to have enabled the southward migration of the visiting elements.

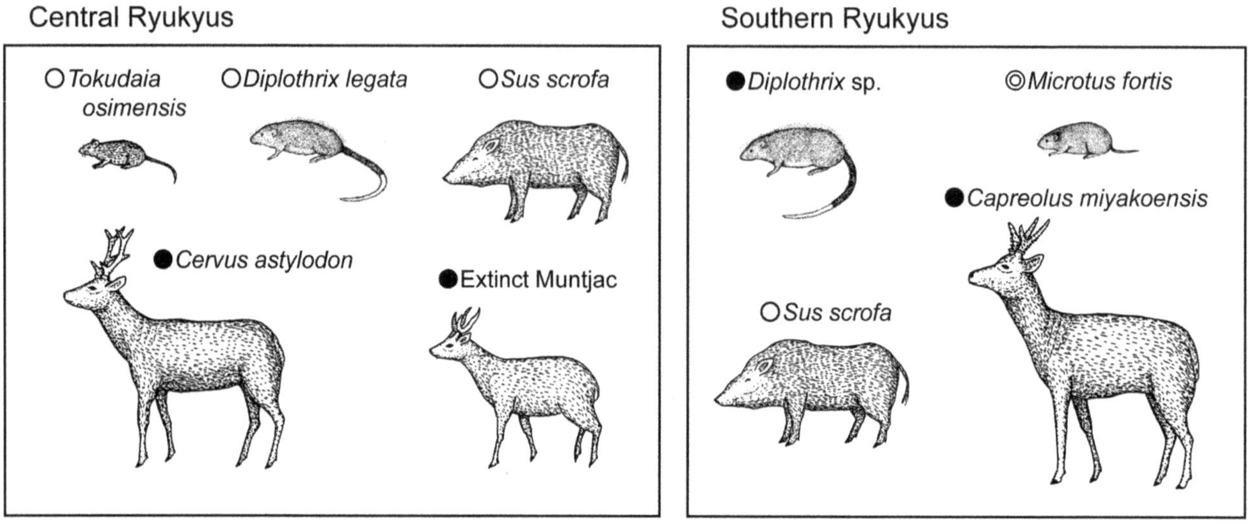

Fig. 6 Representative mammals of OIS 2 and 3 on the Ryukyu Islands. ○extant species now inhabiting the Ryukyu Islands, ●extinct form, ◎ extant species now extinct on the Ryukyu Islands.

The extinctions of mammals occurred in Honshu-Shikoku-Kyushu in the later part of OIS 2. The extinction event was discussed in Kawamura (1992a, 1994, 2007a) and Kawamura and Nakagoshi (1997), and is now being studied vigorously by one of us (Kawamura). The event resulted in a change of faunal composition from OIS 2 to OIS 1, which was possibly related to the cultural change from the Paleolithic to Jomon.

Ryukyu Islands

Terrestrial mammal remains of OIS 3 and OIS 2 occur in cave and fissure sediments on Okinawa and Kume Islands of the central Ryukyus, and on Miyako Island of the southern Ryukyus (Fig. 1). They are relatively rich in number, but much lower in diversity than mammal remains from cave and fissure sediments in Honshu-Shikoku-Kyushu and the adjacent continent (Fig. 6). Moreover, mammals larger than medium-sized deer are absent, but endemic forms of the Ryukyu Islands exist in high numbers. These characters indicate that the faunas of OIS 3 and OIS 2 were of an insular type, as mentioned by Kawamura (1991, 1993, 1994) and others.

1) OIS 3 fauna

On the central Ryukyus, OIS 3 remains are known from Bise Fissure, Hinigusuku Fissure and the Yamashita-cho Cave no.1 site (Fig. 2 and Appendix 3). They comprise only two forms endemic to the Ryukyu Islands, such as *Cervus astylodon* (an extinct deer) and "Muntiacinae" (muntjac) shown in Fig. 6.

On the southern Ryukyus, OIS 3 remains occur in Pinza-Abu Cave and the lower layers of Site A of Mumyono-ana Cave (Fig. 2 and Appendix 3). They are assingned to *Hipposideros* sp. (a leaf-nosed bat), *Microtus fortis* (reed vole), *Diplothrix* sp. (a large rat; probably extinct), *Capreolus miyakoensis* (an extinct roe deer), *Sus scrofa* (wild boar), and others (Fig. 6). Among them, *Diplothrix* sp. and *C. miyakoensis* are endemic to the southern Ryukyus, while *M. fortis*, an extant species now extinct on the Ryukyu Islands, is now distributed widely across the adjacent continent (see Fig. 4 of Kawamura and Nakagawa, 2009). *S. scrofa* is a eurytopic extant species now inhabiting the Ryukyu Islands, as well as Honshu-Shikoku-Kyushu, and the vast area of the Eurasian Continent. Such faunal composition is strikingly different from that of the OIS 3 fauna of the central Ryukyus. This suggests that the central Ryukyus were separated from the southern Ryukyus by the sea during OIS 3.

The OIS 3 faunas of the central and southern Ryukyus are now compared with those of the adjacent continent (north and south China). The former areas are vastly different from the latter in faunal composition, and thus it is inferred that the central and southern Ryukyus were also separated from the adjacent continent during OIS 3. *M. fortis* and *S. scrofa* in the southern Ryukyus are considered to have originated on the continent, but the time of their immigration was probably much earlier than OIS 3 as discussed in Nakagawa *et al.* (2012, this volume).

2) OIS 2 fauna

On the central Ryukyus, OIS 2 remains are known from the Minatogawa site, Chinen Fissure and Shimojibaru Cave (Fig. 2 and Appendix 3). They are *Rhinolophus cornutus* (little Japanese horseshoe bat), *Tokudaia osimensis* (Amami spiny rat), *Diplothrix legata* (Ryukyu rat), *S. scrofa* or *Sus* sp., *C. astylodon*, and "Muntiacinae". *T. osimensis* and *D. legata* are extant species endemic to the central Ryukyus. *S. scrofa* is a eurytopic extant species, and *C. astylodon* and "Muntiacinae" are extinct forms endemic to the Ryukyu Islands.

On the southern Ryukyus, OIS 2 remains are derived from the middle layers of Site A of Mumyono-ana Cave. Most are the same forms as the OIS 3 fauna, and thus the OIS 2 fauna of the southern Ryukyus is greatly different from that of the central Ryukyus, as in OIS 3. The southern Ryukyus are inferred to have separated from the central Ryukyus by the sea even in OIS 2.

The OIS 2 faunas of the central and southern Ryukyus were greatly different from those of the adjacent continent as in OIS 3. It can be inferred that the central and southern Ryukyus were also isolated from the adjacent continent during OIS 2.

In regard to the transition from OIS 3 to OIS 2, Nakagawa *et al.* (2012, this volume) found a remarkable change in the relative abundance of *M. fortis* compared to *Diplothrix* sp. in the southern Ryukyus. The authors considered the shifts to reflect climate and vegetation changes from OIS 3 to OIS 2.

On the central and southern Ryukyus, several mammals became extinct between OIS 2 and OIS 1, as mentioned by Kawamura (1994) and others. They are exemplified by *C. astylodon* and "Muntiacinae" on the central Ryukyus, and by *M. fortis*, *Diplothrix* sp. and *C. miyakoensis* on the southern Ryukyus. The exact time and pattern of their extinction are still unknown, owing to the scarcity of the dated remains.

Conclusion

In Hokkaido, our knowledge of the OIS 3 and OIS 2 faunas is poor. The components of the OIS 3 fauna hitherto known are restricted to *Mammuthus primigenius*, *Palaeoloxodon naumanni* and *Sinomegaceros yabei*, three extinct species of large herbivores (Fig. 3). Among them, *M. primigenius* probably migrated into Hokkaido from the northern part of the Eurasian Continent. *P. naumanni* and *S. yabei* are typical elements of late Middle and Late Pleistocene faunas of Honshu-Shikoku-Kyushu, and seem to have coexisted with *M. primigenius* during OIS 3 under conditions much milder and more wooded than in the northern part of the continent. In regard to the OIS 2 fauna of Hokkaido, *M. primigenius* is known to have existed, but it is unknown whether *P. naumanni* and *S. yabei* survived.

In Honshu-Shikoku-Kyushu, the OIS 3 and OIS 2 faunas are much better understood than in Hokkaido. The OIS 3 fauna of this region was dominated by extant temperate-forest type species now inhabiting the region (Fig. 4). The fauna also included several extinct species, as well as a few extant species now extinct in this region (Fig. 4). They can be regarded as survivors from the Middle and early Late Pleistocene. The OIS 3 fauna of Honshu-Shikoku-Kyushu differed greatly from the contemporaneous faunas of the adjacent Eurasian Continent. It is inferred that Honshu-Shikoku-Kyushu was separated from the continent and Hokkaido during OIS 3.

In Honshu-Shikoku-Kyushu, the OIS 2 fauna was identical to the OIS 3 fauna except for the presence of the visiting elements from the northern part of the Eurasian Continent which were exemplified by *Alces alces* and *Bison priscus* (Fig. 5). The fauna was of the temperate-forest type as a whole, and the visiting elements can be regarded as minor members. The OIS 2 fauna of this region differed greatly from the contemporaneous faunas of the adjacent continent, but most of the visiting elements were also in the OIS 2 fauna of northeast China. It is inferred that Honshu-Shikoku-Kyushu was separated from the continent during OIS 2, but seasonal ice bridges that formed across the Tsugaru Strait in winter enabled the immigration of the visiting elements into Honshu-Shikoku-Kyushu from the continent through Hokkaido.

On the Ryukyu Islands, the faunas of OIS 3 and OIS 2 were characterized by low diversity, lack of large mammals, and a predominance of endemic forms (Fig. 6), and thus considered of an insular type. They showed remarkable regional difference between the central and southern Ryukyus. On the central Ryukyus, extinct and extant forms endemic to this area were prevalent, such as *Cervus astylodon* and *Tokudaia osimensis*. On the southern Ryukyus, the fauna was characterized by extinct forms endemic to the area (*Diplothrix* sp. and *Capreolus miyakoensis*), and by extant species now extinct in the area (*Microtus fortis*). The faunas of the central and southern Ryukyus greatly differed from the contemporaneous faunas of the adjacent continent. It is inferred that the central and southern Ryukyus were separated by the sea from each other, and from the adjacent continent during OIS 3 and OIS 2.

Acknowledgments

We would like to thank Professor S. Matsu'ura (Ochanomizu University) and Mr. H. Taruno (Osaka Museum of Natural History) for reviewing and improving the draft. Thanks are due to Dr. H. Machida (Professor Emeritus of Tokyo Metropolitan University), Dr. M. Oba (Professor Emeritus of Hokkaido University), Professor A. Ono (Meiji University) and Professor H. Yamazaki (Tokyo Metropolitan University), and Associate Professor A. Momohara (Chiba University) for providing valuable information and discussions. We are also grateful to Dr. Y. Kondo (Nojiri-ko Museum), Mr. K. Yasui (Toyohashi Museum of Natural History) and Professor A. A. Vasilevsky (Sakhalin State University) for providing literature. This study was partly supported by a Grant-in-Aid for Scientific Research from Japan Society for the Promotion of Science (project no. 21340145).

References

Abe, Y., 2003, Kaza-ana doketsu nai no sojo to ibutsu no shutsudo-jokyo [Stratigraphy and mode of occurrence of artifacts in the sediments of Kaza-ana Cave]. Dodo, Y., Takigawa, W. and Sawada, J. (eds.) *Search for Japanese*

Pleistocene Human Remains in the Kitakami Mountains: Excavations of the Abakuchi and Kaza-ana Cave Sites in Ohasama, Iwate Prefecture, pp.276-280. Tohoku University Press, Sendai. (in Japanese)

Abe, Y., Nara, T. and Yonekura, K., 2002, Shimokita-hanto ni okeru kyusekkijidai isekikenkyu no juyosei —Iseki-ibutsu to dobutsukaseki no kento kara— [Importance of the researches on Paleolithic sites in the Shimokita Peninsula —Viewed from the examination of sites, artifacts and animal fossils—]. *Shigaku*, **71**, pp.301-319. (in Japanese)

Akagi, S., 1981, *Palaeoloxodon naumanni* (Makiyama) from the sea bottom off the coast of San'in. *The Journal of the Faculty of Education, Tottori University. Natural Science*, **30**, pp.57-64, pls. 1-3. (in Japanese with English abstract)

Akiyama, M., Kamei, T. and Nakai, N., 1988, ^{14}C ages of Naumann's elephant *Palaeoloxodon naumanni* (Makiyama) from the sea bottom off the San'in district in the Sea of Japan by accelerator mass spectrometry —^{14}C age of the Quaternary deposits in Japan (168)—. *Earth Science (Chikyu Kagaku)*, **42**, 29-31. (in Japanese)

Akiyama, M., Kamei, T. and Nakai., N., 1989, Kaiteisan zokaseki no ^{14}C nendai [^{14}C ages of elephant fossils from the bottom of the sea]. *Journal of Fossil Research* (*Fossil Club Bulletin*), **22**, pp.22-23. (in Japanese)

Akiyama, M., Nakamura, T. and Hoshimi, K., 1992, ^{14}C ages of three mammalian fossils from the sea bottom off the the San'in district in the Japan Sea using an accelerator mass spectrometer —^{14}C age of the Quaternary deposits in Japan (175)—. *Earth Science (Chikyu Kagaku)*, **46**, 241-242. (in Japanese)

Anma, T. and Nakagoshi, T., 2001, Excavation at Taishaku-Oburo Cave Site (fifth excavation). *Annual Bulletin of Hiroshima University Taishaku-kyo Sites Research Centre*, **15**, pp.13-26. (in Japanese)

Anma, T., Nakao, A., Yahata, K. and Nakagoshi, T., 2000, Excavation at Taishaku-Oburo Cave Site (third and fourth excavations). *Annual Bulletin of Hiroshima University Taishaku-kyo Sites Research Centre*, **14**, pp.13-34. (in Japanese)

Anma, T., Nakayama, K. and Eguchi, J., 1998, Excavation at Taishaku-Oburo Cave Site (second excavation). *Annual Bulletin of Hiroshima University Taishaku-kyo Sites Research Centre*, **13**, pp.11-18. (in Japanese)

Anma, T., Nakayama, K. and Koizumi, M., 1997, Excavation at Taishaku-Oburo Cave Site (first excavation). *Annual Bulletin of Hiroshima University Taishaku-kyo Sites Research Centre*, **12**, pp.11-17. (in Japanese)

Azuma, Y., 2007, Three new species of fossil terrestrial Mollusca from fissure deposits within the Ryukyu Limestone in Okinawa and Yoron islands, Japan. *Paleontological Research*, **11**, pp.231-249.

Boule, M., Breuil, H., Licent, E. and Teilhard de Chardin, P., 1928. Le Paléolithique de la Chine. *Archives de l'Institut de Paléontologie Humaine*, Mémoire 4, pp.i-viii+1-138, pls.1-30.

Chen, T. M., Hedges, R.E.M. and Yuan, Z. X., 1992, The second batch of accelerator radiocarbon dates for Upper Cave site of Zhoukoudian. *Acta Anthropologica Sinica*, **11**, pp.112-116. (in Chinese with English summary)

Chinzei, K., 1966, The deposits at the Nekata cave-site, Hamakita City, Shizuoka and their geologic age. *Journal of the Anthropological Society of Nippon*, **74**, pp.137-152, 174-175. (in Japanese with English summary)

Chou, P. H., 1965, The fossil remains of vertebrates unearthed from the Palaeolithic cave deposits at Hsiaonanhai in Anyang, Honan Province. *Kaogu Xuebao* (*Chinese Journal of Archaeology*), **1965**, pp.29-49, pls.1-6. (in Chinese with English abstract)

Danzeglocke, U., Jöris, O. and Weninger, B., 2008, CalPal-2007^{online}. http://www.calpal-online.de, accessed 2008-7-27.

Department of Education, Okinawa Prefectural Government (ed.), 1985, *Pinza-Abu: Reports on Excavation of the Pinza-Abu Cave*. 184pp. Department of Education, Okinawa Prefectural Government, Naha. (in Japanese)

Engstrand, L. G. and Östlund, H. G., 1962, Stockholm natural radiocarbon measurements IV. *Radiocarbon*, **4**, pp.115-136.

Fossil Mammal Research Group for Nojiri-ko Excavation, 2000, Vertebrate fossils from the Nojiri-ko Formation during the Lake Nojiri-ko Excavation (1996-1999). *Bulletin of the Nojiri-ko Museum*, no. 8, pp.29-44. (in Japanese with English abstract)

Fossil Mammal Research Group for Nojiri-ko Excavation, 2003, Vertebrate fossils from the Nojiri-ko Formation during the Lake Nojiri-ko Excavation (2000-2002). *Bulletin of the Nojiri-ko Museum*, no. 11, pp.57-77. (in Japanese with English abstract)

Fossil Mammal Research Group for Nojiri-ko Excavation, 2006, Vertebrate fossils from the Nojiri-ko Formation during the Lake Nojiri-ko Excavation (2003-2005). *Bulletin of the Nojiri-ko Museum*, no. 14, pp.55-69. (in Japanese with English abstract)

Fossil Mammal Research Group for Nojiri-ko Excavation, 2008, Vertebrate fossils from the Nojiri-ko Formation during the 16th Lake Nojiri-ko Excavation. *Bulletin of the Nojiri-ko Museum*, no. 16, pp.23-32. (in Japanese)

Gai, P. and Wei, Q., 1977, Discovery of the Late Palaeolithic site at Hutouliang, Hebei. *Vertebrata PalAsiatica*, **15**, pp.287-300, pls.1-2. (in Chinese)

Geology Research Group for Nojiri-ko Excavation, 2004, Geology of the excavation site at Lake Nojiri and its surrounding areas, northern part of Nagano Prefecture —On the redefinition of the Nojiri-ko Formation—. *Bulletin of the Nojiri-ko Museum*, no. 12, pp.1-13. (in Japanese with English abstract)

Guthrie, R. D., 1990, *Frozen Fauna of the Mammoth Steppe: The Story of Blue Babe*. 323pp. The University of Chicago Press, Chicago and London.

Hanaizumi Site Excavation Research Group, 1993, Hanaizumi-iseki [*The Hanaizumi site*]. 161pp. Board of Education of Hanaizumi-cho, Hanaizumi, Iwate Prefecture. (in Japanese)

Hasegawa, Y., 1980, Notes on vertebrate fossils from Late Pleistocene to Holocene of Ryukyu Islands, Japan. *The*

Quaternary Research (*Daiyonki-Kenkyu*), **18**, pp.263-267. (in Japanese with English abstract)

Hasegawa, Y., Sotsuka, T. and Urata, K., 1980, Preliminary report on two fossil baby elephant, *Palaeoloxodon naumanni* from Seiryukutsu-Cave deposits on Hiraodai Karst Plateau, northern Kyushu, Japan. *Bulletin of the Kitakyushu Museum of Natural History*, no. 2, pp.41-47, pl. 5. (in Japanese)

Hasegawa, Y., Tomida, Y., Kohno, N., Ono, K., Nokariya, H. and Uyeno, T., 1988, Quaternary vertebrates from Shiriya area, Shimokita Peninsula, northeastern Japan. *Memoirs of the National Science Museum*, no.21, pp.17-36, pls.1-8. (in Japanese with English summary)

Hemmer, H., 1968, Der Tiger —*Panthera tigris palaeosinensis* (Zdansky, 1924)— im Jungpleistozän Japans. *Neues Jahrbuch für Geologie und Paläontologie, Monatshefte*, **1968**, pp. 610-618.

Hosimi, K. and Morioka, H., 1987, ^{14}C age of *Palaeoloxodon naumanni* (Makiyama) from the sea bottom off the coast of San'in —^{14}C age of the Quaternary deposits in Japan (165)—. *Earth Science (Chikyu Kagaku)*, **41**, 248-250. (in Japanese)

Igarashi, Y., 1975, Palynological study of subsurface geology of the coastal plain along the Ishikari Bay, Hokkaido, Japan. *The Quaternary Research* (*Daiyonki-Kenkyu*), **14**, pp.33-53, pls.1-2.

Igarashi, Y., 1993, Kafunbunseki kara mita Hokkaido no kankyohensenshi [Environmental history of Hokkaido viewed from pollen analysis]. Higashi, S., Abe, H. and Tsujii, T. (eds.) *Hokkaido Viewed from Ecology*, pp.3-21, Hokkaido University Press, Sapporo. (in Japanese)

Igarashi, Y. and Kumano, S., 1981, Vegetational changes during the Last Glacial Age in Hokkaido. *The Quaternary Research* (*Daiyonki-Kenkyu*), **20**, pp.129-141. (in Japanese with English abstract)

Igarashi, Y., Miyata, Y., Noi, H. and Yamada, O., 1990, Fossil pollen and spore assemblages of the Last Glacial Age from the eastern part of the Konsen Plateau, Eastern Hokkaido. *The Quaternary Research* (*Daiyonki-Kenkyu*), **29**, pp.131-138. (in Japanese with English abstract)

Inada, T., 1989, Mode of occurrence of mammalian fossils and Palaeolithic culture in Japan. *Publications of Okayama University, Faculty of Letters*, **2**, pp.1-214, pls. 1-15. (in Japanese)

Inada, T., Kawamura, Y. and Taruno, H., 2005, Harunotsuji-iseki-kaso (Hatahokogawa-so) no kokikoshinsei honyuruikaseki [Late Pleistocene mammals from the Hatahokogawa Formation (underlying sediments of the Harunotsuji archaeological site) on Iki Island, Nagasaki Prefecture, western Japan]. *The Harunotsuji Site : Final Report I*, pp.213-225, Board of Education of Nagasaki Prefecture, Nagasaki. (in Japanese)

Ishikari Lowland Research Group, 1963, A new finding of a molar tooth of *Paleoloxodon* [sic] *namadicus naumanni. Earth Science* (*Chikyu Kagaku*), no.69, p.42. (in Japasene)

Jin, C. Z. and Kawamura, Y., 1996, Late Pleistocene mammal fauna in Northeast China: Mammal fauna including woolly mammoth and woolly rhinoceros in association with Paleolithic tools. *Earth Science* (*Chikyu Kagaku*), **50**, pp.315-330. (in Japanese with English abstract)

Kahlke, H.D., 1961, On the complex of the *Stegodon-Ailuropoda*-Fauna of southern China and the chronological position of *Gigantopithecus blacki* v. Koenigswald. *Vertebrata PalAsiatica*, **1961**, pp.83-108. (in Chinese with English summary)

Kahlke, R.-D., 1994, Die Entstehungs-, Entwicklungs- und Verbreitungsgeschichte des oberpleistozänen *Mammuthus-Coelodonta*-Faunenkomplexes in Eurasien (Groβsäuger). *Abhandlungen der Senckenbergischen Naturforschenden Gesellschaft*, **546**, pp.1-164.

Kamei, T., 1958, Discovery of megacerid deer from Totchu, Nagano-ken, central Japan. *The Journal of Faculty of Liberal Arts and Science, Shinshu University*, no. 8, pp.69-74, pl.1.

Kamei, T., 1967, Nihonkai nanbu kaitei no Naumanzo kaseki [Fossils of the Naumann's elephant from the sea bottom of the southern part of the Japan Sea]. *Tsukumo Chigaku*, no.2, pp.24-31. (in Japanese)

Kamei, T., 1978, On Naumann's elephant, *Palaeoloxodon naumanni* from Churui, Hokkaido. *Monograph of the Association for the Geological Collaboration in Japan*, no.22, pp.345-379, 418. (in Japanese with English abstract)

Kamei, T., 1987, New finds of the wooly mammoths [sic] teeth from the sea bed of the Nemuro strait, Hokkaido, Japan. *Professor Masaru Matsui Memorial Volume*, pp.1-12. (in Japanese)

Kamei, T., 1990, The Japan Sea and elephants. *The Quaternary Research (Daiyonki-Kenkyu)*, **29**, pp.163-172. (in Japanese with English abstract)

Kanto Loam Research Group and Shinshu Loam Research Group, 1962, On the geological age and sedimentary environment of the Hanaizumi Bed. *Earth Science* (*Chikyu Kagaku*), no. 62, pp.1-10; no. 63, pp.10-18. (in Japanese with English abstract)

Kawagoe, T., 1995, ^{14}C data of Taishaku-kyo sites. *Annual Bulletin of Hiroshima University Taishaku-kyo Sites Research Centre*, **10**, pp.109-117. (in Japanese)

Kawamura, Y., 1985, Saishuhyoki iko no nihon no honyudobutsuso no hensen [Succession of the mammalian fauna of Japan since the Last Glacial Period]. *The Earth Monthly*, 7, pp.349-353. (in Japanese)

Kawamura, Y., 1988, Quaternary rodent faunans in the Japanese Islands (Part 1). *Memoirs of the Faculty of Science*, *Kyoto University*, Series of Geology and Mineralogy, **53**, pp.31-348.

Kawamura, Y., 1989, Quaternary rodent faunans in the Japanese Islands (Part 2). *Memoirs of the Faculty of Science*, *Kyoto University*, Series of Geology and Mineralogy, **54**, pp.1-235.

Kawamura, Y., 1990, Nihonretto no honyudobutsuso no oitachi—Tairiku no dobutsuso tono kankei [Origin of the present-day mammalian fauna of the Japanese Islands with special reference to the relationship to continental faunas]. *Mongoloid*, no.5, pp.24-27. (in Japanese)

Kawamura, Y. 1991, Quaternary mammalian faunas in the

Japanese Islands. *The Quaternary Research* (*Daiyonki-Kenkyu*), **30**, pp.213-220.

Kawamura, Y., 1992a, Stratigraphic distribution of mammals in the Taishaku-kyo sites, Hiroshima Prefecture, west Japan. *The Quaternary Research* (*Daiyonki-Kenkyu*), **31**, pp.1-12. (in Japanese with English abstract)

Kawamura, Y., 1992b, Honyudobutsuso no utsurikawari [Historical change of mammalian fauna]. *Science Journal KAGAKU*, **62**, pp.240-244. (in Japanese)

Kawamura, Y., 1993, Higashi-ajia no kokikoshinsei iko no honyudobutsuso [Late Pleistocene and Holocene mammalian faunas in East Asia], *Japanese Scientific Monthly*, **46**, pp. 148-152. (in Japanese)

Kawamura, Y., 1994, Late Pleistocene to Holocene mammalian faunal succession in the Japanese Islands, with comments on the Late Quaternary extinctions. *ArchaeoZoologia*, **6**, pp.7-22.

Kawamura, Y., 1998, Immigration of mammals into the Japanese Islands during the Quaternary. *The Quaternary Research* (*Daiyonki-Kenkyu*), **37**, pp.251-257. (in Japanese with English abstract)

Kawamura, Y., 2003a, Abakuchi doketsu no kokikoshinsei sekitsuidobutsu-itai [Late Pleistocene vertebrate remains from Abakuchi Cave]. Dodo, Y., Takigawa, W. and Sawada, J. (eds.) *Search for Japanese Pleistocene Human Remains in the Kitakami Mountains: Excavations of the Abakuchi and Kaza-ana Cave Sites in Ohasama, Iwate Prefecture*, pp.185-200. Tohoku University Press, Sendai. (in Japanese)

Kawamura, Y., 2003b, Kaza-ana doketsu no kanshinsei oyobi kokikoshinsei no honyurui-itai [Holocene and Late Pleistocene mammalian remains from Kaza-ana Cave]. Dodo, Y., Takigawa, W. and Sawada, J. (eds.) *Search for Japanese Pleistocene Human Remains in the Kitakami Mountains: Excavations of the Abakuchi and Kaza-ana Cave Sites in Ohasama, Iwate Prefecture*, pp.284-386. Tohoku University Press, Sendai. (in Japanese)

Kawamura, Y., 2004, Die Säugetiere der japanischen Inselkette vom Pleistozän zum Holozän. Wieczorek, A., Steinhaus, W. and Sahara, M. (eds). *Zeit der Morgenröte: Japans Archäologie und Geschichte bis zu den ersten Kaisern*, pp.40-44, Reiss-Engelhorn-Museen, Mannheim.

Kawamura, Y., 2007a, Last Glacial and Holocene land mammals of the Japanese Islands: Their fauna, extinction and immigration. *The Quaternary Research* (*Daiyonki-Kenkyu*), **46**, pp.171-177.

Kawamura, Y., 2007b, Recent progress in paleontological studies on the Quaternary mammals of Japan. *Honyurui Kagaku* (*Mammalian Science*), **47**, pp.107-114. (in Japanese)

Kawamura, Y., 2009a, Fossil record of sika deer in Japan. McCullough, D. R., Takatsuki, S. and Kaji, K. (eds.) *Sika Deer: Biology and Management of Native and Introduced Populations*, pp.11-25. Springer, Tokyo, Berlin, Heidelberg, New York.

Kawamura, Y., 2009b, Radiocarbon age of a cheek tooth of a deer from Layer 6 of Taishaku-Oburo Cave Site by accelerator mass spectrometry. *Annual Bulletin of Hiroshima University Taishaku-kyo Sites Research Centre*, **23**, pp.69-75. (in Japanese)

Kawamura, Y., Kamei, T. and Taruno, H., 1989, Middle and Late Pleistocene mammalian faunas in Japan. *The Quaternary Research* (*Daiyonki-Kenkyu*), **28**, pp.317-326. (in Japanese with English abstract)

Kawamura, Y. and Matsuhashi, Y., 1989, Late Pleistocene fissure sediments and their mammalian fauna at Site 5 of Yage Quarry, Inasa, Shizuoka Prefecture, central Japan. *The Quaternary Research* (*Daiyonki-Kenkyu*), **28**, pp.95-102. (in Japanese with English abstract)

Kawamura, Y., Matsuhashi, Y. and Matsu'ura, S., 1990, Late Quaternary mammalian faunas at Suse Quarry, Toyohashi, central Japan, and their implications for the reconstruction of the faunal succession. *The Quaternary Research* (*Daiyonki-Kenkyu*), **29**, pp.307-317. (in Japanese with English abstract)

Kawamura, Y. and Nakagawa R., 2009, Quaternary small mammals from Site B of Mumyono-ana Cave on Miyako Island, Okinawa Prefecture, Japan. *Bulletin of Aichi University of Education (Natural Sciences)*, **58**, pp.51-59.

Kawamura, Y., Nakagawa, R. and Fujita, M., 1996, Preliminary report on the Quaternary vertebrates from the Naumann Branch of Seiryukutsu Cave. *Tokakuji-Shugendo Historic Sites*, pp.106-118, The Kanda-machi Board of Education, Kanda, Fukuoka Prefecture. (in Japanese)

Kawamura, Y. and Nakagoshi, T., 1997, Late Quaternary mammalian extinctions in central and western Honshu, and related problems. *Annual Bulletin of Hiroshima University Taishaku-kyo Sites Research Centre*, **12**, pp.155-168. (in Japanese)

Kirillova, I. V. and Tesakov, A. S., 2008, New mammalian elements of the Ice Age assemblage on the Sakhalin Island. *Mammal Study*, **33**, pp.87-92.

Kobayashi, K., 1965, Radiocarbon date of Totchu Conifer Bed, Akashina-cho, Nagano Prefecture —^{14}C age of the Quaternary deposits in Japan XXV—. *Earth Science* (*Chikyu Kagaku*), no. 81, pp.44-45. (in Japanese)

Kobayashi, H., Hirose, T., Sugino, M. and Watanabe, N., 1974, University of Tokyo radiocarbon measurements V. *Radiocarbon*, **16**, pp.381-387.

Kobayashi, H., Matsui, Y. and Suzuki, H., 1971, University of Tokyo radiocarbon measurements IV. *Radiocarbon*, **13**, pp.97-102.

Kondo, M. and Matsu'ura, S., 2005, Dating of the Hamakita human remains from Japan. *Anthropological Science*, **113**, pp.155-161.

Koyama, Y., Saito, M., Shimai, Y., Totani, M. and Yamada, O., 1977, The measurement of ^{14}C in limestone and its application to ^{14}C dating. *Acta Humanistica et Scientifica Universitatis Sangio Kyotiensis*, Natural Science Series, **6**, pp.90-95. (in Japanese with English abstract)

Kuzmin, Y. V., 1992, Palaeoenvironment of the Late Palaeolithic of Primorye (the former Far East U. S.S.R.). *Man and Environment*, **17**, pp.11-20.

Li, Y. X. and Cai, H. Y., 1986, A Paleolithic site at Puding, Guizhou, *Acta Anthropologica Sinica*, **5**, pp.162-171, pls. 1-2. (in Chinese with English abstract)

Lister, A. and Bahn, P., 1994, *Mammoths*. 168pp. Macmillan, New York.

Machida, H. and Arai, F., 2003, *Atlas of Tephra in and around Japan* (revised edition). 336pp. University of Tokyo Press, Tokyo. (in Japanese)

Matsumoto, H., Mori, H., Marui, K. and Ozaki, H., 1959, On the discovery of the Upper Pliocene fossiliferous and culture-bearing bed at Kanamori, Hanaizumi Town, Province of Rikuchu. *Bulletin of the National Science Museum*, **4**, pp.287-324, pls. 25-48.

Matsu'ura, S., 1981, Asparatic acid racemization dating of fossil bones from the Layer 25 of Taishaku-Kannondo Cave Site. *Annual Bulletin of Hiroshima University Taishaku-kyo Sites Research Centre*, **4**, pp.89-94. (in Japanese)

Matsu'ura, S., 1999, A chronological review of Pleistocene human remains from the Japanese Archipelago. Omoto K. (ed.), *"Interdisciplinary Perspectives on the Origins of the Japanese" International Symposium No. 11-B*, pp.181-197, International Research Center for Japanese Studies, Kyoto.

Matsu'ura, S. and Kondo, M., 2003a, Abakuchi doketsu no kai-taisekibutsu ni kansuru nendaibunseki [Chronological analyses on the lower sediments of the Abakuchi cave site]. Dodo, Y., Takigawa, W. and Sawada, J. (eds.) *Search for Japanese Pleistocene Human Remains in the Kitakami Mountains: Excavations of the Abakuchi and Kaza-ana Cave Sites in Ohasama, Iwate Prefecture*, pp.49-53. Tohoku University Press, Sendai. (in Japanese)

Matsu'ura, S. and Kondo, M., 2003b, Kaza-ana doketsu daiyonso no taisekinendai ni kansuru yobitekibunseki [Preliminary analyses on the age of Layer 4 of the Kaza-ana cave site]. Dodo, Y., Takigawa, W. and Sawada, J. (eds.) *Search for Japanese Pleistocene Human Remains in the Kitakami Mountains: Excavations of the Abakuchi and Kaza-ana Cave Sites in Ohasama, Iwate Prefecture*, pp.281-283. Tohoku University Press, Sendai. (in Japanese)

Matsuzawa, I. and Kosaka, T., 1987, On the Quaternary System in the Cape Erimo District —with special reference to the horizon of *Mammonteus primigenius* (M^2)—. *Professor Masaru Matsui Memorial Volume*, pp.71-78. (in Japanese)

Minato, M., 1955, Zu den Mammonteusfaunen Hokkaidos. *Japanese Journal of Geology and Geography*, **26**, pp.105-113, pl.7.

Nakai, N., Arita, Y., Nakamura, T., Kamei, T., Akiyama, M. and Sawada, K., 1991, AMS radiocarbon ages of mammal fossils from Lake Nojiriko, Nagano Pref. and environmental changes during the last glacial age. *Summaries of Researches Using AMS at Nagoya University*, **2**, pp.26-39. (in Japanese with English abstract)

Nakagawa, R., Yoneda, M., Uno, H. and Shibata, Y., 2007, AMS ^{14}C dating of mammalian remains from Naumann Branch of Seiryukutsu Cave, Hiraodai Karst Plateau, Fukuoka Prefecture, Japan. *Journal of the Speleological Society of Japan*, **32**, pp.35-41.

Nakamura, T., Ohta, T., Miyamoto, M., Minami, M., Oda, H. and Ikeda, A., 1998, AMS ^{14}C age of collagen separated from a molar fossil of Naumann's elephant collected from the Uwa-sea, Ehime Prefecture. *Summaries of Researches Using AMS at Nagoya University*, **9**, pp.286-297. (in Japanese with English abstract)

Nakashima, R., Itoh, M., Kaneko, N., Taru, H., Toshimitsu, S., Nakazawa, T. and Isobe, I., 2004, A fossil elephantoid molar of *Palaeoloxodon naumanni* (Makiyama) collected from the latest Pleistocene deposits of the Hanamurogawa River, Tsukuba City, Ibaraki, Japan. *The Quaternary Research* (*Daiyonki-Kenkyu*), **43**, pp.225-230. (in Japanese with English abstract)

Naora, N., 1959, On the fossils found in Hanaizumi, Iwate Prefecture. *The Quaternary Research* (*Daiyonki-Kenkyu*), **1**, pp.118-124. (in Japanese with English abstract)

Nasu, T., 1991, Naumanzo wo meguru kokankyo [Paleoenvironments of the Naumann's elephant]. Kamei, T. (ed.) *Japanese Proboscidean Fossils*, pp. 170-179, Tsukiji-shokan, Tokyo. (in Japanese)

Niwa, R. and Kawamura, Y., 2001, Quaternary mammals from Taishaku-Oburo Cave Site, Jinseki, Hiroshima Prefecture: Paleontological study on mammalian remains obtained by fine mesh screening (Part 2). *Annual Bulletin of Hiroshima University Taishaku-kyo Sites Research Centre*, **15**, pp.115-133. (in Japanese)

Nojo, A., Takahashi, Y., Tanaka, N., Takanashi, S., Hashimoto, C. and Iwamizawa Research Group, 2002, Stratigraphy and paleoenvironmental changes of the Upper Pleistocene in the eastern end area of the Ishikari Low Land, Hokkaido, Japan. *Earth Science* (*Chikyu Kagaku*), **56**, pp.253-268. (in Japanese with English abstract)

Okumura, K., Ishida, S., Kawamura, Y., Kumada, M. and Tamiya, S., 1982, Latest Pleistocene mammalian assemblage of Kumaishi-do Cave, Gifu Prefecture and the significance of its ^{14}C age. *Earth Science* (*Chikyu Kagaku*), **36**, pp.214-218. (in Japanese with English abstract)

Ono, Y., 1991, Kita no rikkyo [Northern land bridge of Japan]. *Mongoloid*, no.10, pp.37-44. (in Japanese)

Ooi, N., Tsuji, S., Danhara, T., Noshiro, S., Ueda, Y. and Minaki, M., 1997, Vegetation change during the early last Glacial in Haboro and Tomamae, northwestern Hokkaido, Japan. *Review of Palaeobotany and Palynology*, **97**, pp.79-95.

Oshiro, I. and Nohara, T., 2000, Distribution of Pleistocene terrestrial vertebrates and their migration to the Ryukyus. *Tropics*, **10**, pp.41-50.

Otsuka, H., 2002, Ryukyu-retto no kosekitsuidobutsuso to sono kigen [Paleovertebrate fauna of the Ryukyu Islands and its origin]. Kimura, M. (ed.) *The Formation of the Ryukyu Arc and Migration of Biota to the Arc*, pp.111-127, Okinawa Times Co., Naha. (in Japanese)

Otsuka, H. and Nagafuji, T., 1994, Quaternary geology of the Ohno-cho area, Oita Prefecture, Japan, with special reference to the Naumann's elephant-bearing bed. *Reports of the Faculty of Science, Kagoshima University*

(Earth Sciences and Biology), no. 27, pp.133-156. (in Japanese with English abstract)

Otsuka, H., Nakamura, T. and Ota, T., 2004, ^{14}C ages of vertebrate fossil beds in Okinawa Islands, the Ryukyus. *Summaries of Researches Using AMS at Nagoya University*, **15**, pp.41-51. (in Japanese with English abstract)

Pei, W. C., 1940, The Upper Cave fauna of Choukoutien. *Palaeontologia Sinica*, New Ser. C, no.10, pp.1-101.

Qiu, Z. D., Li, C. K. and Hu, S. J., 1984, Late Pleistocene micromammal fauna of Sanjiacun, Kunming. *Vertebrata PalAsiatica*, **22**, pp.281-293. (in Chinese with English abstract)

Shen, G. J. and Fang, Y. S., 2001, Re-dating Late Pleistocene *Ailuropoda-Stegodon* fauna and its implications. Deng, T. and Wang, Y. (eds.). *Proceedings of the Eighth Annual Meeting of the Chinese Society of Vertebrate Paleontology*, pp.143-148, China Ocean Press, Beijing. (in Chinese with English abstract)

Shikama, T. and Tsugawa, S., 1962, Megacerid remains from Gunma Prefecture, Japan. *Bulletin of the National Science Museum*, **6**, pp.1-13. pls. 1-6.

Shiomi, H., 1999, *Taishakukyo-isekigun [The Taishaku-kyo Sites]*. 155pp. Kibijin Shuppan, Okayama. (in Japanese)

Stuiver, M., Deevey, E. S. and Gralenski, L. J., 1960, Yale natural radiocarbon measurements V. *American Journal of Science Radiocarbon Supplement*, **2**, pp.49-61.

Suzuki, H. and Tanabe, G., 1982, Introduction, Suzuki, H. and Hanihara, K. (eds), *The Minatogawa Man: The Upper Pleistocene Man from the Island of Okinawa*. pp.1-5, University of Tokyo Press, Tokyo.

Takahashi, K., Izuho, M., Soeda, Y. and Chang, C. H., 2005, The chronological record of the woolly mammoth (*Mammuthus primigenius*) in Japan, and its new findings. *Journal of Fossil Research* (*Fossil Club Bulletin*), **38**, pp.116-125. (in Japanese with English abstract)

Takahashi, K., Kitagawa, H., Soeda, Y. and Oda, H., 2008. Reexamination of the Churui specimen of *Palaeoloxodon naumanni* (Proboscidea) from Churui, Hokkaido, Japan. *Fossils*, no.84, pp.74-80. (in Japanese with English abstract)

Takahashi, K., Shimaguchi, T. and Kamiya, H., 2006, *Palaeoloxodon naumanni* fossils from Shikkari, Higashi-doori, Shimokita-gun, Aomori Prefecture, Japan and its AMS ^{14}C dating. *Journal of Fossil Research* (*Fossil Club Bulletin*), **39**, pp.21-27. (in Japanese with English abstract)

Takahashi, K., Soeda, Y., Izuho, M., Aoki, K., Yamada, G. and Akamatsu, M., 2004, A new specimen of *Palaeoloxodon naumanni* from Hokkaido and its significance. *The Quaternary Research* (*Daiyonki-Kenkyu*), **43**, pp.169-180.

Takai, F., 1975, Fossil deer from the Yamashita-cho cave no. 1. *Journal of the Anthropological Society of Nippon*, **83**, pp.280-293. (in Japanese with English abstract)

Takai, F. and Hasegawa, Y., 1966, Vertebrate fossils from the Gansuiji Formation. *Journal of the Anthropological Society of Nippon*, **74**, pp.155-167, 176. (in Japanese with English summary)

Takai, F. and Hasegawa, Y., 1978, Giant deer remains of Taishaku-Mawatari Rock-shelter Site. *Annual Bulletin of Hiroshima University Taishaku-kyo Sites Research Centre*, **1**, pp.68-71, pl.10. (in Japanese)

Takigawa, W., Nara, T. and Abe, Y., 2003, Abakuchi doketsu nai no sojo to ibutsu no shutsudo-jokyo [Stratigraphy and mode of occurrence of artifacts in the sediments of Abakuchi Cave]. Dodo, Y., Takigawa, W. and Sawada, J. (eds.) *Search for Japanese Pleistocene Human Remains in the Kitakami Mountains: Excavations of the Abakuchi and Kaza-ana Cave Sites in Ohasama, Iwate Prefecture*, pp.28-32. Tohoku University Press, Sendai. (in Japanese)

Taruno, H.and Majima, N., 1996, Mizuumi shuhen no dobutsuso [Fauna around the lake]. *Urban Kubota*, no. 35, pp.30-43. (in Japanese)

Tong, H. W., Shang, H., Zhang, S. Q., Liu, J. Y., Chen, F. Y., Wu, X. H. and Li, Q., 2006, Mammalian biostratigraphy of Tianyuan Cave, compared with that of Upper Cave at Zhoukoudian (Choukoutien). *Acta Anthropologica Sinica*, **25**, pp.68-81. (in Chinese with English abstract)

Uma-oi Collaborative Research Group, 1987, Late Pleistocene stratigraphy and paleogeography of the eastern marginal area of the Ishikari Lowland, Central Hokkaido, Japan. *Earth Science* (*Chikyu Kagaku*), **41**, pp.303-319. (in Japanese with English abstract)

Vereshchagin, N. K., 1979, Pochemu Vymerli Mamonty [*Why the mammoths died out*]. 195pp. Nauka, Leningrad. (in Russian)

Wang, X. Q., Ding, J. P. and Tao, F. H., 1983, Microliths from Xueguan, Puxian County, Shanxi. *Acta Anthropologica Sinica*, **2**, pp.162-171, pls.1-2. (in Chinese with English abstract)

Wu, R. K., Wu, X. Z. and Zhang, S. S. (eds.), 1989. *Early Humankind in China*. 437pp., 8 pls., Science Press, Beijing. (in Chinese)

Xue, X. X. and Zhang, Y. X., 1991, Quaternary mammalian fossils and the fossil human beings in China. Zhang, Z. H. and Shao, S. X. (eds.) *The Quaternary of China*, pp.307-374, China Ocean Press, Beijing.

Yamada, G., Akamatsu, M., Nakaya, H. and Kumasaki, N., 1996, AMS-^{14}C date of mammoth molar found on off the coast of Rausu, eastern part of Hokkaido. *Bulletin of the Historical Museum of Hokkaido*, no.24, pp.1-8. (in Japanese with English abstract)

Yamasaki, F., Hamada, T. and Fujiyama, C., 1966, RIKEN natural radiocarbon measurements II. *Radiocarbon*, **8**, pp.324-339

Yano, M., 1991, On the occurrence of mammalian fossils from the eastern part of Ishikari lowland in Hokkaido. *The Annual Report of the Historical Museum of Hokkaido*, no.19, pp.9-21. (in Japanese with English abstract)

Yasui, K., Kusuhashi , N. and Matsuoka, H., 2004, The new specimens of *Palaeoloxodon naumanni* (Makiyama) from the Kumaishi-do Cave, Gujo City, Gifu Prefecture, and it's [sic] ^{14}C age. *Abstracts with Programs, the 2004 Annual Meeting, the Palaeontological Society of Japan*, p.48. (in Japanese)

Yasui, K. and Matsuoka, H., 2002, Gifu-ken Hachiman-cho Kumaishi-do kara sanshutsushita okami-kaseki to sono igi [Wolf remains from Kumaishi-do Cave, Hachiman-cho, Gifu Prefecture, and their significance]. *Abstracts with Programs, the 151th Regular Meeting, the Palaeontological Society of Japan*,p. 21. (in Japanese)

Yokoyama, Y., 1992, Gendai-jinrui no kigen wo saguru. Hihakaibunseki niyoru kojinkotsu no nendaisokutei [Quest for the origins of modern humans: Dating of ancient human bones by nondestructive analysis]. *Science Journal KAGAKU*, **62**, pp.195-196. (in Japanese)

Yoneda, M., 2003, Abakuchi doketsu yoji-jinkotsu no nendai bunseki to shokuseifukugen [Chronological analysis and reconstruction of paleodiet of infant human bones from Abakuchi Cave]. Dodo, Y., Takigawa, W. and Sawada, J. (eds.) *Search for Japanese Pleistocene Human Remains in the Kitakami Mountains: Excavations of the Abakuchi and Kaza-ana Cave Sites in Ohasama, Iwate Prefecture*, pp.95-102. Tohoku University Press, Sendai. (in Japanese)

Zhang, S. S., 1988, A brief report of the tentative excavation in Ma'anshan paleolithic site. *Acta Anthropologica Sinica*, **7**, pp.64-74, pl.1. (in Chinese with English summary)

Appendix 1 Detailed data on the localities yielding mammal remains of OIS 2 and OIS 3 in Hokkaido. Calendar dates are calibrated using Danzeglocke *et al.* (2008).

Locality	*Mammal forms obtained*	*OIS allocated*	*Radiocarbon date with calendar date in parenthesis*	*Material dated by radiocarbon method*	*Remarks*	*Literature*
1. Sea-bottom off Rausu (off Shiretoko)	*Mammuthus primigenius*	OIS 3	23,816±884 yBP (28.5±1.1 cal ka) 25,010±120 yBP (30.0±0.3 cal ka)	Molars of *M. primigenius*		Kamei (1987), Nakai *et al.* (1991), Takahashi *et al.* (2005)
2. Sea-bottom 16km off Rausu	*M. primigenius*	OIS 3	38,920±760 yBP (43.2±0.7 cal ka)	Molar of *M. primigenius*		Yamada *et al.* (1996)
3. Sea-bottom north off Notsuke-saki	*M. primigenius*	OIS 2	20,243±670 yBP (24.2±0.8 cal ka) 20,770±120 yBP (24.8±0.2 cal ka)	Molars of *M. primigenius*		Kamei (1987), Akiyama *et al.* (1989), Takahashi *et al.* (2005)
4. Higashi-baro, Yubetsu	*Palaeoloxodon naumanni*	OIS 3	30,520±220 yBP (34.7±0.3 cal ka)	Molar of *P. naumanni*	Sediments yielding the molar uncertain.	Takahashi *et al.* (2004)
5.Bansei,Churui, Makubetsu	*M.primigenius*	OIS3	42,850±510 yBP (46.4±1.2 cal ka)	Molar of *M. primigenius*	*P.naumanni* described by Kamei (1978) occurred from older sediments of the same locality.	Takahashi *et al.* (2008)
6. Maruyama, Kuriyama	*P. naumanni*	OIS 3 or OIS 5			Dated by stratigraphy including tephrochronology; Nojo *et al.* (2002) insisted the earlier age.	Ishikari Lowland Research Group (1963), Uma-oi Collaborative Research Group (1987), Yano (1991), Nojo *et al.* (2002)
7. Yamamasu, Yuni	*M. primigenius*	OIS 3	45,110±480 yBP (48.4±1.6 cal ka)	Molar of *M. primigenius*	A wood from the horizon near that of the molar was dated to 58,450±160 yBP; Takahashi *et al.* (2005) called the locality Higashi mikawa instead of Yamamasu.	Ono (1991), Yano (1991), Takahashi *et al.* (2005)
8. Higashimikawa, Yuni	*M. primigenius*	OIS 3	37,400±250 yBP (42.1±0.3 cal ka)	Molar of *M. primigenius*	Exact position of the site yielding the molar is unknown.	Takahashi *et al.* (2005)
9. Iwanai or Higashimikawa, Yuni	*Sinomegaceros yabei*	OIS 3 ?			Geologic age inferred from stratigraphic correlation with the sediments yielding the molar of *M.primigenius* at Yamamasu.	Ono (1991), Yano (1991)
10. Ogoshi, Erimo	*M. primigenius*	OIS 2	19,580±80 yBP (23.4±0.3 cal ka)	Molar of *M. primigenius*	A wood fragment from the bed possibly yielding the molar was also dated to 21,810±150 yBP by the radiocarbon method.	Minato (1955), Matsuzawa and Kosaka (1987), Takahashi *et al.* (2005)

Appendix 2 Detailed data on the localities yielding mammal remains of OIS2 and OIS 3 in Honshu-Shikoku-Kyushu. Calendar dates are calibrated using Danzeglocke *et al.* (2008).

Appendix 2-1

Locality	*Mammal forms obtained*	*OIS allocated*	*Radiocarbon date with calendar date in parenthesis*	*Material dated by radiocarbon method*	*Remarks*	*Literature*
11. Shikkari (Loc.4), Higashidori	*Palaeoloxodon naumanni* and *Sinomegaceros yabei*	OIS 3	23,570±130 yBP (28.5±3.7 cal ka) 25,920±120 yBP (30.9±3.4 cal ka) 31,470±160 yBP (35.4±4.0 cal ka)	Molars of *P. naumanni*	This locality was also called Loc. 4 (Hasegawa *et al.* 1988).	Hasegawa *et al.* (1988), Abe *et al.* (2002), Takahashi *et al.* (2006)
12. Abakuchi site (Layers VIb to VII-4), Hanamaki	*Selenarctos thibetanus*, ?*Ursus* sp. and other 10 forms	OIS 2 to OIS 3	48,260±1100 yBP (52.3±2.2 cal ka)	Metacarpus of a bear from Layer VII-3	Layers I to VIa are assigned to the Holocene.	Takigawa *et al.* (2003), Yoneda (2003), Matsu'ura and Kondo (2003a), Kawamura (2003a)
13. Kaza-ana site (Layers 4 and 5), Hanamaki	Elephantid, *Alces alces* and other 30 forms	OIS 2	18,140±60 yBP (21.8±0.3 cal ka)	Femur of a elephantid	Layers 1 to 3 are assigned to the Holocene.	Abe (2003), Matsu'ura and Kondo (2003b), Kawamura (2003b)
14. Hanaizumi site, Ichinoseki	*P. naumanni, S. yabei, A. alces, Bos primigenius, Bison priscus* and other 2 forms	OIS 2	18,470±660 yBP (22.1±0.8 cal ka) 21,430±1260 yBP (25.8±1.6 cal ka)	Molar of a bovid Molar of *P. naumanni*	Many radiocarbon dates on woods were obtained from various horizons of the sediments of this site (Stuiver *et al.* 1960; Engstrand and Östlund, 1962; Yamasaki *et al.* 1966).	Matsumoto *et al.* (1959), Naora (1959), Kanto Loam Research Group and Shinshu Loam Research Group (1962), Hanaizumi Site Excavation Research Group (1993)
15.Hanamurogawa, Tsukuba	*P. naumanni*	OIS 3	27,340±860 yBP (32.0±0.8 cal ka)	Wood		Nakashima *et al.*(2004)
16. Nojiri-ko (Tategahana site), Shinano	*P. naumanni, S. yabei* and other 5 forms	OIS 3	34,500±670 yBP (39.7±1.0 cal ka) to 48,800±1950 yBP (53.5±3.3 cal ka): range of ^{14}C dates from successive horizons yielding mammal remains	Molars of *P. naumanni*	These dates are considered to be reliable, instead of many radiocarbon dates on plant materials obtained before.	Nakai *et al.* (1991), Taruno and Majima(1996), Fossil Mammal Research Group for Nojiri-ko Excavation (2000, 2003, 2006, 2008), Geology Research Group for Nojiri-ko Excavation (2004)
17. Totchu, Azumino	*S. yabei*	OIS 2	15,750±390 yBP (19.0±0.4 cal ka)	Wood		Kamei (1958), Shikama and Tsugawa (1962), Kobayashi (1965)

Appendix 2-2

Locality	***Mammal forms obtained***	***OIS allocated***	***Radiocarbon date with calendar date in parenthesis***	***Material dated by radiocarbon method***	***Remarks***	***Literature***
18. Kumaishi-do Cave, Gujo						
·F_4	*P. naumanni, S. yabei, A.alces* and other 24 forms	OIS 2	16,720±880 yBP (20.1±1.0 cal ka)	Bone fragments of *P. naumanni, S. yabei*, and *C. praenipponicus*	F_4 is the main site with abundant remains in well-statified sediments. F_3 is the other site near F_4, whose sediments yield mammal remains somewhat earlier than F_4 (Kawamura, 1988).	Okumura *et al.* (1982), Kawamura (1988)
·Site other than F_4 andF_3	*Canis lupus*	OIS 3	31,010±320 yBP (35.1±0.4 cal ka)	Ulna of *C. lupus*	The remains occurred from differnt sediments from those of F_4 and F_3.	Yasui and Matsuoka (2002)
·Site other than F_4 and F_3	*P. naumanni*	OIS 3	23,960±200 yBP (28.8±0.4 cal ka)	Molar of *P. naumanni*	This site differs from the above mentioned site, and is situated about 20m away from F_4. The only specimen from this site was found on the cave floor (not in sediments).	Yasui *et al.* (2004)
19. Negata (Nekata) site, Hamakita, Hamamatsu (*Homo* bed and *Felis* bed)	*Homo sapiens, Panthera* cf. *pardus* and other 5 forms	OIS 2	13,860±50 yBP (17.1± 0.2 cal ka) 13,970±90 yBP (17.2±0.2 cal ka) 14,050±50 yBP (17.3±0.2 cal ka) 14,200±50 yBP (17.4±0.2 cal ka) 17,910±70 yBP (21.5± 3.4 cal ka)	Human parietal Human occipital Human humerus Metapodial of a deer Phalanx of *P.* cf. *pardus*	Hemmer (1968) regarded *P.* cf. *pardus* from this site as *P. tigris*.	Chinzei (1966), Takai and Hasegawa (1966), Kondo and Matsu'ura (2005)
20. Yage Quarry (Loc. 5), Hamamatsu	*Ursus* cf.*arctos, Cervus* sp. and other 13 forms	OIS 2	18,040±990 yBP (21.6±1.2 cal ka)	Limb bones of *Cervus* sp. and indeterminate mammal bones		Kawamura and Matsuhashi (1989)

Appendix 2-3

Locality	***Mammal forms obtained***	***OIS allocated***	***Radiocarbon date with calendar date in parenthesis***	***Material dated by radiocarbon method***	***Remarks***	***Literature***
21. Suse Quarry (East Fissure), Toyohashi	*Cervus praenipponicus*, and other 13 forms	OIS 2	14,710±670 yBP (17.9±0.7 cal ka)	Long bone fragments of large mammals		Kawamura *et al.* (1990)
22. Mawatari site (Layer 4), Shobara	*S. yabei*	OIS 2	12,080±100 yBP (14.1±0.3 cal ka)	Fresh water shell		Takai and Hasegawa (1978), Kawagoe (1995), Shiomi (1999)
23. Kannondo site (Horizons M to P), Jinseki-kogen	*Ursus* sp., *P.* cf. *pardus*, elephantid and other 39 forms	OIS 2 to OIS 3			These horizons intercalate SUk and AT tephras, which are dated to 20-21 and 26-29 cal ka respectively (Machida and Arai, 2003). A radiocarbon date of calc tufa and an amino acid racemization date of a bone fragment were also obtained from the horizons (Koyama *et al.* 1977: Matsu'ura, 1981)	Kawamura (1988, 1992a), Shiomi (1999)
24. Oburo site (Lower part of Layer 5, and Layer 6), Jinseki-kogen	*Microtus epiratticepoides*, *Lemmus* or *Myopus* sp. and other 11 forms	OIS 2 to OIS 3	31,700±900 yBP (36.2±1.3 cal ka)	Tooth of a deer		Anma *et al.* (1997, 1998, 2000), Anma and Nakagoshi (2001), Niwa and Kawamura (2001). Kawamura (2009b)
25. Off San'in	*P. naumanni*, *M. primigenius*	OIS 3	23,680±880 yBP (28.4±1.2 cal ka) 29,000±300 yBP (33.5±0.4 cal ka) 35,560±1300 yBP (40.2±1.4 cal ka) 38,500±600 yBP (43.0±0.7 cal ka) 48,870±1770 yBP (53.5±3.1 cal ka)	Molar of *M. primigenius* Incisor of *P. naumanni* Incisor of *P. naumanni* Incisor of *P. naumanni* Incisor of *P. naumanni*	Kamei (1990) inferred that the molar of *M.primigenius* was derived from a carcass transported from north China.	Kamei (1967, 1990), Akagi (1981), Hosimi and Morioka (1987), Akiyama *et al.* (1988, 1992),

Appendix 2-4

Locality	***Mammal forms obtained***	***OIS allocated***	***Radiocarbon date with calendar date in parenthesis***	***Material dated by radiocarbon method***	***Remarks***	***Literature***
26. Off Natori-Kajitanihana, Ikata	*P. naumanni*	OIS 3	29,200±870 yBP (33.5±0.8 cal ka)	Molar of *P. naumanni*		Nakamura *et al.* (1998)
27. Seiryukutsu Cave (Naumann Branch), Kanda	*P. naumanni*, *M.epiratticepoides* and other 16 forms	OIS 3	45,500±2200 yBP (49.3±2.9 cal ka)	Limb bone fragment of ?*Cervus* sp.		Hasegawa *et al.* (1980), Kawamura *et al.* (1996), Nakagawa *et al.* (2007)
28. Dainoharu, Bungo-ono	*P. naumanni*	OIS 3	37,250±1880 yBP (41.5±1.9 cal ka)	Wood	Otsuka and Nagafuji, (1994) reported a radiocarbon date (16,890±130 yBP) of "carbonized woods" from the same sediments. This date is unreliable, because it contradicts the tephrochronology of the sediments.	Inada (1989)
29. Hatahokogawa Formation, Iki	*P. naumanni*, *S. yabei*	OIS 3	39,920±440 yBP (43.7±0.6 cal ka)	Wood	Mammal remains were obtained from the uppermost part of the formation.	Inada *et al.* (2005)

Appendix 3 Detailed data on the localities yielding mammal remains of OIS 2 and OIS 3 on the Ryukyu Islands. Calendar dates are calibrated using Danzeglocke *et al.* (2008).

Appendix 3-1

Locality	*Mammal forms obtained*	*OIS allocated*	*Radiocarbon date with calendar date in parenthesis*	*Material dated by radiocarbon method*	*Remarks*	*Literature*
30. Bise Fissure, Motobu	*Cervus astylodon*	OIS 3	23,050±70 yBP (27.6±0.4 cal ka)	Land shell		Otsuka (2002)
31. Hinigusuku Fissure, Kitanakagusuku	*Cervus astylodon* and "Muntiacinae"	OIS 3	28,160±90 yBP (32.6±0.3 cal ka) 20,890±770 yBP (25.1±1.0 cal ka)	Land shell Land shell		Otsuka (2002), Otsuka *et al.* (2004)
32. Yamashita-cho Cave no.1 site, Naha	*Cervus astylodon*	OIS 3	32,100±1000 yBP (36.6±1.4 cal ka)	charcoal		Kobayashi *et al.* (1971), Takai (1975)
33. Minatogawa site, Yaese	*Sus scrofa, Cervus astylodon,* "Muntiacinae" and other 3 forms	OIS 2	18,250±650 yBP (21.8±0.8 cal ka) 16,600±300 yBP (19.9±0.4 cal ka) 10,470±50 yBP (12.4±0.2 cal ka) 9865±35 yBP (11.2±0.0 cal ka)	Charcoal Charcoal Land shell Land shell	Yokoyama (1992) reported a U-Pa date of a human skull (19,200±1800 years ago)	Kobayashi *et al.* (1974), Hasegawa (1980), Suzuki and Tanabe (1982), Otsuka *et al.* (2004)
34. Chinen Fissure, Nanjo	*Tokudaia osimensis* and *Sus* sp.	OIS 2	19,093±325 yBP (22.9±0.4 cal ka)	Land shell		Azuma (2007)
35. Shimojibaru Cave, Kumejima	*Diplothrix* sp., *Cervus astylodon* and "Muntiacinae"	OIS 2	15,200±100 yBP (18.3±0.3 cal ka)	Crab		Matsu'ura (1999), Oshiro and Nohara (2000)
36. Pinza-abu cave, Miyakojima	*Sus scrofa, Capreolus miyakoensis* and other 5 forms	OIS 3	26,800±1300 yBP (31.5±1.2 cal ka) 25,800±900 yBP (30.6±0.9 cal ka)	Charcoal Charcoal		Department of Education, Okinawa Prefectural Government (1985)

Appendix 3-2

Locality	***Mammal forms obtained***	***OIS allocated***	***Radiocarbon date with calendar date in parenthesis***	***Material dated by radiocarbon method***	***Remarks***	***Literature***
37. Site A of Mumyono-ana Cave, Miyako-jima	*Microtus fortis*, *Diplothrix* sp. and other 5 forms	OIS 2 to OIS 3	20,700±110 yBP (24.7±0.2 cal ka) to 23,670±140 yBP (28.6±0.4 cal ka)	Humerus of *Diplothrix* sp.		Nakagawa *et al.* (2012)

A New OIS 2 and OIS 3 Terrestrial Mammal Assemblage on Miyako Island (Ryukyus), Japan

Ryohei Nakagawa

Mie Prefectural Museum
Komei-cho 147-2, Tsu-shi, Mie, 514-0006 Japan
E-mail: nakagr00@pref.mie.jp

Yoshinari Kawamura

Aichi University of Education,
Hirosawa 1, Igaya-cho, Kariya, 448-8542 Japan

Shin Nunami

Asakura Publishing Co., Ltd,
Ogawa-cho 6-29, Shinjuku-ku, Tokyo, 162-8707 Japan

Minoru Yoneda

Department of Integrated Biosciences, Graduate School of Frontier Sciences, The University of Tokyo,
Kashiwanoha 5-1-5, Kashiwa-shi, Chiba, 277-8562 Japan

Motomasa Namiki

Okinawa Prefecture Archeological Center,
Uehara 193-7, Nishihara-cho, 903-0125 Japan

Yasuyuki Shibata

Environmental Chemistry Division, National Institute for Environmental Studies,
Onogawa 16-2, Tsukuba, Ibaraki, 305-8506 Japan

Abstract: *Quaternary mammal remains have been recovered from several localities on Miyako Island, the southern Ryukyus. Among the localities, Pinza-abu Cave had been the only locality yielding rich mammalian remains until we excavated a new locality in a cave called Mumyono-ana. The mammal assemblage of this cave contains at least 11 forms including the reed vole (Microtus fortis), an endemic large rat (Diplothrix sp.) and the wild boar (Sus scrofa). The relative abundance of each form can be analyzed in accordance with the stratigraphic sequence of the cave sediments because the remains are rich enough for the analysis.*

Bones of mammals from the four layers of the sediments were dated by the AMS radiocarbon method. The ages of the bones from the lower part (Layers D and C), middle part (Layer B) and upper part (Layer Z) are approximately 28,700 to 25,300, 24,700 and 10,100 calBP, respectively. The ages of the lower and middle layers indicate that the former and latter layers are assignable to the later part of OIS 3 and early part of OIS 2, respectively. Layer A is unable to be dated due to the scarcity of bones suitable for dating, although it is considered to be almost synchronous with Layer Z because of the lithologic similarity between the two layers. The faunal change of terrestrial mammals from OIS 3 to OIS 2 on Miyako Island is reconstructed based on the assemblages of these layers. The observed faunal change indicates that the vegetational change such as expansion of grasslands with reduction of forests occurred between the two stages.

Introduction

Miyako Island, one of three large islands in the southern Ryukyus, is situated about 300 km southwest of Okinawa Island and about 350 km east of Taiwan (Fig. 1). This island has a triangular outline, and is characterized by low and flat topography with the highest point of only 114.5 m above sea level (Fig. 2). Such topography has enabled the most part

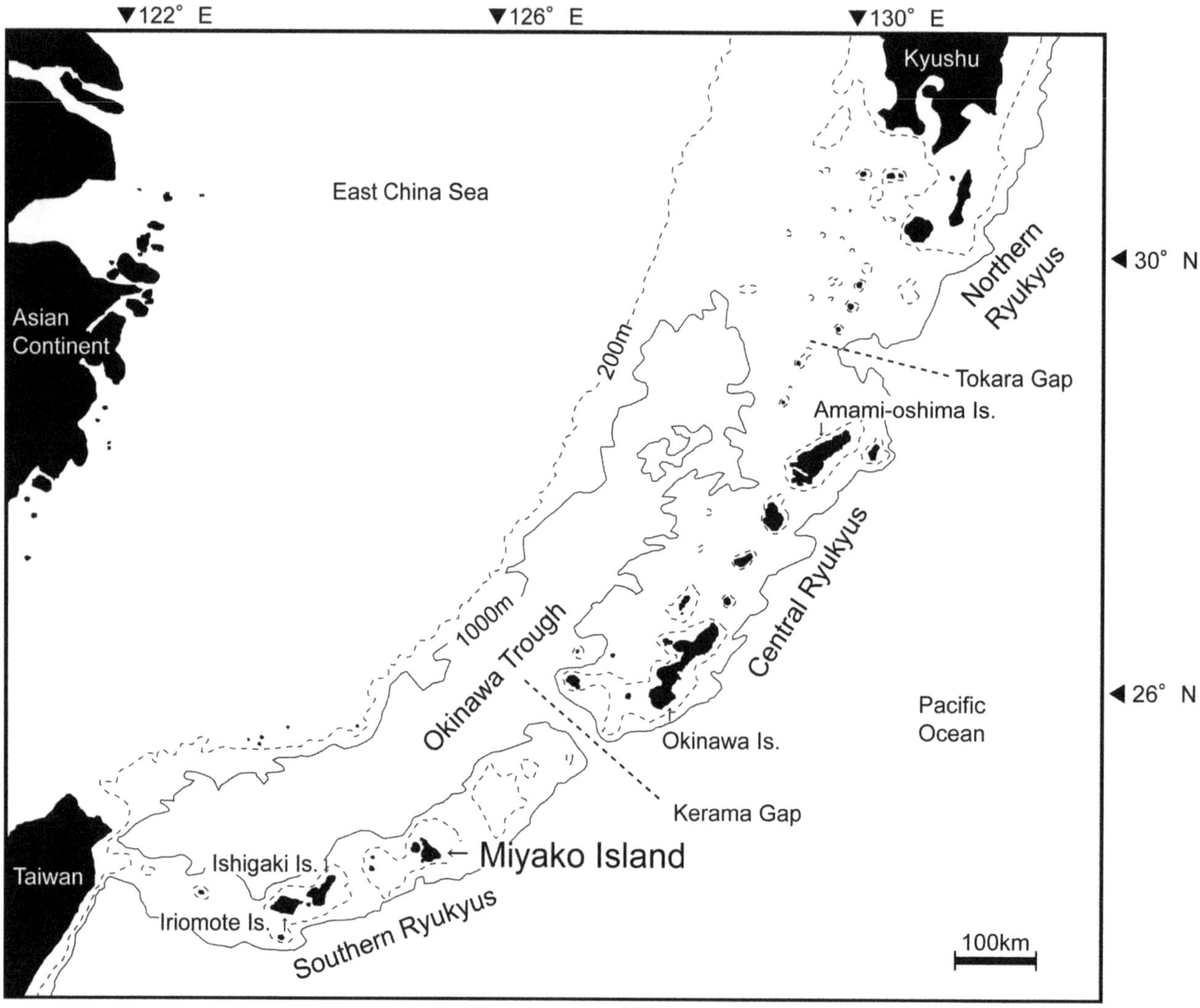

Fig. 1. Map of the Ryukyu Islands showing the location of Miyako Island. Around the islands, bathymetric contours are also shown (200 and 1000m).

of the island to be used for agriculture, which results in a present-day absence of non-commensal terrestrial mammals on the island. From a geological point of view, this island is mostly occupied by the Ryukyu Group consisting of Pleistocene limestone, which overlies Pliocene mudstone and sandstone of the Shimajiri Group. Several caves are formed in the limestone, some of which have yielded Quaternary terrestrial mammal remains.

The first record of terrestrial mammal remains from this island appeared in Tokunaga (1936), who briefly mentioned the occurrence of deer antlers from "old coral reef". The antlers seem to have occurred from the limestone itself, but not from caves. On the other hand, deer remains reported by Tokunaga and Takai (1938) were obtained from three caves on this island and an adjacent small island. Soon after these reports, Tokunaga (1940) and Otuka (1941) described elephant molars from Tanabaru Cave (Fig. 2). The latter author allocated them to *Palaeoloxodon*, although a recent revision regarded them as *Mammuthus trogontherii* (Taruno and Kawamura, 2007). Furthermore, Otuka (1941) described deer remains from the same cave, and referred them to a new species of roe deer, *Capreolus tokunagai*. He also stated that bones of similar roe deer occurred from ten other localities on this island and adjacent small islands, where the bones occurred similarly to those of Tanabaru Cave. Subsequently Hasegawa *et al.* (1973) reported remains of a large rat (*Diplothrix* cf. *legata*) and deer including a roe deer newly named *C. miyakoensis* (this name probably preoccupied by *C. tokunagai*) from three caves including Tanabaru Cave.

Through these works, Miyako Island was known to be important for understanding Pleistocene terrestrial mammal faunas of the southern Ryukyus, because mammal remains were few on other islands of the same area. However, the remains reported in these works were restricted to the large rat, elephant and deer, and were insufficient in specimen number. They also lacked reliable stratigraphic and age controls.

A subsequent discovery of abundant and diversified remains from Pinza-Abu Cave (Fig. 2) greatly increased the knowledge on the Pleistocene terrestrial mammal fauna

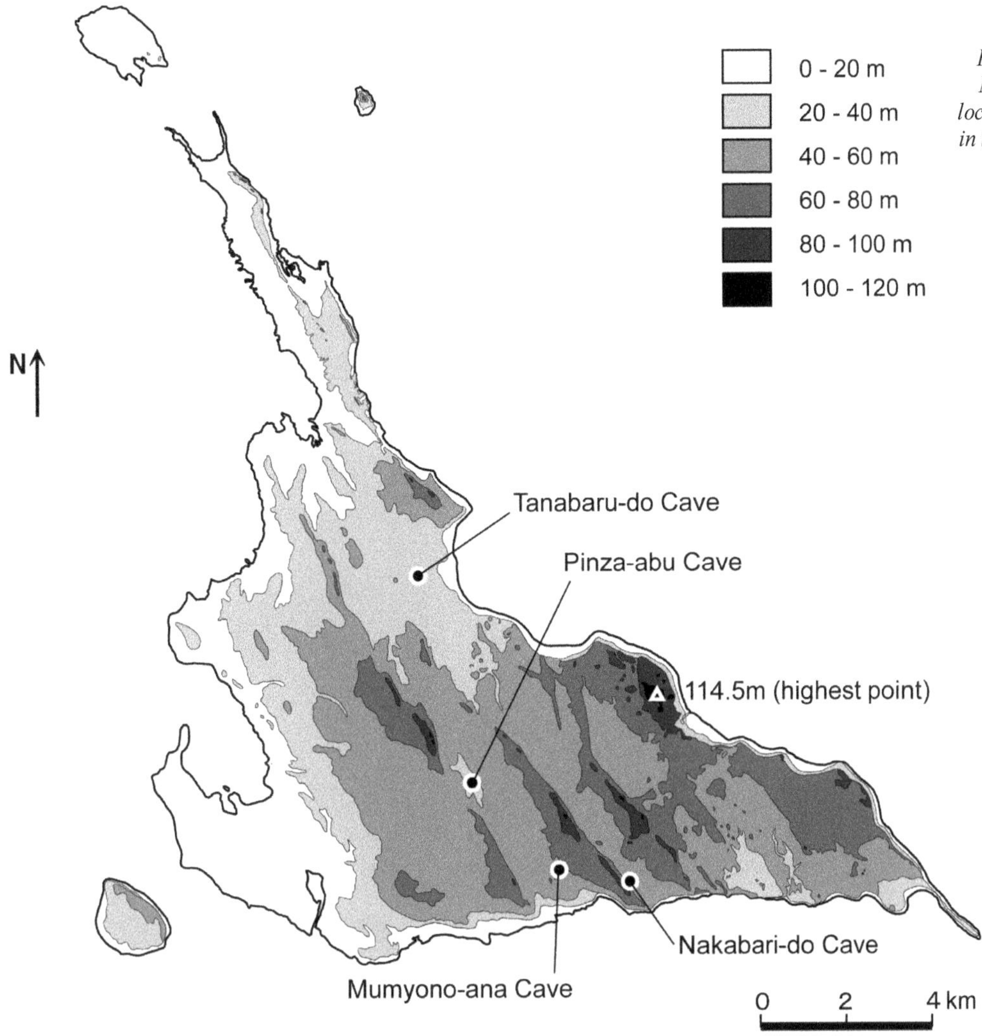

Fig. 2. Topographic map of Miyako Island showing the locations of the caves discussed in the text and the highest point of the island.

of Miyako Island (The Department of Education, Okinawa Prefectural Government, 1985). The remains examined in this work were collected from several sites in the cave, and included 11 forms of terrestrial mammals, most of which were new to the Pleistocene fauna of the island. Two radiocarbon dates of charcoal indicative of the Late Pleistocene were reported, but unfortunately, the sampling sites and horizons of the charcoal were not recorded. Moreover the contemporaneity between the charcoal and mammal remains was uncertain.

In February 2005, two of us (Nakagawa and Nunami) found abundant mammal remains in Mumyono-ana Cave situated near Pinza-abu Cave (Fig. 2). Mumyono-ana with the meaning of an unnamed cave in Japanese was named and described by The Society of Scientific Expedition, Ehime University (1977). After the discovery of the remains, we carefully excavated the sediments of a single site in the cave (Site A in Fig. 3), in accordance with the stratigraphic sequence. The excavations resulted in collecting a large number of diversified mammal remains with detailed stratigraphic records. Some of the remains collected from several horizons were dated by the AMS radiocarbon method, and most of the ages obtained fell into OIS 2 and OIS 3. We believe that the remains provide the best record of the mammal fauna during OIS 2 and OIS 3 among those hitherto known from Miyako Island, because the remains were abundant in all the horizons of the single site which were well-dated by the radiocarbon method. In this paper, we present the stratigraphic and chronological data obtained from the cave, and describe its mammal assemblage. We also discuss the faunal transition from OIS 3 to OIS 2 and its environmental background, as well as the paleogeography around Miyako Island.

Cave description and excavation

Mumyono-ana Cave (Fig. 3), probably formed by the collapse of a doline, opens on the base of the southern wall of an elliptical depression with a length of about 45 m, a width of about 35 m and a depth of about 5 m. The bedrock of the cave is referable to the Tomori Limestone of the Ryukyu Group (Yazaki and Oyama, 1980) or to the middle part of the Miyakojima Limestone of the same group (Nakamori, 1982)

Mumyono-ana Cave is a small inclined cave with a maximum length of about 18 m, and has a keyhole-shaped plan (Fig.3). Its entrance is broad, and measures about 18 m in width and about 6 m in height. From this entrance, the

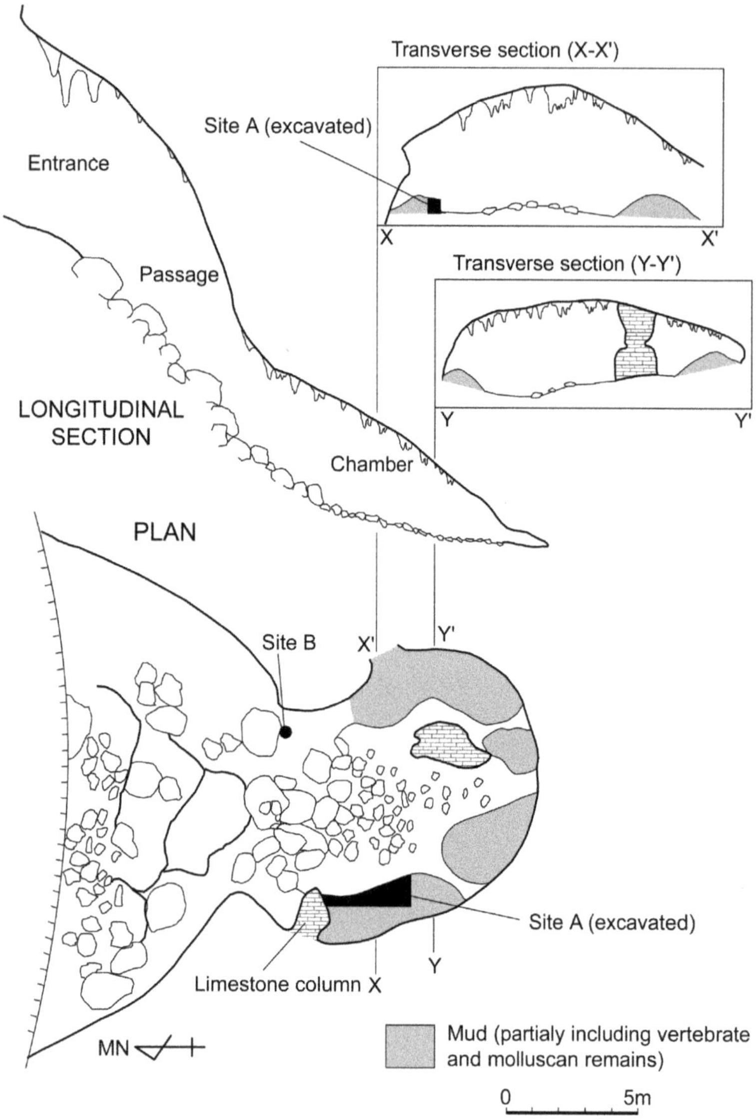

Fig. 3. Plan and sections of Mumyono-ana Cave showing the location of the sites yielding mammal remains.

cave inclines steeply and narrows toward the inside to be a passage with a length about 10 m. Fallen blocks and large breccias pile up on its floor. The passage grades into a small chamber with a round plan whose diameter is about 10 m. Smaller limestone breccia and reddish brown brecciated mud cover the floor of the chamber. The former and latter are distributed in the central and peripheral parts of the chamber, respectively. This suggests that the brecciated mud was originally deposited all over the floor, and subsequent water inflow from the entrance of the cave washed away finer sediments of the brecciated mud from the central part. The ceiling of the chamber is ornamented with stalactite. Two large columns are present in the chamber.

In February 2005, our reconnaissance to this cave revealed that a number of small mammal remains were included in the sediments collected from Sites A and B (Fig. 3). Among them, Site A is more suitable for excavations because thick brecciated mud is exposed much more extensively at this site. In April and July of the next year, the first and second excavations were carried out at Site A in order to collect vertebrate remains as well as to record the stratigraphic sequence of the sediments. These excavations were conducted in the area with the maximum length of 3m, maximum width of 1.5m and maximum depth of 70 cm from the surface of the sediments (Fig. 3).

Stratigraphy

The stratigraphic sequence of the sediments at Site A was recorded during the excavations, as shown in Fig. 4. The sediments were divided into seven layers, although the boundaries between the layers are unclear. The layers overlap imbricately toward the inside of the cave. This suggests that the sediments of each layer were transported in the direction from the entrance to chamber. The sediments of each layer are described as follows:

1) Layer D

This layer is more than 20 cm in thickness and is observed at the northern end of Site A. It consists of brown sticky mud with pebble- to cobble-sized breccias. This layer yields land shell and mammal remains.

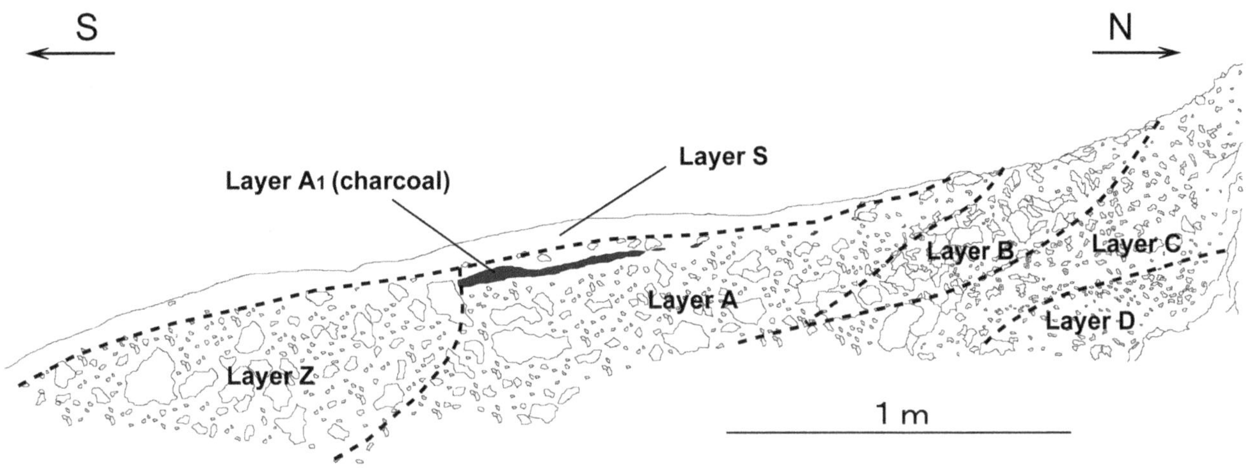

Fig. 4. Section of the sediments observed at Site A.

2) Layer C

This layer is 10 to 30 cm in thickness, and inclines southward at angles of 20° to 30° . It consists of brown sticky mud with pebble- to cobble-sized breccias. The breccias increase in size southward, and their maximum diameters attain 20 cm at the southern most part of the layer. Abundant mammal bone fragments and land shells occur from this layer. The abundant occurrence of land shells, distinguishes this layer from Layer B.

3) Layer B

This layer is about 20 cm in maximum thickness. It thins off southward and westward, and then disappears. Thus its distribution is limited to a small area with a length of about 1 m in north-south direction and a width of about 0.4 m in east-west direction. This layer consists of cobble-sized breccias with brown mud as the matrix, which contains mammal bone fragments and land shells.

4) Layer A

This layer is about 50 cm in maximum thickness. It gradually thins off northward and then disappears. It consists of brown mud with breccias of 1 to 5 cm in diameter. The diameters of the breccias increase to about 20 cm southward and downward. This layer is distinguishable from Layer B by the large amount of charcoal contained.

5) Layer A_1

A black band of charcoal is intercalated in the uppermost part of Layer A. The band named Layer A_1 is very thin along the north-south section of Site A (Fig. 4), but gradually thickens westward up to 30 cm.

6) Layer Z

Layer Z is distributed in the southern part of Site A. It consists of brown mud with breccias of less than 20 cm in diameter, and contains a lot of charcoal. Its lithofacies is similar to that of Layer A. But this layer is distinguished from Layer A because it shows indications of disturbance after depositon, and because its lower surface clearly cuts Layer A_1 (Fig. 4).

7) Layer S

This layer gradually decreases in thickness northward and disappears near the northern end of Site A (Fig. 4). But it thickens westward to be more than 30 cm in width. This layer consists of brown mud with granule-sized breccias. Two pairs of a thin sand band and mud band are intercalated in the layer. The thin sand bands contain charcoal. Shells of a giant snail were obtained from the lowest part of Layer S. Giant snails were introduced into the Ryukyu Islands in the twentieth century. This fact suggests that this layer was deposited in the twentieth century.

Technique for collecting mammal remains

About 1.2 metric tons of sediments were sampled from the layers (150 kg from Layer D, 285 kg from Layer C, 130 kg from Layer B, 115 kg from Layer A, 155 kg from Layer A_1, 280 kg from Layer Z and 110 kg from Layer S), and transported to some places suitable for screen-washing on the island. The sediments were washed in water by using 0.5 mm mesh screens at the places. Residues on the screens were dried up naturally, and then all vertebrate remains were picked up from the residues with the naked eyes.

^{14}C dating by AMS measurement

The ^{14}C dating method adopted in this study is based on Longin (1971) and Yoneda *et al.* (2002), and is described in detail in Nakagawa *et al.* (2007). For the purpose of extracting pure collagen, we first washed the samples with 1M NaOH, and eliminated mineral matter from them with 1M HCl in a cellulose tube. Then the residue was heated

Table 1. ^{14}C age, state of preservation of collagen, and calendric age calibrated by CalPal-2007online (Danzeglocke et al., 2008).

Layer	*Sample no.*	*Laboratory code*	*Sample(taxon / portion)*	*Contents ratio (%)*	*C/N ratio (mol)*	^{14}C age (yBP)	*Calendric age (calBP)*
Z	KAZ01	TERRA-050707b	? *Sus scrofa*, limb bone fragment	0.38	3.37	8,914± 49	10,050±100
B	KAB01	TERRA-091306a	*Diplothrix* sp., humerus	0.79	3.40	20,700±110	24,690±240
C	KAC02	TERRA-052207a	*Diplothrix* sp., humerus	0.63	3.90	21,180± 91	25,340±340
C	KAC03	TERRA-091306a	*Diplothrix* sp., humerus	0.81	3.38	23,420±130	28,200±200
D	KAD01	TERRA-091306a	*Diplothrix* sp., humerus	0.53	3.47	23,670±140	28,650±400
D	KAD02	TERRA-091306a	*Diplothrix* sp., humerus	1.95	3.89	22,130±110	26,650±460
D	KAD03	TERRA-052207a	*Diplothrix* sp., humerus	0.10	4.14	23,210±100	27,960±210

Table 2. Number of the specimens (total of maxillae, mandibles and cheek teeth) allocated to each mammalian form in each layer.

Specimens/Layer	*S*	*Z*	A_1	*A*	*B*	*C*	*D*
Order Insectivora							
Suncus sp.	-	2	1	1	-	-	-
Order Chiroptera							
Pteropus sp.	-	4	7	5	3	3	-
Hipposideros sp.	1	22	51	40	--	12	4
Rhinolophus sp.	-	9	19	19	-	4	-
Miniopterus sp.	-	-	-	1	-	-	-
Order Rodentia							
Microtus fortis	8	43	50	72	29	34	28
Diplothrix sp.	3	14	3	4	-	148	124
Ruttus sp.	1	2	4	-	-	-	-
Mus sp.	1	1	-	-	-	-	-
Order Artiodactyla							
Sus scrofa	-	14	7	3	-	2	-
Cervidae, gen. et sp. indet.	-	-	-	-	-	-	1

in water at 90°. The gelatinized solution obtained was regarded as collagen.

The collagen was converted into CO_2 by an elemental analyzer (Elemental Vario EL; Yoneda *et al.*, 2004), and the carbon and nitrogen ratio was measured. The CO_2 was reduced by hydrogen with iron catalysts into graphite (Kitagawa *et al.*, 1993). The $^{14}C/^{12}C$ of the graphite ratio was measured by an AMS system at NIES-TERRA (Tanaka *et al.*, 2000).

The results of the dating are shown in Table 1. Remains of the rat (*Diplothrix* sp.) and wild boar (?) were used for the samples of the present dating. On the other hand, those of the vole (*Microtus fortis*) were not used, because they are too small to extract enough amount of the collagen.

The samples were not in so good range of collagen content and carbon-nitrogen ratio, but the rat remains from the Layer D were dated to almost the same age. The age deviations due to contamination of the samples can be regarded as negligibly small, although they show low content of collagen or varied carbon-nitrogen ratio.

The samples from Layers C and D and from Layer B were dated to approximately 28,700 to 25,300 calBP and 24,700 calBP, respectively. Thus these layers are considered to be deposited in the later part of OIS 3 and the early part of OIS 2. The sample from Layer Z was dated to approximately 10,100 calBP, which indicates to be of early Holocene age (within OIS 1).

Mammal assemblage

Among the remains obtained from Site A, our taxonomic study was conducted exclusively for maxillae, mandibles and isolated teeth of mammals. The study has revealed that they include at least 11 forms belonging to 10 genera as shown in Table 2. This table also shows the number of the identified specimens from each layer.

1) Insectivores

Only four remains are allocated to insectivores. All of them are obtained from the upper layers of Holocene age (Layers A, A_1 and Z), and are identified as *Suncus* sp. They show the following characteristics in the mandibles and lower molars: horizontal ramus large and M_3 not so reduced as in genus *Anourosorex*. Compared with an extant specimen of *Suncus murinus* from Ishigaki Island (Fig. 1), the remains are similar to the extant specimen in mandibular and dental morphology, but the remains are too incomplete to determine their specific position. *Suncus* sp. possibly immigrated into Miyako Island in association with humans during the Holocene, because some species of *Suncus* are now commensal .

2) Chiropterans

All the layers yield chiropteran remains, which increase in number toward the upper layers. Four forms are recognized in the remains. Among the four, *Pteropus* sp. is a large-sized form. Its remains are relatively few, and are concentrated especially in the upper layers (Layers A_1, A and Z). Molars of *Pteropus* sp. show a simple shape with two large cusps. Compared with an extant specimen of *Pteropus dasymallus* from Ishigaki Island, the remains are slightly different from the extant specimen in the degree of the coalescence of the two cusps. Further examination is required in order to determine whether the difference is intraspecific or interspecific.

Hipposideros sp. is a medium-sized form. Its remains are common in the assemblage, and are predominant especially in the upper layers (Layers A_1, A and Z). They show the following characteristics: ascending ramus of mandible relatively low; only two premolars present, of which P_2 is smaller than P_4. In comparison with an extant specimen of *Hipposideros turpis* from Ishigaki Island, the remains are slightly different from the extant specimen in molar patterns. Further examination is needed to identify the remains at a species level.

Rhinolophus sp. is a small-sized form in the chiropterans. Its remains are common in the assemblage, and possess the following characteristics in the maxillae and mandibles: ascending ramus low; P^2 and P_3 extremely small. The remains were compared with extant specimens of *R. pumulus* and *R. cornutus*. Some morphological differences are observed between the remains and the specimens of *R. pumulus* and *R. cornutus*. Recently Hirasawa *et al.*, (2006) reported *Rhinolophus* sp. from Nakabari-do Cave near Mumyono-ana Cave (Fig. 2), but its taxonomic relationship to the present remains is still unclear. We need more remains including radii for precise identification.

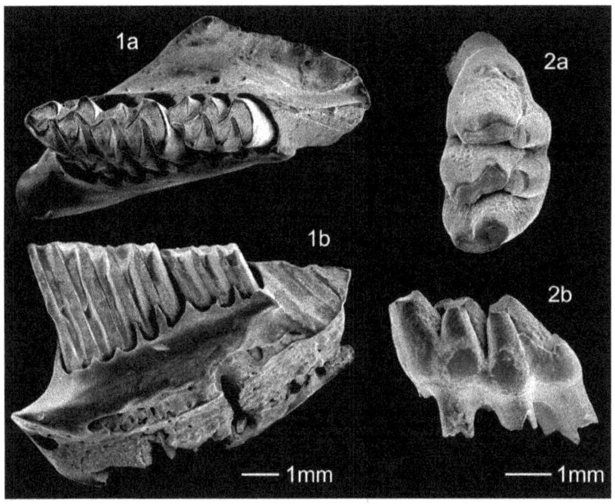

Fig. 5. Microtus fortis (1) and Diplothrix sp. (2) from Mumyono-ana Cave. 1a: occlusal view of the right mandible with M_1 and M_2 from Layer C (MKN-467), 1b: lingual view of the same specimen, 2a: occlusal view of the upper right M^1 from Layer D (MKN-984), 2b: buccal view of the same specimen.

Miniopterus sp. is a medium- to small-sized form. It comprises only one fragmental mandible from Layer A. Its horizontal ramus is elevated posteriorly.

3) Rodents

Arvicolid remains are abundant in the assemblage. They are highly predominant especially in the upper and middle layers (Layers A and B). The mandibles and molars of the remains show the characteristics well coincident with those of *Microtus* (Fig. 5); namely the molars without roots but with rich crown cementum; their enamel well differentiated in thickness and showing the positive or *Microtus*-type differntiation; their salient angles sharply pointed; M^3 and M_1 with three and five triangles respectively, usually well separated from each other; and mandible deeply pocketed between the molar row and ascending ramus. The present remains show considerable homogeneity in molar morphology, where a magnitude of variation seems to be comparable to that of a single living *Microtus* species. Thus they are referable to a single species of *Microtus*. In the present remains, M^3 has a C- or F-shaped posterior loop, and its Is4 is often open (for terminology of arvicolid molars, see Fig. 75 of Kawamura, 1988). M_1 also has a relatively simple anterior loop, where LRA5 is shallow or faint, and BRA4 is lacking or faint. These features as well as their size agree with those of *M. fortis*, and thus the remains are allocated to this species. Department of Education, Okinawa Prefectural Government (1985) and Kaneko and Hasegawa (1995) examined arvicolid remains from Pinza-Abu Cave, and referred some of them to *M. fortis*, *M. oeconomus* and

M. sp. However, all the remains seem to be conspecific with the present remains, judging from the descriptions and figures of these authors. Further investigation is required to dissolve this specific discrepancy.

Large murid remains are abundant in the assemblage. In contrast to the remains of *M. fortis*, however, they are highly predominant especially in the lower layers (Layers C and D). Their maxillae, mandibles and molars resemble those of *Diplothrix legata*, the only species of the genus *Diplothrix*, from Okinawa Island (Fig. 5), but differs from the genus *Rattus* in having larger size and more complicate molar patterns. In spite of the similarity to *D. legata*, the present remains are different from *D. legata* in the anterior border of M^1 locating anteriorly to the posterior end of the incisive foramen; in having larger molars; better developed metacone in M^1 and simpler cusp patterns of the upper molars. We consider that these differences are sufficiently specific, and thus the remains are assigned to *Diplothrix* sp.

Medium-sized murid remains are few, and occur exclusively from the upper layers (Layers S, A_1 and Z). M_2 of the remains possess three roots. This character distinguishes the remains from the genus *Niviventer*, but coincide with the character of the genus *Rattus*. The coincidence of other dental characters with *Rattus* indicates the allocation of the remains to *Rattus*, but the specific determination cannot be made owing to the scarcity of the remains. *Rattus* sp. is possibly immigrated in association with humans, because several species of *Rattus* are now commensal.

Only two remains of a small-sized murid occurred from the upper layers (Layers S and Z). The remains are assigned to *Mus* sp., because M_1 has neither medial anteroconid nor buccal accessory cusps. Present commensal habit of *Mus* species suggests that *Mus* sp. immigrated with human as *Rattus* sp. did.

4) Artiodactyls

Remains of *Sus scrofa* are common in the assemblage, and more abundant in the upper layers (Layers A_1 and Z). The sizes of the remains are as small as those of the living Ryukyu wild boar, *Sus scrofa riukiuanus*, from Ishigaki, Iriomote and Okinawa Islands. Remains of *Sus scrofa* from Pinza-abu Cave (Department of Education, Okinawa Prefectural Government, 1985) are considerably larger than the present remains.

Remains allocated to Cervidae, gen. et sp. indet. comprise only two specimens (one incisor and one molar), and might be referred to "*Capreolus miyakoensis*" previously reported from this island.

Discussion

1) Temporal faunal change

The above-mentioned taxonomic study has shown remarkable faunal differences between the assemblages of the lower and upper layers. In the lower layers (Layers D and C), *Diplothrix* sp. is predominant, while *M. fortis* is much more abundant than *Diplothrix* sp. in the overlying layers (Layers B and A). This indicates that *Diplothrix* sp. prevailed on this island in the later part of OIS 3, but it decreased to be a minor element in OIS 2 and the early Holocene. During the latter periods, *M. fortis* was predominant on the island. *Sus scrofa* are few in the lower layers (Layers D and C), while it is abundant in the upper layers (Layers Z and A). The upper layers also yield *Suncus* sp. and two murid forms (*Rattus* sp. and *Mus* sp.). Furthermore chiropterans are abundant in these layers, although this seems to reflect the increase of suitable habitats for chiropterans in this cave. Layer S, the uppermost layer, yielded only few mammals. *Microtus fortis* and *Diplothrix* sp. from this layer may be secondarily derived from the underlying layers, because there is no evidence that these forms survived until the later Holocene on this island.

2) Paleogeographic and paleoenvironmental implications

This study has revealed that *Diplothrix* sp. decreased remarkably between the later part of OIS 3 and the early part of OIS 2, while *Microtus* steadily increased from the later part of OIS 3 to the early Holocene. *Diplothrix legata* is now a forest dweller on Amami-oshima and Okinawa Islands. *Diplothrix* sp. seems to live in the same habitat, because it resembles *D. legata* in dental morphology and there is no evidence indicative of a different habitat from that of *D. legata*. On the other hand, the genus *Microtus* is generally considered to be the voles highly adapted to grass-eating habit. *M. fortis* prefers to inhabit grasslands along rivers and swamps rather than dry steppes. The remarkable change in the relative abundance of *Diplothrix* sp. against *M. fortis* suggests that forests reduced remarkably between OIS 3 and OIS 2 in relation to the expansion of relatively wet grasslands on Miyako Island.

On the Ryukyu Islands, the localities yielding rich and diversified mammalian remains of the later part of OIS 3 and the early part of OIS 2 are restricted to the Minatogawa site on Okinawa Island (Hasegawa, 1980), and Pinza-abu Cave (Department of Education, Okinawa Prefectural Government, 1985) and Mumyono-ana Cave on Miyako Island. As regards to cervids, the Minatogawa site yields *Cervus astylodon* and Muntiacinae, gen. et sp. indet., while Pinza-abu Cave yields "*Capreolus miyakoensis*". As regards to rodents, *Diplothrix legata* and *Tokudaia osimensis* occur from the Minatogawa site. On the other hand, Pinza-abu and Mumyono-ana Caves yield *Microtus fortis* but no *Tokudaia*. It is likely that *Diplothrix* from Pinza-abu and Mumyono-ana Caves are specifically different from *D. legata,* although remains of *Diplothrix* from the former cave are originally identified as *D.* cf. *legata* (Department of Education, Okinawa Prefectural Government, 1985). Thus the fauna of Okinawa Island is quite different from that of Miyako Island in the later part of OIS 3 and the early part of OIS 2, and it is inferred that Okinawa Island was already separated from Miyako Island by the sea during this period.

On the other hand, Department of Education, Okinawa Prefectural Government (1985) inferred that a land bridge was formed between Miyako Island and the adjacent Asian Continent during the "Würm Glacial Period" (roughly corresponding to OIS 3 and OIS 2), through which "northern elements" such as *Capreolus*, *Microtus* and *Sus* immigrated southward into Miyako Island. This scenario cannot be accepted for the following reasons:

1. The OIS 3 and OIS 2 faunas of Mumyono-ana and Pinza-Abu Caves are of insular type, because non-flying terrestrial mammals of the faunas are composed of smaller forms, and are much less diversified than Late Pleistocene continental faunas of southern China.
2. "*Capreolus miyakoensis*" and *Diplothrix* sp. from Mumyono-ana and Pinza-Abu Caves are considered to be endemic forms of Miyako Island. Endemism on an island generally requires long duration of its isolation from the adjacent continent or other islands.
3. The sea between Miyako Island and Taiwan is too deep to have been dried up by the sea level drop generally estimated for OIS 3 and OIS 2, even though tectonic movements are taken into consideration (more than 500 m in depth).

During the later part of OIS 3 and the early part of OIS 2, Miyako Island was probably separated by the sea from Taiwan which connected to the adjacent continent, and the non-flying terrestrial mammals including *Microtus*, *Diplothrix* and *Capreolus* are considered to have immigrated into the island much earlier.

Conclusion

The excavations of Mumyono-ana Cave have revealed the stratigraphic sequence of the sediments yielding abundant mammalian remains. The sediments can be divided into seven layers such as Layers D, C, B, A, A_1, Z and S in ascending order. Bone samples from Layers D and C were dated to the later part of OIS 3 (ca 28,700-25,300 calBP) by the AMS radiocarbon method (Table 1). Those from Layer B and Layer Z were also dated to ca 24,700 calBP and ca 10,100 calBP, respectively by the same method. Thus Layer B and Layer Z are assignable to the early part of OIS 2 and the early Holocene, respectively.

The mammal remains from these layers are referred to 11 forms (Table 2), of which four are flying mammals (chiropterans). Among the non-flying forms, the large rat (*Diplothrix* sp.) is highly predominant in the lower layers (Layers D and C), but it becomes much fewer in the middle and upper layers (Layers B, A, A_1, Z and S). On the other hand, the reed vole (*Microtus fortis*) is highly predominant in these layers, but its frequency is much lower in the lower layers. The stratigraphic change in the relative abundance of these two forms suggests that forests reduced remarkably between OIS 3 and OIS 2 in relation to the expansion of relatively wet grassland.

It is inferred that Miyako Island was already separated from Okinawa Island by the sea during the later part of OIS 3 and the early part of OIS 2, because the non-flying mammal fauna of Mumyono-ana and Pinza-abu Caves on Miyako Island are quite different from that of the Minatogawa site on Okinawa Island. The separation of Miyako Island from Taiwan during the same period is also inferred from the faunal characters of the assemblages from Mumyono-ana and Pinza-abu Caves including the occurrences of endemic forms, as well as the depth of the sea between Miyako Island and Taiwan.

Acknowledgements

We would like to express our appreciation to Mr. H. Yamauchi (Okinawa Caving Association) for providing useful information about the cave, to Mr. S. Hirasawa (Kyoto University) for assisting in the excavations and Dr. H. Uno (National Institute for Environmental Studies) for helping in the ^{14}C dating. Parts of this research were supported by Grants-in-Aid for Scientific Research from the Society for the Promotion of Science (project number 21340145 to Kawamura and 19310011 to Yoshimura).

References

Danzeglocke, U., Jöris, O. and Weninger, B., 2008, CalPal-2007^{online}. http://www.calpal-online.de, accessed 2008-7-27.

Department of Education, Okinawa Prefectural Government, 1985, *Pinza-Abu* [Report on Excavation of the Pinza-Abu Cave]. 184pp. The Department of Education, Okinawa Prefectural Government, Naha. (in Japanese)

Hasegawa, Y., 1980, Notes on vertebrate fossils from Late Pleistocene to Holocene of Ryukyu Islands, Japan. *Quaternary Research* (*Daiyonki-Kenkyu*), vol. 18, pp.263-267.

Hasegawa, Y., Otsuka, H. and Nohara, T., 1973, Fossil vertebrates from the Miyako Island (Studies of the palaeovertebrates fauna of Ryukyu Islands, Japan. Part I). *Memoir of the National Science Museum*, no.6, pp.39-52, pl. 6-7. (in Japanese with English summary).

Hirasawa, S., Nakagawa, R. and Nunami, S., 2006, On the dentary of *Rhinolophus* specimens from the Nakabari Cave (Abucha), Miyako Island, Okinawa Prefecture, Japan (preliminary report). *Journal of Speleological Society Japan*, vol.31, pp.42-48. (in Japanese with English summary).

Kaneko, Y. and Hasegawa, Y., 1995, Some fossil arvicolid rodents from the Pinza-Abu Cave, Miyako Island, the Ryukyu Islands, Japan. *Bulletin of the Biogeographical Society of Japan*, vol. 50, no.1, pp.23-37.

Kawamura, Y., 1988, Quaternary rodent faunas in the Japanese Islands (Part 1). *Memoirs of the Faculty of Science, Kyoto University*, Series of Geology and Mineralogy, vol. 53, nos.1,2, pp.31-348.

Kitagawa, H., Masuzawa, T., Nakamura T. & Matsumoto, E., 1993, A batch preparation method for graphite targets with low-background for AMS C-14 measurements. *Radiocarbon*, vol. 35, pp.295-300.

Longin, R., 1971, New method of collagen extraction for radiocarbon dating. *Nature*, vol. 230, pp.241-242.

Nakagawa, R., Yoneda, M., Uno, H. and Shibata, Y., 2007, AMS ^{14}C dating of mammalian remains from Naumann Branch of Seiryukutsu Cave, Hiraodai Karst Plateau, Fukuoka Prefecture, Japan. *Journal of Speleological Society of Japan*, vol. 32, pp.35-41.

Nakamori, T., 1982, Geology of Miyako Gunto, Ryukyu Islands, Japan. *Contributions from the Institute of Geology and Paleontology, Tohoku University*, no. 84, pp.23-39. (in Japanese with English abstract)

Otuka, Y., 1941, On the stratigraphic horizon of *Elephas* from Miyako Is., Ryukyu Islands, Japan. *Proceeding of the Imperial Academy* (*of Japan*), vol.17, no.2, pp.43-47.

Society of Scientific Expedition, Ehime University, 1977, *Tanken no, 6* [Exploration no. 6. —Report on the Survey of Caves on Miyako, Yonaguni and Ishigaki Islands, and on the Kii Peninsula—], 84pp. The Society of Scientific Expedition, Ehime University, Matsuyama. (in Japanese)

Tanaka, A., Yoneda, M., Uchida, M., Uehiro, T., Shibata, Y. and Morita, M., 2000, Recent advances in C-14 measurement at NIES-TERRA. *Nuclear Instruments and Methods in Physics Research*, B, 172, pp.107-111.

Taruno, H. and Kawamura, Y., 2007, Mammoths of East Asia: Revisions of their taxonomy, chronospatial distribution, evolution, and immigration into Japan. *Jubilee Publication in Commemoration of Prof. Kamei Tadao's 80th Birthday*, pp.59-78. (in Japanese and English abstract)

Tokunaga, S., 1936, Fossil land mammals from the Riukiu Islands. *Proceeding of the Imperial Academy* (*of Japan*), vol. 7, no. 8, pp. 255-257.

Tokunaga, S., 1940, A fossil elephant tooth discovered in Miyakojima, an island of the Ryukyu Archipelago, Japan. *Proceeding of the Imperial Academy* (*of Japan*), vol.16, no.3, pp.122-124.

Tokunaga, S. and Takai, F., 1938, Deer fossils found on the Ryukyu Islands. *The Journal of the Geological Society of Japan*, vol. 45, no. 537, pp. 470-471. (in Japanese)

Yazaki, K. and Oyama, K., 1980, *Geology of the Miyako-jima District*. Quadrangle Series, Scale 1:50,000. 87pp. Geological Survey of Japan, Tsukuba. (in Japanese with English abstract)

Yoneda, M., Shibata, Y., Tanaka, A., Uehiro, T., Morita, M., Uchida, M., Kobayashi, T., Kobayashi, C., Suzuki, R., Miyamoto, K., Hancock, B., Dibden, C. & Edmonds, J. S., 2004, AMS 14C measurement and preparative techniques at NIES-TERRA. *Nuclear Instruments and Methods in Physics Research*, B, 223-224, pp.116-123.

Yoneda, M., Tanaka, A., Shibata, Y., Morita, M., Uzawa, K., Horita, M., and Uchida, M., 2002, Radiocarbon Marine Reservoir Effect in Human Remains from the Kitakogane Site, Hokkaido, Japan. *Journal of Archaeological Science*, 29, pp.529-536.

Taphonomy of Vertebrate Remains from Funakubu Second Cave in Okinawa Island, Japan

Shin Nunami

Asakura Publishing Co., Ltd,
Shin-ogawamachi 6-29, Shinjuku-ku, Tokyo, 162-8707 Japan.
E-mail: nunami@asakura.co.jp

Ryohei Nakagawa

Mie Prefectural Museum
Koumei-cho 147-2, Tsu, Mie, 514-0006 Japan.
E-mail: nakagr00@pref.mie.jp

Abstract: *A vertebrate fossil accumulation was recovered from Funakubu Second Cave in Okinawa Island, Japan. The cave was formed between the Shinzato Formation in the Shimajiri Group (the Late Pliocene - Early Pleistocene) and the Naha Formation in the Ryukyu Group (the Middle Pleistocene). All the vertebrate remains were unearthed from a reddish-brown muddy sediment. The remains include marine and terrestrial vertebrates together. Marine material consists of fine-sized fragmented teeth of sharks and bony fishes, and they are highly rounded, dissolved and colored. Terrestrial material, however, consists of relatively coarse-sized fragmented teeth and bones of terrestrial vertebrates. These can be separated into two groups: one has suffered heavy rounding, dissolution and coloration, like the marine material, while the other shows little evidence of these characteristics. These two groups, named Land-A and Land-B, could have been individually transported into the cave from distinct sources during different time periods. In addition, 14C measurements and collagen extraction were carried out, and an absolute age of 10,540 ± 340 years BP. was obtained from the fragmented teeth of deer. However, the collagen could have been highly altered because of its low content and the sample's extraordinarily high C/N ratio, which would cast doubt on the reliability of the age of the fossil.*

Introduction

The evidence of human activity is occasionally recorded in cave deposits as skeletal material or artifacts. Although these have been reported from limited deposits so far, fossils of vertebrate animals contemporary with humans are often recovered from many of the cave deposits. Resolution of the change of non-human vertebrate distribution should provide us with further understanding about the history of human activity. However, there is a possibility that vertebrate assemblages in cave deposits include fossils of different periods because of mixing. Therefore, it is difficult to date fossils and deposits without considering the process of fossilization.

A number of Pleistocene vertebrate fossils have been recovered from the Ryukyu Islands located in the southwestern part of Japan. Since Matsumoto (1926) reported a deer fossil from limestone on the Shimajiri coastline of Okinawa Island, a variety of reports and taxonomic studies about these remains have been done; for example, Ie-mura Kyoiku Iinkai (1977, 1978) reported 12 species of vertebrate remains including a human mandible fragment and a large amount of deer remains from Gohezu-do Cave in Ie Island; Suzuki and Hanihara (1982) made careful studies about "the Minatogawa Man" recovered from the Minatogawa fissure on Okinawa Island; Okinawa-ken Kyoiku Iinkai (1985) reported over 40 vertebrate species including a fragment of human skull from Pinza-abu Cave in Miyako Island. Most of these studies dealt with vertebrate remains from caves and fissures which developed in limestone of the Ryukyu Group distributed broadly over the Ryukyu Islands. In addition, these remains include mammals which are now extinct among the Ryukyu Islands. Therefore, they could have significant implication for reconstructing the Pleistocene fauna of Japan and in resolving the relationship between faunal change and human occupation.

In order to understand the human influence accurately, it is necessary to calculate the absolute age of the remains recovered from fossil localities and archaeological sites. Some dates are from charcoal or terrestrial mollusk fossils from caves and fissures on the Ryukyu Islands (Takai, 1975; Suzuki and Hanihara, 1982; Azuma, 2007 and so on). However, we should be careful about these dates in the case of cave or fissure deposits, because there is a possibility that the remains have totally different postmortem histories, even if they were unearthed from the same deposit or layer. An evaluation of the effectiveness of absolute ages would also be needed with physically and chemically damaged samples.

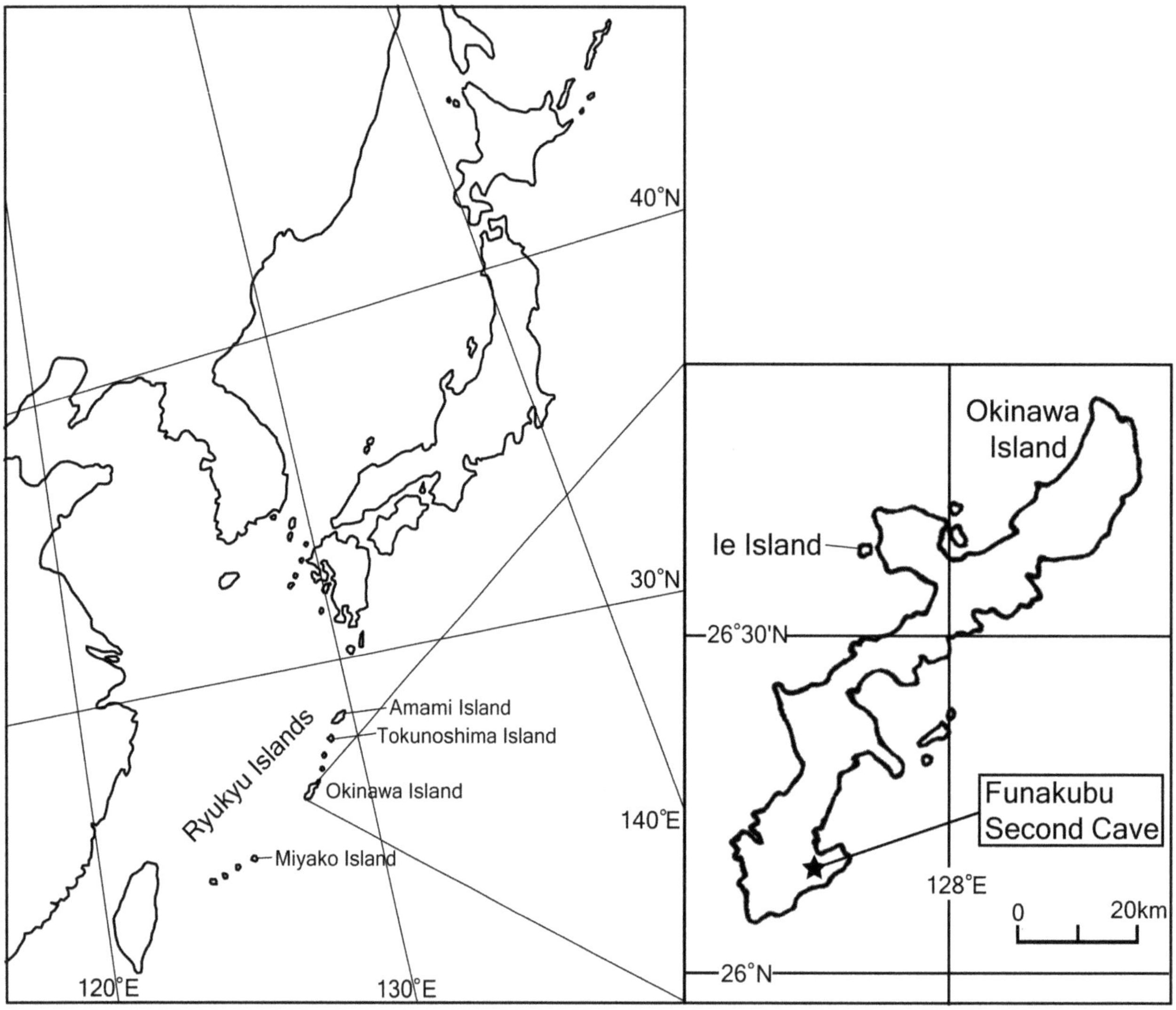

Fig. 1 Location of the Ryukyu Islands and Funakubu Second Cave.

This paper presents taphonomic information about the vertebrate accumulation from Funakubu Second Cave located on Okinawa Island, Japan (Fig. 1). Each specimen was observed in detail and described for characteristics such as degree of edge rounding, existence of cracks or fractures, dissolution, staining, coloration and maximum length. Specimens of terrestrial vertebrates were categorized into two groups according to differences in these traits. Then, the postmortem histories of marine materials and two groups of terrestrial materials were examined. In addition, collagen extraction and subsequent ^{14}C age measurements were carried out on the damaged vertebrate material.

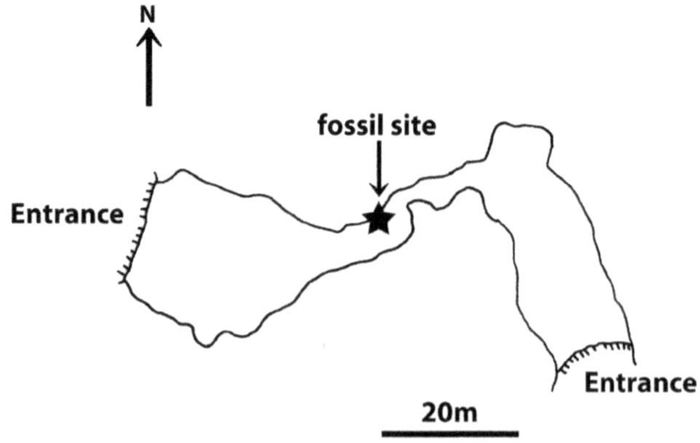

Fig. 2 Plan of Funakubu Second Cave (modified from Nakachi, 1992).

Location and Settings

Funakubu Second Cave is located at Nanjo-shi, southeastern Okinawa Island, and belongs to the Funakubu Cave System which is composed of six caves formed in the large scale doline about 5 km wide from east to west. The cave was developed transversally by chemical processes of ground water along the surface of unconformity between the Shinzato Formation in the Shimajiri Group (the Late Pliocene - Early Pleistocene) and the Naha Formation in the Ryukyu Group (the Middle Pleistocene) (Kaneko and Ujiie, 2006). The inner wall and floor of the cave

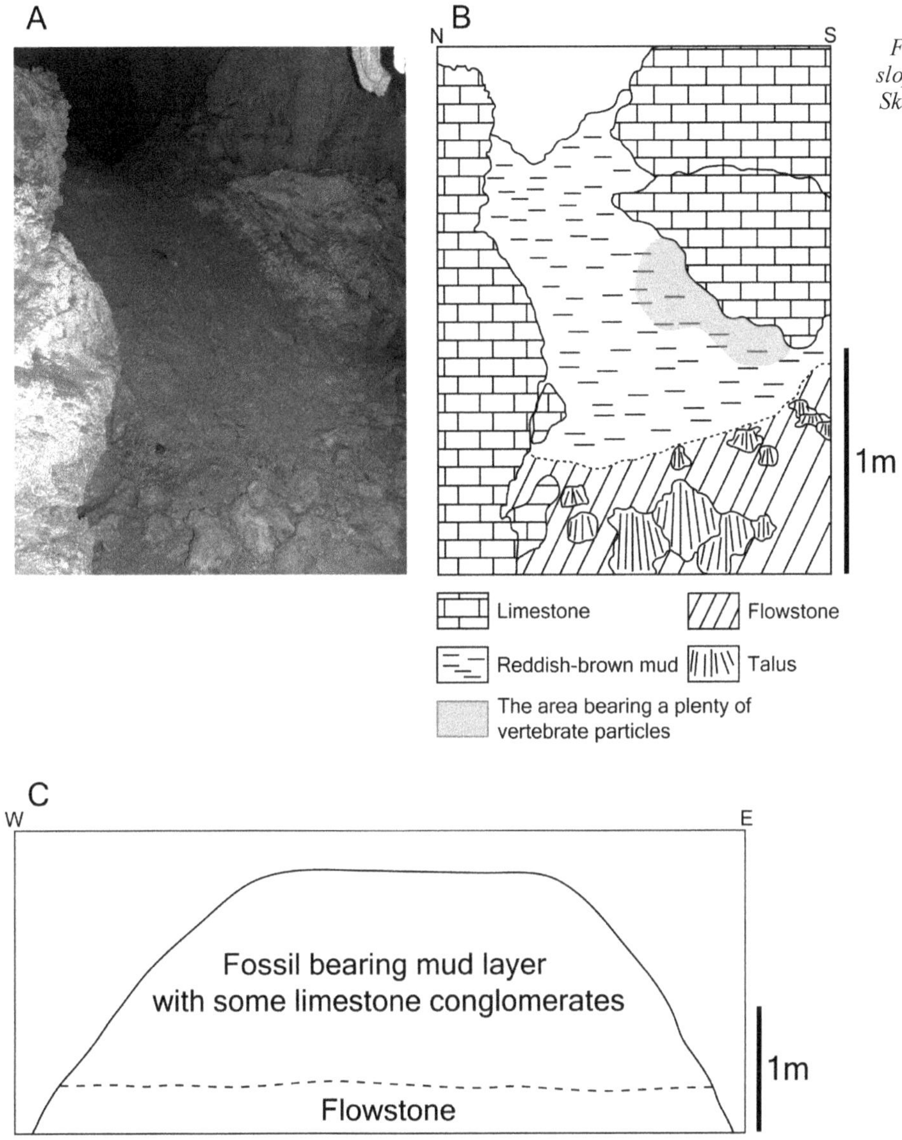

Fig. 3 A: Photograph of the western slope of the fossil-bearing sediment. B: Sketch of A. C: Pattern diagram of the mound-like sediment.

are covered with secondary products, such as columns, stalactites and flowstones. Numerous boulders of limestone about 1 meter in diameter lie everywhere in the cave, and muddy sediments are deposited on the floor. The cave has two entrances; the western one is 3 meters in height and 15 meters in width, while the eastern one is 5.5 meters in height and 12 meters in width. Funakubu Second Cave is 84 meters in length (Fig. 2), the longest cave in the Funakubu Cave System (Nakachi, 1992).

The fossil bearing sediment (Fig. 3) is about 40 meters from the western entrance. This sediment is mainly composed of reddish-brown calcareous mud with some unweathered limestone conglomerates of less than 100 millimeters in diameter. Most of the sediment is massive and very poor sorted, although there might be an area bearing plenty of vertebrate particles near its boundary with the southern limestone (Fig. 3B). The sediment is shaped like a mound (Fig. 3C), about 5 m wide from east to west and is composed of a mud layer of about 1.7 m thick on a calcareous flowstone. Vertebrate remains are found throughout the sediments together with fragments of invertebrate fossils, such as brachiopods, bivalves and bryozoans.

Methods

All specimens in this study were collected from the western slope of the sediment mound. For collecting vertebrate remains without omission, the screening method described below was adopted. About 60 kg of cave deposits were collected and wet sieved through 0.5 mm mesh. After drying, fossil remains were removed.

Systematics and terminology for vertebrate identification were complied following Ueno *et al.* (1974), Sakai & Hanamura (1976), Ie-mura Kyoiku Iinkai (1977, 1978), Hasegawa & Nohara (1978), Sakai *et al.* (1978), Okinawa-ken Kyoiku Iinkai (1985), Cappetta (1987), Ueno & Sakamoto (1999), Abe (2000), and Ikeda (2007). After measuring of the remains with calipers and recording their maximum lengths, their taphonomic characteristics, such as degree of edge rounding, existence of cracks or fractures, degree of fragmentation, dissolution, as well as staining and coloration, were described.

Important fossil samples were photographed with a digital camera or scanning electron microscope (JEOL, JSM-6060B) in the Kyoto University Museum.

Fossils

A total of 2,459 vertebrate remains were found in the deposits from Funakubu Second Cave. About 39% of them were taxonomically identified to at least order level, and 8% to genera or species level. Precise identification of vertebrate remains, both marine and terrestrial, revealed fauna consisting of 11 orders, 10 families, 12 genera and 1 species (Fig.4 and Table 1). All samples are housed in the Kyoto University Museum.

1) Sharks and Rays

Of the 225 fossil teeth of sharks and rays, approximately 66% were identified. The sharks are *Carcharhinus* sp., *Heterodontus* sp., *Isurus* sp., *Carcharodon carcharias*, *Galeocerdo* sp., *Sphyrna* sp., and *Squalus* sp. As for rays, only *Dasyatis* sp. was found. Ueno *et al.* (1974) reported various fish remains from the Chinen Sand (correlated to the Chinen Formation in the Ryukyu Group of Kaneko and Ujiie, 2006) which outcrops about 2 km to the west of Funakubu Second Cave, and documented species such as *Heterodontus* sp., *Isurus* sp., *Carcharhinus* sp. and *Galeocerdo* sp. from both localities.

2) Bony fishes

Of the 882 fossil teeth of bony fishes, about 81% could be identified. Remains are mainly fragmented teeth with a wide variety of shapes; cobblestone type, tusk type, keyhole-like type, pea-shaped type, pot-shaped type, dome type, etc. They are identifiable to Order Perciformes when compared to modern samples, and might include Family Scaridae, Sparidae and Labridae, but accurate identification about teeth of bony fishes is quite difficult so far because any detailed comparison between recent species and fossils has not been fully done. The rest are identified as *Diodon* sp. Ueno *et al.* (1974) described Order Perciformes as "Teleostei, Gen et sp. indet." from the Chinen Sand.

3) Reptiles

Two fossil vertebrae were recovered. One of them is identified as *Trimeresurus* sp., and the other as *Elaphe* sp., based on their shapes and characteristics as typified by the existence of hypapophysis. These identified taxa are extant on Okinawa Island.

4) Mammals

Of 1,350 recovered mammal fossils, only 7% could be

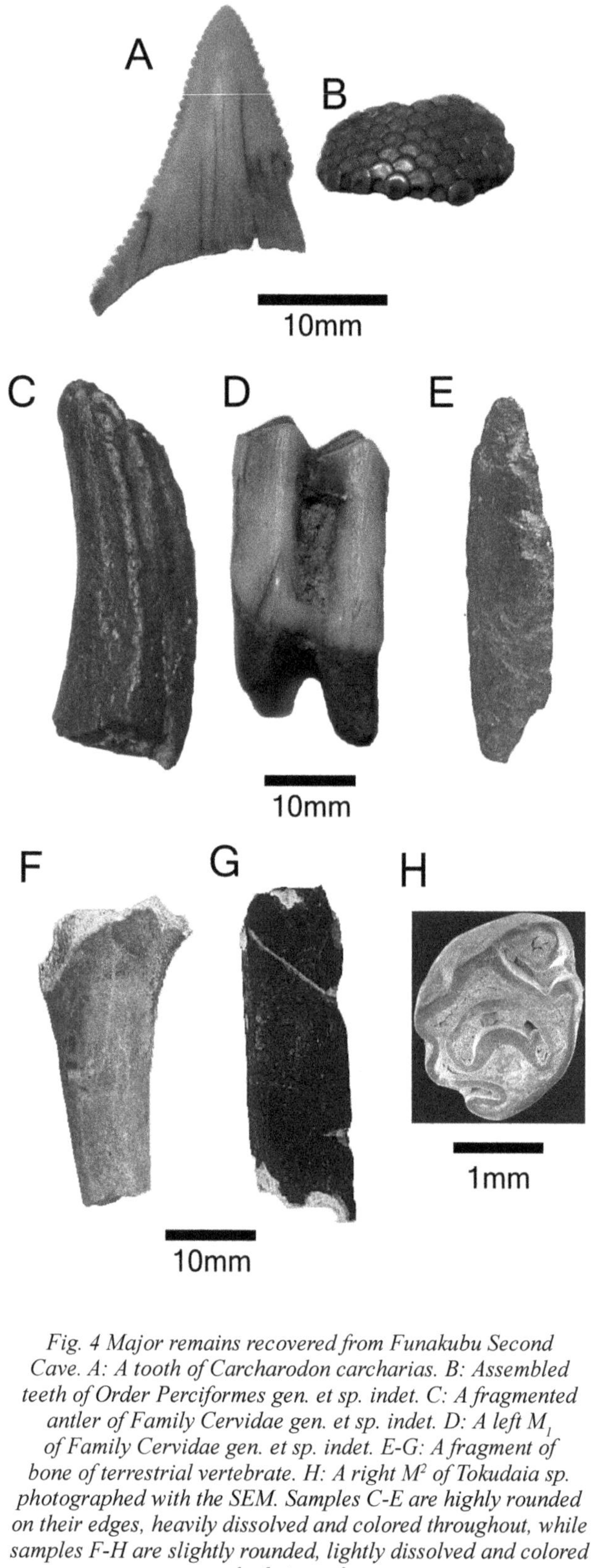

Fig. 4 Major remains recovered from Funakubu Second Cave. A: A tooth of Carcharodon carcharias. B: Assembled teeth of Order Perciformes gen. et sp. indet. C: A fragmented antler of Family Cervidae gen. et sp. indet. D: A left M_1 of Family Cervidae gen. et sp. indet. E-G: A fragment of bone of terrestrial vertebrate. H: A right M^2 of Tokudaia sp. photographed with the SEM. Samples C-E are highly rounded on their edges, heavily dissolved and colored throughout, while samples F-H are slightly rounded, lightly dissolved and colored only their surfaces.

Table 1 A list of vertebrate remains from Funakubu Second Cave.

Taxa	*No. of specimen*	*Taxa*	*No. of specimen*
Sharks & Rays		Reptiles	
Charcharhinus sp.	75	*Elaphe* sp.	1
Galeocerdo sp.	2	*Trimeresurus* sp.	1
Sphyrna sp.	2		
Heterodontus sp.	49	Mammals	
Carcharodon carcharias	4	Family Cervidae gen. et sp. indet.	90
Isurus sp.	13	*Tokudaia* sp.	4
Squalus sp.	1	Order Chiroptera gen. et sp. indet.	4
Dasyatis sp.	2	indet. teeth	33
		indet. bones	1219
Bony Fishes			
Order Perciformes gen. et sp. indet.	667	Total	2459
Diodon sp.	35		
indet. teeth of fishes	257		

identified. Indeterminate bone fragments were grouped by shape as thick, robust, or compact.

More than 90% of the identified remains represent fragmented antlers, molars, premolars and incisors of Family Cervidae, gen. et sp. indet. Two kinds of extinct deer, *Cervus astylodon* (Ryukyu-jika) and Subfamily Muntiacinae gen. et sp. indet. (Ryukyu-mukashikyon), have been reported from Pleistocene sediments on the Ryukyu Islands (for example, Takai, 1975 and Ie-mura Kyoiku Iinkai, 1977, 1978). The deer remains of Funakubu Second Cave are also assignable to these two taxa on the basis of molar and premolar morphology. However, these taxa are usually identified and described by the shape of their antlers. Furthermore, the differences in their cranial parts, except for canines, are still unclear. The unearthed antlers are so fragmentary that it is difficult to make a taxonomic assignment. Therefore, the deer fossils are best identified as Family Cervidae in this study.

Four molars (right M^1, two right M^2 and right M_3) were identified as *Tokudaia* sp. with their characteristic occulusal surfaces and relatively higher crowns than those of other members of Family Muridae. *Tokudaia osimensis* is an endemic species of the Ryukyu Islands, and they are believed to inhabit Amami Island, Tokunoshima Island and the northern part of Okinawa Island, while a recent genetic study indicates that rats of each island belong to distinct species, *Tokudaia osimensis* in Amami Island, *Tokudaia tokunoshimensis* on Tokunoshima Island and *Tokudaia muenninki* on Okinawa Island (Kaneko, 2001 and Endo and Tsuchiya, 2006). Recovered molars seem to be *Tokudaia muenninki*, but in order to identify them more accurately, further comparative morphological studies on the teeth of the three species need to occur.

The existence of bat is denoted by 1 molar, 1 tibia and 2 cochlear ducts, although these remains are inadequate for comparison and identification because of high degrees of rounding and fragmentation. They are thus treated as Order Chiroptera gen. et sp. indet.

Taphonomic description

The collection is dominated by fragmented terrestrial vertebrate bones, followed by fish teeth and teeth of terrestrial vertebrates, whereas bony elements of marine vertebrates are absent.

1) Surface characteristics

Almost all of the fish teeth are moderately to highly rounded and polished on their surfaces, which are comparable to the subangular to subrounded categories in sedimentology. They also suffer heavy fragmentation, so samples do not fully retain their original shape. All sharks and most of the bony fishes are represented by isolated teeth (Fig. 4A), but some bony fish teeth remain intact (Fig. 4B). Furthermore, the roots of most of the teeth were lost except for those which are highly dissolved and darkly stained roots. The coloration completely permeates the crowns of teeth and varies from black, to dark brown, to ochre, orange, blue-white, milky white and so on.

Most of the teeth and bones of terrestrial vertebrates are also moderately to highly rounded, have no longitudinal cracks or fractures on their surfaces, and suffer heavy dissolution (Fig. 4C-E). Except for some teeth, they also suffer heavy fragmentation except for some teeth (Fig. 4D). All examples are isolated teeth and disarticulated bones. The brown, black or ochre coloration completely permeates the crowns of teeth like with the marine remains. However, their roots and bones are still unstained in contrast.

There are some samples in the terrestrial material which show slight rounding, have no longitudinal cracks or fractures on their surfaces and suffer no or slight dissolution (Fig. 4F-H). The bones also suffer fragmentation, but the teeth retain their original shapes (Fig. 4F). All the samples are isolated teeth and disarticulated bones. They retain white color or are grey to light brown on the surfaces.

2) Size distribution

The distribution of maximum lengths of the marine remains is shown in Fig. 5-A, and of terrestrial remains in Fig. 5-B. Marine remains are relatively smaller (1-8 mm), while terrestrial remains are comparatively larger (4-16 mm or larger).

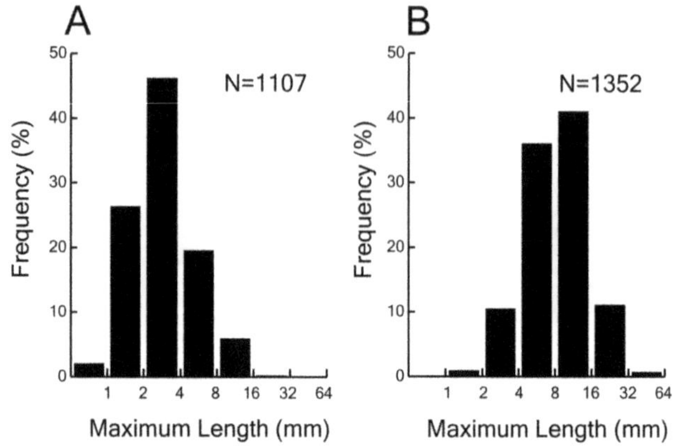

Fig. 5 Histograms showing the distribution of maximum lengths. A: Marine remains. B: Terrestrial remains.

^{14}C measurements

^{14}C measurements and collagen extraction were carried out using the NIES-TERRA system at the National Institute of Environmental Studies, Tsukuba, Japan. A radiocarbon date of 10,540 ± 340 years BP was obtained from a highly rounded, dissolved and dark colored fragmented deer tooth (TERRA-071106b, FNA 002). Its collagen content was 0.09 wt%, and the C/N ratio was 18.17. It is the first radiocarbon date directly measured on deer remains from Okinawa Island.

Discussion

1) Postmortem history

Marine vertebrate fossils are well rounded, heavily dissolved and colored, and only include fragmented teeth, which lack their roots. These characteristics would demonstrate the following processes occurred; rounding and fragmentation would indicate that samples suffered physical damage in some postmortem movement such as transport, falling, or reworking. Dissolution would result from chemical processes in the acidic ground water; coloration would occur during burial in deposit bearing enough metal ion, for example, manganese or iron (Andrews, 1990). At the same time, the bony parts of the marine vertebrates could have been destroyed by postmortem processes. In addition, the remains suffered from chemical alteration while buried and exposed to slightly acidic ground water that bore plenty of metal ions.

Most of the terrestrial materials are also well rounded, heavily dissolved and colored. They have no longitudinal cracks or fractures on their surfaces and include a number of fragmented teeth and bones, which are isolated or disarticulated. Behrensmeyer (1978) suggested that longitudinal cracks or fractures are produced by subaerial exposure for a certain period of time. Consequently, their exposure was limited, but suffered plenty of physical damage before they were buried. They also were affected by chemical processes while buried in sediments containing slightly acidic, ion-rich ground water.

Some terrestrial fossils lack high rounding, heavy dissolving or coloring. The teeth maintain their original shapes. They seem to have undergone less postmortem processes than the well-rounded and fragmented fossils. Besides, since they suffered little chemical damage, they could have been added to the fossil record relatively recently. Thus it would be possible to divide terrestrial materials into two groups according to site taphonomy—Land-A (Fig. 4C-E; high rounding, fragmentation, dissolving and coloration) and Land-B (Fig. 4F-H; low rounding, dissolving and coloration).

At this time it is impossible to explain the coexistence of the marine and terrestrial material. As a matter of course, there are some deposits from which both marine and terrestrial remains are found together like archaeological shell middens. However, most fossils from these sites are generally well-preserved, slightly rounded and not much dissolved. And each part of carcasses is preserved relatively abundantly. The pattern of breakage and fragmentation of remains are characteristic, such as a spiral fracture. Furthermore, human fossils and their artifacts are sometimes recovered together. Their influences are occasionally remained on bone material as a trace of hunting and manufacturing. Considering that the vertebrate accumulation of Funakubu Second Cave doesn't apply to these characteristics, it is suggested that they are different from remains of shell mounds.

2) ^{14}C chronology

The resultant absolute age of 10,540 ± 340 years BP is the youngest among the dates of deer fossils reported from sites in the Ryukyu Islands. However, deer tooth is included in the Land-A group on the basis of physical damage and chemical corrosion. Additionally, the collagen content was low and the C/N ratio was far from a normal value (refer to DeNiro, 1985). It is thus considered that the collagen was altered and the reliability of the age of the fossil is questioned (just for reference, ^{14}C ages tend to err younger when collagen is altered).

Conclusions

Taphonomic information about the vertebrate fossil accumulation in Funakubu Second Cave was examined. It is obvious that there are two groups in the terrestrial mammals, which suffered different degrees of physical and chemical damage and were deposited in the cave at different times. A damaged deer tooth with highly altered collagen produced a questionable ^{14}C age of 10,540 ± 340 years BP.

Consequently, careful taphonomic analysis and observation of fossils about whether mixing of several discrete groups occur are needed when dealing with fossils from caves and fissures. It is also necessary for the accurate evaluation of dates to demonstrate the contemporaneity of fossils and charcoal even if they were recovered from the same deposits. Lastly, it is important as well to assess the physical damage and chemical corrosion on fossils before meaningful conclusions can be drawn.

Acknowledgement

The authors would like to express our appreciation to members and former members of the Biosphere Group, Kyoto University, Drs. T. Setoguchi, H. Kamiya, T. Ohno, H. Matsuoka, H. Naruse, N. Sakakura, K. Sato, and T. Nishimura for their valuable advice throughout the investigation. We also would like to express our appreciation to Dr. Y. Kawamura (Aichi University of Education) for his valuable instruction and encouragement throughout this work. We thank Dr. D. Matsumoto (AIST) for his assistance during the survey of the Funakubu Second Cave and valuable discussion about this work. We also thank Drs. Y. Ota (Kitakyushu Museum of Natural History and Human History) and H. Maeda (Kyoto University) for their critical review to this manuscript. We wish to express thanks to Mr. H. Yamauchi, representative of the Okinawa Caving Association who gave lodging and information about caves with kindness throughout the survey and provided us a glimpse into the fossils found in caves on the Ryukyu Islands. During writing this paper we were greatly inspired by these materials.

References

Abe, H., 2000, *Nihonsan honyurui tokotsu zusetsu* [Illustrated skulls of Japanese mammals]. Hokkaido Daigaku Tosyo Kankokai, Sapporo.

Andrews, P., 1990, *Owls, Caves and Fossils.* University of Chicago Press, Chicago.

Azuma, Y., 2007, 'Three new species of fossil terrestrial Mollusca from fissure deposits within the Ryukyu Limestone in Okinawa and Yoron Islands, Japan'. *Paleontogical Research*, 11, pp. 231-249.

Behrensmeyer, A. K., 1978, 'Taphonomic and ecologic information from bone weathering'. *Paleobiology*, 4, pp. 150-162.

Cappetta, H., 1987, *Handbook of Paleoichthyology 3B, Chondrichthyes II Mesozoic and Cenozoic Elasmobranchii.* Gustav Fischer Verlag, Stuttgart and New York, Stuttgart.

DeNiro, M. J., 1985, 'Postmortem presavation and alterration of *in vivo* bone collagen isotope ratios in relation to palaeodietary reconstruction'. *Nature*, 317, pp. 806-809.

Endo, H. and Tsuchiya, K., 2006, 'A new species of Ryukyu spiny rat, *Tokudaia* (Muridae: Rodentia), from Tokunoshima Island, Kagoshima Prefecture, Japan'. *Mammal Study*, 31, pp. 47-57.

Hasegawa, Y. and Nohara, T., 1978, *Ishigaki-shi Ishisukuyama doubutsu igai gunsyu no gaiyou* [Outline of vertebrate assemblages from Ishisuku-yama at Ishigaki city]. In Okinawa-ken Kyoiku Iinkai (ed.) *Ishisukuyama*, pp. 44-98. Okinawa-ken Kyoiku Iinkai, Okinawa.

Ie-mura Kyoiku Iinkai, 1977, *Okinawa-ken Ie-jima Gohezu-do no chosa –dai 1ji gaiho-* [Investigation of the Gohezu-do Cave at Ie Island, Okinawa Prefecture - primary report -]. Ie-mura Kyoiku Iinkai, Okinawa.

Ie-mura Kyoiku Iinkai, 1978, *Okinawa-ken Ie-jima Gohezu-do no chosa –dai 2ji gaiho-* [Investigation of the Gohezu-do Cave at Ie Island, Okinawa Prefecture - secondary report -]. Ie-mura Kyoiku Iinkai, Okinawa.

Ikeda, T., 2007, 'A comparative morphological study of the vertebrae of snakes occurring in Japan and adjacent regions'. *Current Herpetology*, 26, pp. 13–34.

Kaneko, Y., 2001, 'Morphological discrimination of the Ryukyu spiny rat (genus *Tokudaia*) between the islands of Okinawa and Amami Oshima, in the Ryukyu Islands, southern Japan'. *Mammal Study*, 26, pp. 17-33.

Kaneko, N. and Ujiie, H., 2006, *Itoman oyobi Kudaka-jima chiiki no chisitsu* [Quadrangle Series, 1:50,000, Geology of the Itoman and Kudaka-jima District]. Chisitsu Chosa Sogo Center, Tsukuba.

Matsumoto, H., 1926, 'On some new fossil cervicorns from Kazusa and Liukiu'. *Science Reports of the Tohoku Imperial University, 2nd series, Geology*, 10, pp. 17-26.

Nakachi, M., 1992, *Funakubu Cave System no gaiyo ni tsuite* [Outline of the Funakubu cave system]. In Gyokusen-do Cave Festival jimukyoku (ed.) *Gyokusen-do Cave System*, pp. 83-87. Tamagusuku-mura Kyoiku Iinkai and Nanto World Kabushiki Gaisya, Okinawa.

Okinawa-ken Kyoiku Iinkai, 1985, *Pinza-abu* [Reports on Excavation of the Pinza-Abu Cave]. Okinawa-ken Kyoiku Iinkai, Naha.

Sakai, T. and Hanamura, H., 1976, *Yokusyumoku no ha no keitaigaku teki kenkyu I. Kikugashirakomori ka* [A morphological study on the dentition of chiroptera. I. Rhinolophidae]. *Shika Kiso Igakukai Zassi*, 18, pp. 442-445.

Sakai, T., Hanamura, H. and Toda, Y., 1978, *Yokusyumoku no ha no keitaigaku teki kenkyu II. Hinakomori ka* [A morphological study on the dentition of chiroptera. II. Vespertilionidae]. *Shika Kiso Igakukai Zassi*, 20, pp. 738-755.

Suzuki, H. and Hanihara, H., 1982 ed., *The Minatogawa Man. The Upper Pleistocene man from the Island of Okinawa.* University of Tokyo Press, Tokyo.

Takai, F., 1975, *Yamashita-cho Dai1-do hakken no shika kaseki* [Fossil deer from the Yamashita-cho Cave No. 1]. *Jinruigaku Zassi*, 83, pp. 280–293.

Ueno, T., Nohara, T., and Hasegawa, Y., 1974, *Okinawa-jima san gyorui kaseki ni tsuite (Ryukyu syoto no kosekitsui dobutsuso – sono4)* [Fish fossils from Okinawa Island. (Paleo-vertebrate fauna of Ryukyu Islands – part IV)]. *Kokuritsu Kagaku Hakubutsukan Senpo*, 7, pp. 53-60.

Ueno, T. and Sakamoto K., 1999, *Sakana no bunrui no zukan* [The picture book of fish taxonomy]. Tokai University Press, Tokyo.

Formative History of Terrestrial Fauna of the Japanese Islands during the Plio-Pleistocene

Keiichi Takahashi

Scientific Research Department
Lake Biwa Museum, Kusatsu, Shiga, 525-0001, Japan.
Email: takahasi@lbm.go.jp

Masami Izuho

Archaeology Laboratory, Faculty of Social Sciences and Humanities, Tokyo Metropolitan University
1-1, Minami Osawa, Hachioji City, Tokyo 192-0397, Japan.
Email: izuhom@tmu.ac.jp

Abstract: *This paper presents a framework of the evolution of the mammalian fauna in the Plio-Pleistocene in the Japanese Islands in order to better understand the correlation between fossil and paleoenvironment records of the Late Pleistocene, as well as provide a model of large mammal extinctions.*

Based on the paleoclimatic records and the presence of land bridges connected to the Asian continent, the history of terrestrial mammals of the Japanese archipelago during the Plio-Pleistocene can be divided into four periods; First Period (5.3-3.5Ma), Second Period (3.5-2.4Ma), Third Period (2.4-1.7Ma), and Fourth Period (1.7-0.01Ma). Late in the fourth period, the Palaeoloxodon- sinomegaceroides complex and mammoth fauna inhabited the Paleo-Honshu Island and the Paleo-Sakhalin/Hokkaido/Kurile peninsula, respectively. The boundary of these two faunal groups that migrated back and forth north-south matched the range of vegetation that shifted from climatic fluctuations as indicated by radiocarbon dates and the geographical distribution of the fossil records. Also, the dates and distributions of fossil records demonstrate that formation processes of the present faunal community in the Japanese Islands are basically explained by climatic changes and associated changes in landscapes and flora.

Finally, the archaeological record, as well as direct evidence of fossil records and inferred mammalian behavior in Eurasia and the Japanese Islands, implies that the large mammal extinctions were not caused by human factors such as mass killings. Rather, the fragmentary data examined in this paper support the hypothesis that climate-induced vegetation changes were the main causes.

Keyword:

Introduction

The formative history of the mammalian fauna in the Plio-Pleistocene (5.3 to 0.01Ma) of the Japanese Islands is one of the major topics in paleo-vertebrate research in Japan (Shikama, 1962; Hasegawa, 1977; Kamei *et al.*, 1988a, b; Kawamura, 2003a). Despite the increased discovery of fossil records and the steady progress in their studies, many questions are still unanswered because of the sporadic nature of the fossil record in the Japanese Islands. For example, a relatively large number of fossil specimens have been recovered from cave deposits dating later than the mid Middle Pleistocene, while fossil records prior to this are rare. Detail composition of mammal faunas that inhabited the Japanese Islands during the Pliocene to early Middle Pleistocene and their relationship to the faunal communities of mainland Asia are still ambiguous.

Some recent paleoenvironmental studies provide new evidence, however, allowing us to better understand the formative history of the mammalian fauna of the Japanese Islands. The Ocean Drilling Program (ODP) conducted a boring survey in the Sea of Japan in 1989. The analysis of organic carbon content and diatom abundance in sediment cores clearly showed changes in the relative influx of the East China Sea coastal water into the Sea of Japan (Koizumi, 1992; Tada, 1994, 1999). These results have shed light on the timing of the land bridge exposure in the southern strait area. In addition, identifications of widely spread tephras from numerous localities on the Japanese Islands allow us to establish a reliable timescale for reconstructing an evolutionary history of the mammal and plant records (Machida and Arai, 2003).

Not enough attention, however, has been paid to reports of mammalian fauna in the Japanese Islands, and newly obtained environmental records are seldom shared among researchers. In this paper, we present a framework of the evolution for the mammal fauna in the Plio-Pleistocene in the Japanese Islands. Furthermore, we discuss the

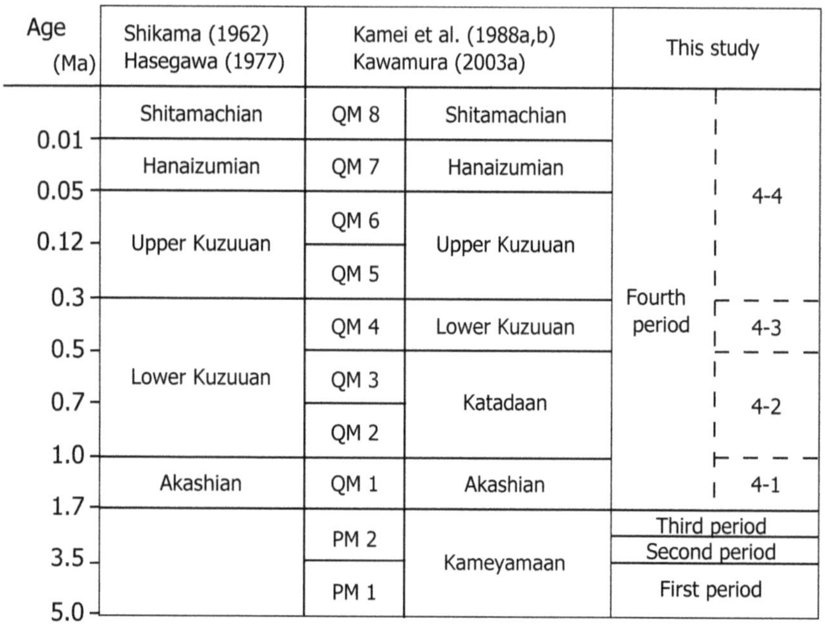

Figure 1. Comparison of divisions for the history of Pleistocene terrestrial mammals on the Japanese Islands.

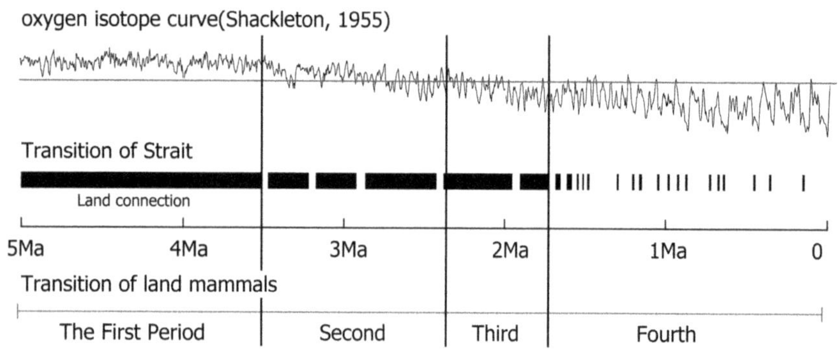

Figure 2. Oxygen isotope curve, appearance of land bridges, and history of large mammals on the Japanese Islands.

correlation between fossil and paleoenvironmental records of the Late Pleistocene in the Japanese Islands, and provide a model of large mammalian extinctions.

Formative history of the mammalian fauna in the Plio-Pleistocene of the Japanese Islands

The formative history and evolution of land mammals in the Japanese Islands have been discussed in Shikama (1962), Hasegawa (1977), Kamei *et al.*, (1988a), and Kawamura (2003a). Shikama (1962) and Hasegawa (1977) divided the history of Japanese Pleistocene land mammals into four periods (Figure 1), while Kamei *et al.*, (1988b) and Kawamura (2003a) divided the Pliocene into seven periods. Based on the paleoclimatic records and the presence of land bridges to the Asian continent, the history of terrestrial mammals of the Japanese archipelago are here divided into four periods for the Plio-Pleistocene (Figure 1, 2).

1) First Period (5.3-3.5Ma)

During this period, a land bridge was present across the current southern strait (Koizumi, 1992; Kitamura and Kimoto, 2004) (Figure 3), and the climate was relatively warmer than at present. Subtropical plant fossils in this period extended their distribution northwards in continental Asia, and they moved across the southern land bridge into the Japanese Archipelago (Suzuki and Manabe, 1982; Sohma, 1986; Momohara, 1994).

Numerous sub-tropical faunal species have been reported from Ajimu, Oita Prefecture in Kyushu (Takahashi and Kitabayashi, 2001). The fossil assemblage, consisting of Sambar (*Cervus unicolar*), Rhinocerotidae, four species of turtles including the big-headed variety (*Platysternon megacephalum*), the Chinese striped-necked turtle (*Ocadia sinesis*), and the Kwangtung River Turtle (*Chinemys* cf. *nigricans*), two species of crocodilians including *Alligator sinensis*, the giant salamander, and four bird genera representing four families in four orders, are all subtropical or tropical extant inhabitants of southern China and southeast Asia. Additionally, fossils of a large extinct stegodont, *Stegodon miensis*, have been recovered from this locality.

Although it is difficult to reconstruct the habitat of *S. miensis*, its fossils are often associated with the mammals mentioned above, suggesting that it inhabited a subtropical environment. The oldest, northern-most example of this species was found at Sendai City, Miyagi Prefecture.

Figure 3. Map of the Japanese Islands and surrounding regions showing present straits between continent and islands.

This implies that subtropical mammalian species, such as *S. miensis*, moved southwards as the Pliocene climate gradually became colder.

While the Ajimu fauna in the southern Japanese Islands was principally composed of subtropical species, the fossil flora is said to resemble that of warm, broad-leaf, deciduous forests (Iwanai and Hase, 1986, Yamakawa, 2001). This may indicate that subtropical animal species that migrated northwards across the southern land bridge during warm periods later adapted to cooler climates in the Japanese Archipelago.

2) Second Period (3.5 – 2.4Ma)

The land bridge across the southern strait between continental Asia and the Japanese archipelago submerged at 3.5Ma, allowing a warm current to flow into the Sea of Japan. Since this event, land connections between the mainland and Japan occurred several times during the Second Period, specifically at 3.2Ma, 2.9Ma and 2.4Ma (Kitamura and Kimoto, 2006). Fossils of the Osaka Group demonstrate that subtropical species of flora (e.g., *Fortunearia sinensis, Keteleeria davidiana,* and *Liquidambar* sp.) became extinct, and those of temperate coniferous and cool temperate forests appeared and expanded (Momohara, 1994).

In mainland Asia, the subtropical mammalian community that extended its range to northern China during warm periods began to migrate southwards as cooling occurred. In contrast, a large number of mammal species in Japan, unable to migrate southwards due to the isolated nature of the Japanese Archipelago, became extinct. *Stegodon miensis*, which colonized the Japanese Islands during the First Period, may have survived the Second Period climatic cooling climate in Japan by decreasing its body size. This scenario is supported by the following: (1) no fossils of *S. miensis* younger than 3Ma have been found, (2) the smaller stegodont *S. aurorae*, which evolved from *S. miensis*, appeared in the fossil record around 2.5Ma, and (3) *S. aurorae* has only been found in the Japanese Archipelago (Taruno, 1991, 1999) .

3) Third Period (2.4 – 1.7Ma)

As with the Second Period, the land connection across the southern strait to the Asian continent was discontinuous during the Third Period (Kitamura and Kimoto, 2006). The climate, however, was cooler compared with the Second Period, causing subtropical mammals to venture southwards, while temperate mammals that inhabited northern China, migrated to the Japanese Islands when land bridges were present. Thus, the Japanese mammalian fauna of this period was composed of subtropical mammals that survived in the cooler climate along with those that migrated from the Asian continent.

Mammals of this period have been grouped as the *S. aurorae-Elaphurus* fauna (Shikama, 1962; Hasegawa, 1977). In addition to *Elaphurus*, it included a large canid (*Canis falconeri*), and Kazusa deer (*Cervus kazusensis*). Reliable fossil evidence of *S. aurorae* first occurs in the Mitsumatsu Tephra of the Osaka Group dated at 1.8Ma. *S. aurorae*-like archaic species are recovered from layers between the upper level of the Gauss Normal Chron and the lower level of the Matsuyama Reverse Chron (Taruno, 1999). Aiba *et al.* (2010) proposed a new species under the name of *A. protoaurorae* for the *S. aurorae*-like archaic species. Fossils of *Elaphurus, C. kazusensis* and *Canis falconeri* first occur in levels dated around 1.8Ma. The Kuchinotsu Group in Kyushu Island yielded *S. aurorae* along with various species of deer, such as *C. kazusensis, C. japonicas,* and *C. shimabarensis*. Deer similar to these species have been found from Chinese Early Pleistocene sites, such as Shansi Hsihoutu, Shansi Lantian, and Hebei Nihewan. Fossils of *Canis falconeri* have been reported from the Wushan site in Sichuan Province dated to 2 – 1.8Ma (Koizumi, 2003). Although these dates from the Chinese sites still have problems, it is probable that these temperate mammals in the Asian Continent migrated to the Japanese islands during this period.

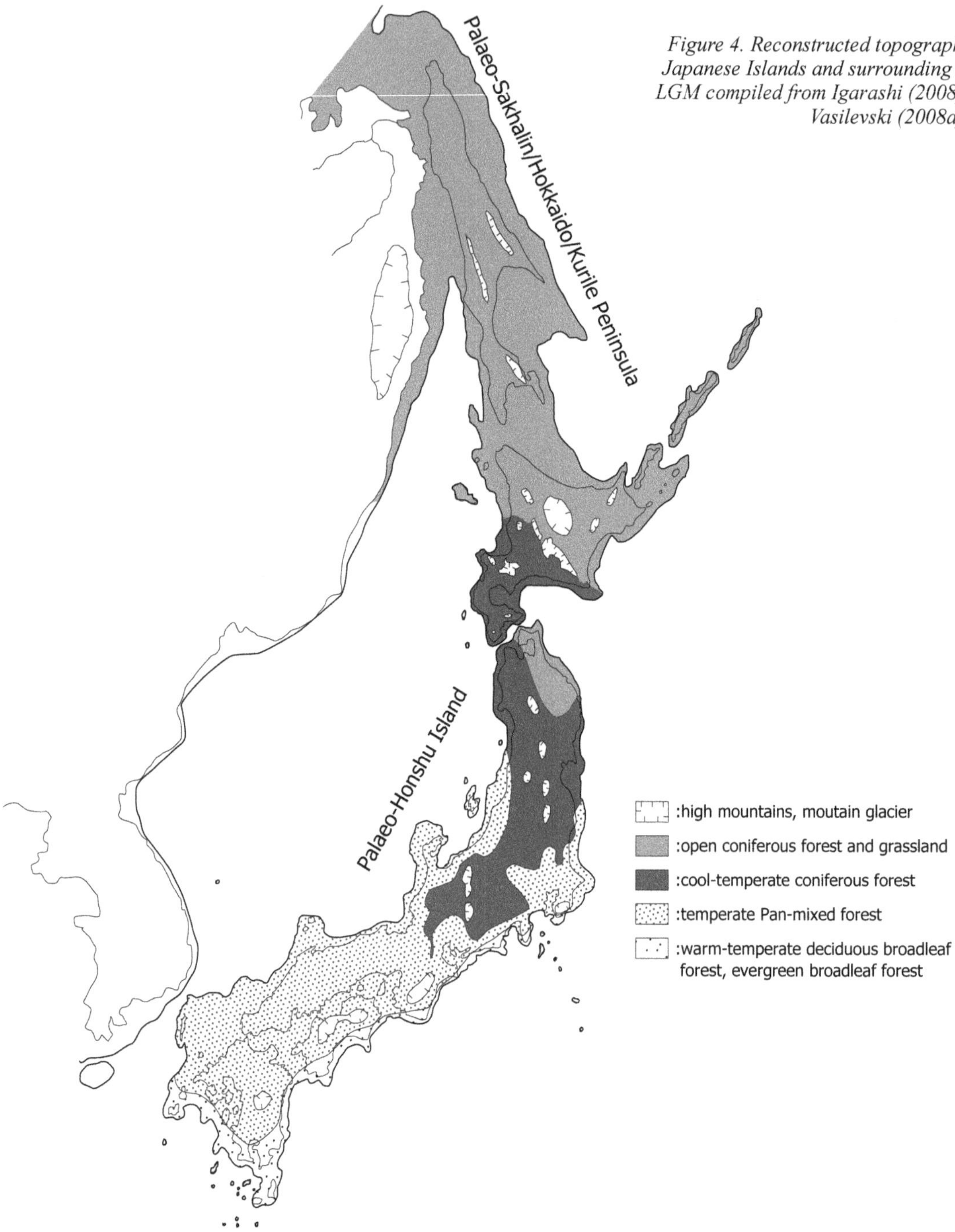

Figure 4. Reconstructed topography and flora of the Japanese Islands and surrounding regions during the LGM compiled from Igarashi (2008), Tsuji (2004), and Vasilevski (2008a)

4) Fourth Period (1.7 – 0.01Ma)

This period is quite distinct from the rest; glacioeustatic sea level oscillations resulted in the presence of land bridges across the southern strait for geologically short durations when the sea level lowered. (Tada and Irino, 1994; Kitamura *et al.*, 2001). During the existence of the land bridges, faunal migration occurred from the northern area of China. Since the last land bridge at the Korea and Tsushima Straits area was present approximately 130ka, the formation of the present Japanese fauna, including endemic species, accelerated after this time. Later in this paper, we will more fully describe the formation history of mammals in this period.

Mammalian fauna of the Late Pleistocene

1) Appearances of land bridges during the Late Pleistocene

Figure 4 shows a reconstructed topography of the Japanese Islands and surrounding regions during the Last Glacial Maximum (LGM) (ca. 24 – 19 ka). The land masses consisting of the Japanese Islands were composed of two main regions. Hokkaido was a part of the Paleo-Sakhalin/Hokkaido/Kurile Peninsula by connection with a land bridge between the Sakhalin and Kurile Islands (Kunashiri and Shikotan Islands). Honshu was connected to Shikoku and Kyushu to form Paleo-Honshu Island. This island would not have been connected to the Paleo-Sakhalin/Hokkaido/

Kurile Peninsula during the Last Glacial, although distances across the straits were shortened to only a few to a dozen kilometers. Since the Sea of Japan was rarely influenced by currents flowing from the Pacific Ocean, it was a lake-like enclosed sea. Besides the fact that there was no inflow of warm currents onto the Sea of Japan, major currents, including the Kuroshio, were a long way from the coasts. This resulted in the entire archipelago becoming influenced by a continental climate, with colder and drier conditions, unlike the present Holocene warm and humid climate. Although small changes in the terrestrial areas did occur in response to short climatic fluctuations, no dramatic changes related to the Asian continent happened until the onset of the Holocene.

The flora in this period was conditioned by these climates, and therefore they were composed of plant species that were more tolerable to cold climates than the present flora (Igarashi, 2008). In northern and eastern Hokkaido, a mosaic of cold grassland and coniferous forest was present. The region of western Hokkaido and eastern Paleo-Honshu was covered with cold temperate coniferous forest, and the western half of Paleo-Honshu was dominated by temperate mixed broad-leaf and coniferous forest. The warm deciduous broad-leaf forest presently distributed in southwestern Japan was restricted to the southern shore of Paleo-Honshu Island (Tsuji, 2004) (Figure 4).

2) Two faunas in the Japanese Islands

During the Late Pleistocene, the land bridges that greatly impacted the distribution of Japanese flora and fauna disappeared, although the northern land bridge that connected Hokkaido and eastern Siberia lasted until the end of the Pleistocene (Ono, 1991). This northern land bridge, however, was narrow and the north entrance was situated in the extremely cold region of the Amur River mouth. Moreover, climatic zones and vegetation distribution varied from north to south on this land bridge. These conditions imply that mammal migrations to the Japanese Archipelago via the northern land bridge made a smaller contribution to the Japanese fauna than those of the southern land bridge.

The paleogeographical history indicates that the Late Pleistocene fauna was formed by cessation of migration via the western land bridge during the Middle Pleistocene (130 ka or earlier), coupled with population isolation in the Japanese Archipelago (Kamei *et al.*, 1988a; Kawamura, 1998; Kawamura *et al.*, 1989).

The Late Pleistocene fauna can be divided into two groups—one with Naumann's elephant (*Palaeoloxodon naumanni*), and another with mammoths. The former is composed of *P. naumanni*, giant deer (*Sinomegaceros yabei*), extinct cervid (*Cervus praenipponicus*), Sika deer (*C. nippon*), brown bear (*Ursus arctos*), marten (*Martes melampus*), least weasel (*Mustera nivalis*), European badger (*Meles meles*), raccoon dog (*Nycterutes procyonoides*), Japanese monkey (*Macaca fuscata*), wolf (*Canis lupus*), and fox (*Vulpes vulpes*). Hasegawa (1972) called this group the *Palaeoloxodon- sinomegaceroides* complex. All middle to large-sized fossil mammals found on the Japanese Islands comprise only these species, and most of them migrated from continental Asia across the western land bridge during the mid Middle Pleistocene (ca. 0.4-0.3 Ma), or 0.13Ma. Notable Chinese sites, such as Zoukoudian Locality 1 (mid Middle Pleistocene), the Dali site (mid Middle Pleistocene), the Diangcun site (early Late Pleistocene), and the Xujiayao site (early Late Pleistocene), have yielded large to middle sized mammals, including Naumann's elephant, tigers, wolves, foxes, raccoons, black bears, brown bears, hyenas, cats, horses, rhino, boars, giant deer, grey deer, red deer, gazelles, bovids, sheep, and camels (Wu *et al.*, 1989).

In the Japanese Islands, it was only later than MIS 10 (ca. 0.36-0.34 Ma) when these mammals including Naumann's elephant appeared (Taruno and Kamei, 1993). A small number of mammal species, such as Naumann's elephant, tigers, wolves, horses, grey deer, and boars, have been found dating to this time, and they are similar to the *P. naumanni* faunal group of continental China. This similarity clearly suggests that the mammalian species in the Middle Pleistocene and early Late Pleistocene of continental China migrated to the Japanese Islands via the western land bridge. The characteristics of these mammals are: (1) their principal habitat was in temperate climates, (2) the majority was forest-dwelling animals, and (3) they have a high percentage of endemic species (Kawamura, 1991).

The mammoth fauna group, mostly consists of species that lived in grassland. In Siberia, it includes medium to large animals, including steppe bison (*Bison oriscus*), European ass (*Equus hydruntinus*), red deer (*Cervus elaphus*), giant deer (*Megaroceros giganteus*), musk ox (*Ovibos moschatus*), woolly rhinoceros (*Coelodonta antiquitatis*), wolverine (*Gulo gulo*), polar fox (*Alopex tagopus*), cave hyena (*Crocuta spelaea*), woolly weasel (*Mustela putorius*), reindeer (*Raingifer tarandus*), cave lion (*Pantera soelea*), snowshoe rabbit (*Lepus timidus*), camargue horse (*Equus caballus*), cave bear (*Ursus spelaeus*), saiga antelope (*Saiga tatarica*), European ground squirrel (*Spermophilus citellus*), wolf (*Canis lupus*), Eurasian badger (*Meles meles*), brown bear (*Ursus arctos*), beaver (*Castor fiber*), and moose (*Alces alces*). Among these species, the fossils associated with mammoths in Siberia and Alaska are bison, horse, and reindeer (Lister and Burn, 2007).

Sakhalin Island on the northern part of the Paleo-Sakhalin/Hokkaido/Kurile Peninsula has yielded species of the mammoth faunal assemblage such as mammoths, brown bear, Siberian musk deer, horse, moose, snow sheep, *Panther* sp., wolf, and arctic fox (Kirillova, 2003; Kuzmin *et al.*, 2005; Vasilevski, 2008a, 2008b). The mammoth fauna in the Japanese Islands includes mammoths and bison only from localities in Hokkaido, and moose from Hanaizumi Town (Iwate Prefecture), Kaza-ana Cave (Iwate Prefecture), Lake Nojiri (Nagano Prefecture), and Kumaishido (Gifu Prefecture) (Hasegawa and Matsushima, 1968; Okumura *et*

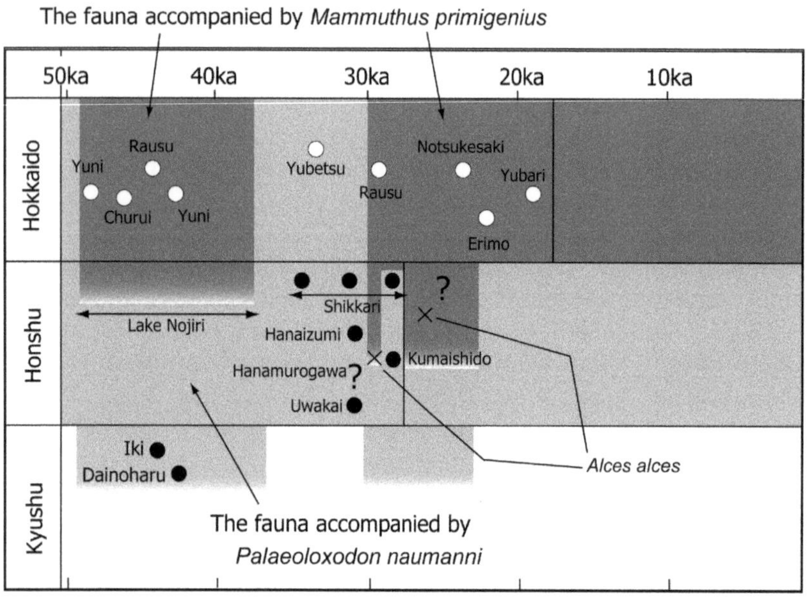

Figure 5. Repeated migrations of the Mammoth fauna and the Naumann's elephant fauna across the Japanese Islands after 50 ka. Age of Alces alces is estimated position.

al., 1978; Takakuwa and Fossil Mammal Research Group for Nojiriko Excavation, 2009; Kawamura, 2003b).

Among the localities bearing moose, the fourth layer of Kaza-ana Cave, Ohazama (Iwate Prefecture) has yielded the Shinto Shrew (*Sorex Shinto*), Japanese shrew-mole (*Dymecodon pilirostris*), and Japanese mountain mole (*Euroscaptor mizura*), the latter of which now lives in the high mountains of Honshu, and were obviously members of the mammoth fauna that migrated into Honshu during previous periods. These small to middle-sized animals in the mammoth fauna are thought to have migrated to the south according to their habitat and temperature preferences. Hence, it is hypothesized that the basic components of the mammoth fauna reduced in number while migrating south.

All Japanese fossil records of mammoth have been found in Hokkaido, and none in the Honshu Islands. This suggests that the grassland in which mammoths could survive was limited to Hokkaido in the Japanese Archipelago. In fact, the pollen records in the Kenbuchi Basin, northern Hokkaido have demonstrated that the basin was extensively covered with grassland during the period approximately 25 to 16 ka (Igarashi *et al.*, 1993).

The present distribution of moose extends from northern North American to between 85-65° N in Eurasia. Their habitat is coniferous forest dominated by birch and spruce, river deltas, shrub tundra, and coastal forest. They suffer stress when winter temperatures are above -5°C and summer temperatures are above 14°C, and do not survive when temperatures above 27°C last for a long duration (Rodgers, 2001). During the LGM, the sub cold-temperate coniferous forest extended to the southern Japanese Islands (Morita *et al.*, 1998), in which moose likely inhabited. In addition, moose typical of cold climates, and *P. naumanni* and *S. yabei*, which both lived in temperate conditions, were found from in Hanaizumi (Iwate Prefecture) and Kumaishido (Gifu Prefecture), although their contemporaneity is questioned as we discuss below.

3) Repeated migration of two faunal groups

In the Late Pleistocene, the fauna that lived in the Japanese Islands consisted of the two groups—one with Naumann's elephant (*P. naumanni*) and giant deer that lived in temperate conditions, and the other with mammoths and moose that lived in cold einvironments. How did these faunal groups change their habitats in response to drastic short-cycles of climatic shifts? Studies have recently progressed with respect to this question. In Hokkaido, twelve specimens of mammoths and four specimens of *P. naumanni* have been reported. For the mammoth, AMS radiocarbon dates were successfully obtained from seven of nine specimens including two that were already dated (Table 1) (Takahashi *et al.*, 2004, 2006).

The oldest date obtained is 45,110±480RCYBP and the youngest date is 16,320±90RCYBP for mammoths from the Japanese Islands. During these times in Hokkaido, vegetation was dominated by coniferous forest, which now has retreated to Sakhalin Island, and grassland, similar to the climate and vegetation in Siberia where mammoths lived.

Among the four specimens of *P. naumanni* reported, the Kuriyama and Churui specimens are dated to MIS 5 (Nojo *et al.*, 2002; Takahashi *et al.*, 2010), and the Yubetsu specimen is dated to 30,520±220RCYBP (Takahashi *et al.*, 2004). A short warm period is recognized between 34 – 26 ka in MIS3 of Hokkaido, when cool temperate forests, mainly consisting of *Betula* sp., *Alnus* sp., *Ulmus* sp., *Corylus* sp., as well as *Tsuga* sp., extended its distribution (Igarashi, 2008). During this period, mammoths retreated north to the southern limit of the cool temperate forest, and *P. naumanni* extended north to areas where mammoths once

Table 1. Radiocarbon dates for terrestrial mammals on the Japanese Islands.

Specific name	Material	Locality	^{14}C age			Reference
			Lab Code	Method	(BP)	
Mammuthus primigenius	Right upper third molar	Yubari City, Hokkaido but exact point unknown	Beta-187606	AMS	16,320±90	Makiyama(1938), Takahashi *et al.*(2005, 2006a)
Mammuthus primigenius	Right upper second molar	Ogoshi, Erimo-cho, Horoizumi-gun, Hokkaido	Beta-188519	AMS	19,580±80	Minato(1955, 1967), Matsuzawa and Kosaka(1987)
Mammuthus primigenius	Right lower third molar	North of Notsukewsaki in the Nemuro Channel, Hokkaido	not reported Beta-184269	AMS	20,248±670 20,770±120	Akiyama *et al.*(1989), Takahashi *et al.*(2005, 2006a)
Mammuthus primigenius	Right lower third molar	6 miles point from the Rausu Port toward the Kunashiri Island, Hokkaido	not reported Beta-185830	AMS	23,816±884 25,010±120	Nakai *et al.*(1989), Takahashi *et al.*(2005, 2006a)
Mammuthus primigenius	Right lower third molar	30 km off Onsentsu, Shimane Pref.	NUTA-633	AMS	23,680±880	Kamei(1990) Akiyama *et al.*(1992)
Mammuthus primigenius	Right upper first molar	Higashi-mikawa, Yuni-cho, Yubari-gun, Hokkaido	IAAA-32222	AMS	45,110±480	Ono(1991), Nakaya *et al.*(1992) , Takahashi *et al.*(2005, 2006a)
Mammuthus primigenius	Right lower third molar	16 km southeast off the Rausu Port, Hokkaido	Beta-85090	AMS	38,920±760	Yamada *et al.*(1996)
Mammuthus primigenius	Right lower first molar	1 km off the Notsukesaki, Hokkaido	Beta-184935	AMS	>43,100	Takahashi *et al.*(2005,2006a)
Palaeoloxodon naumanni	Right upper second molar	Higashibaro, Yubetsu-cho, Monbetsu-gun, Hokkaido	Beta-134800	AMS	30,520±220	Takahashi *et al.*(2004)
Shinomegaceros yabei	Antler	Higashi-mikawa, Yuni-cho, Yubari-gun, Hokkaido	not reported	AMS	49,250±740	Ono(1991) Nakaya *et al.*(1992)
Palaeoloxodon naumanni	Right upper second or third molar	Shikkari, Higashi-douri, Shimokita-gun, Aomori Prefecture	IAAA-52429	AMS	31,470±160	Takahashi et al(2006b)
Palaeoloxodon naumanni	Right upper second molar	Shikkari, Higashi-douri, Shimokita-gun, Aomori Prefecture	IAAA-53431	AMS	23,570±130	Takahashi *et al.*(2006b)
Palaeoloxodon naumanni	Right upper second or third molar	Shikkari, Higashi-douri, Shimokita-gun, Aomori Prefecture	IAAA-52430	AMS	25,920±120	Takahashi *et al.*(2006b)
Palaeoloxodon naumanni	Molar fragment	Hanaizumi, Hanaizumi-cho, Nishiiwai-gun, Iwate Prefecture	GaK-15893	β-ray	21,430±1,260	Hanaizumi Site Excavation Research Group(1993)
Palaeoloxodon naumanni	Piece of wood under the molar yielding horizon	River bed of Hanamuro River, Tsukuba City, Ibaragi Prefecture	Gak-13276	AMS	27,340±860	Nakajima *et al.*(2004)

Specific name	**Material**	**Locality**	**^{14}C age**			**Reference**
			Lab Code	**Method**	**(BP)**	
Palaeoloxodon naumanni	Piece of wood from the molar yielding horizon	Ubafutokoro, Odanaka, Nakano City, Nagano Prefecture	Gak-3774 Gak-3775 Gak-3776	β-ray	19,350±600 ～31,700+3,500-2,400	Fossil Elephant Research Group (1973), Tomizawa (1979)
Palaeoloxodon naumanni	Molar and tusk	Lake shore of Shinanomachi, Kamiminochi-gun, Nagano Prefecture	31 dates obtained	AMS	48,799±1950 ～31,920±700	Sawada *et al.*(1992)
Palaeoloxodon naumanni	Right upper third molar	Kumaishido cave, Miyama, Gujyo, Gifu Prefecture	not reported	AMS	23,960±200	Yasui *et al.*(2004)
Palaeoloxodon naumanni	Left tusk	Off Hinomisaki, Taisya-cho, Izumo, Shimane Prefecture	NUTA-477	AMS	29,000±300	Kamei(1967) Akiyama *et al.*(1988)
Palaeoloxodon naumanni	Left tusk	Off Hinomisaki, Taisya-cho, Izumo, Shimane Prefecture	NUTA-478	AMS	38,500±600	Akagi(1981) Akiyama *et al.*(1988)
Palaeoloxodon naumanni	Left tusk	Off Hinomisaki, Taisya-cho, Izumo, Shimane Prefecture	Gak-12898	β-ray	35,560±1,300	Hoshimi and Morioka(1987)
Palaeoloxodon naumanni	Left lower third molar	Off Natori, Misaki-cho, Nishiuwa-gun, Ehime Prefecture	NUTA-5722	AMS	29,200±870	Nakamura *et al.*(1998)
Palaeoloxodon naumanni	Left lower third molar	Off Morishima Islands, Towa-cho, Ohshima-gun, Yamaguchi Prefecture	not reported	AMS	24,280±190	Hitagawa *et al.*(2006)
Palaeoloxodon naumanni	Piece of wood from the horizon	Dainoharu, Ohno-cho, the Ohita Prefecture,	not reported	β-ray	37,250±1,880	Inada(1989)
Bison sp.	Right horn	Seashore of Yakumo-cho, Yamakoshi-gun, Hokkaido	Beta-87674	amino acid?	Ca.18,000	Kimura(2004)
Ursus thibetanus japonicus	Left forth meta- carpal bone	Abakuchi cave site, Ohohazama- machi, Hienuki-gun, Iwate Prefecture	Beta-117393	AMS	46,260±1100	Dodo *et al.*(2003)
Elephantidae gen.et sp.indet.	Fragment of femur	Kaza-ana cave site, Ohohazama- machi, Hienuki-gun, Iwate Prefecture	Beta-117392	AMS	18,140±60	Dodo *et al.* (2003)
Bovidea gen.et sp.indet	Right lower second molar	Hanaizumi, Hanaizumi-cho, Nishiiwai-gun, Iwate Prefecture	GaK-15798	β-ray	18,470±660	Hanaizumi Site Excavation Research Group(1993)
Unknown	Bone fragment	Suse quarry, Kuzu-machi, Aso-gun, Tochigi Prefecture	Gak-11495	β-ray	14,710±670	Kawamura(1988)
Unknown	Bone fragment	Yage quarry, Hamamatsu City, Shizuoka Prefecture	Gak-11494	β-ray	18,040±990	Kawamura(1988)
Shinomegaceros yabei	Piece of wood 5.4m under Left antler	Tochu, Akashinamachi, Higashi-chikuma-gun, Nagano Prefecture	Gak-161	β-ray	15,750±390	Kobayashi (1965)
Shinomegaceros yabei	Antler and limb bones	Lake shore of Shinanomachi, Kamiminochi-gun, Nagano Prefecture	6 dates obtained	AMS	41,250±1190 ～30,583±1291	Sawada *et al.*(1992)
Palaeoloxodon naumanni, *Shinomegaceros yabei*, *Cervus* (*Nipponicervus*) *praenipponicus*	Bone fragment	Kumaishido cave, Miyama, Gujyo, Gifu Prefecture	Gak-7007	B-ray	16,720±880	Okumura *et al.*(1982)

Specific name	*Material*	*Locality*	^{14}C *age*			*Reference*
			Lab Code	*Method*	*(BP)*	
Canis lupus	Right ulna	Kumaishido cave, Miyama, Gujyo City, Gifu Prefecture	not reported	AMS	31,010±320	Yasui *et al.*(2002)
Some kinds of mammal	Tephra, Tufa	Kannondo cave site, Jinsekikougen-cho, Hiroshima Prefecture		β-ray amino acid	>Ca.16000-31,900,	Kawamura(1988)
Large size deer	Tibia	"Kuchimino-se", east of Amami- oshima Island, Kagoshima Prefecture	NUTA-3144	AMS	25,740±220	Nakamura *et al.* (1996)

lived (Takahashi *et al.*, 2006). Consequently, the boundary of these two faunal groups that migrated back and forth north to south matched the range of vegetation that shifted with climatic fluctuations (Figure 5).

Formation of the present Japanese fauna

In the Japanese Islands, the extinct large to middle sizes mammals are mammoths, Naumann's elephant, and giant deer. We will discuss the newest absolute dates (Table 1).

The youngest AMS dates for *P. naumanni* are around 23,000RCYBP— 23,570±130RCYBP from the Shikkari specimen, Aomori Prefecture (Takahashi *et al.*, 2006), and 23,960±200RCYBP from Kumaishido, Gifu Prefecture. Although it is well quoted, the date of 16,720±880RCYBP for *P. naumanni* from Kumaishido (Gifu) was in fact obtained from combined specimens of *P. naumanni, S. yabei*, and *C. praenipponicus*, not only *P. naumanni* itself (Okumura *et al.*, 1982). Therefore, we exclude this date from our discussion. We also omit the dates measured by the Beta method, since it skews dates a few thousand years too young (Sawada *et al.*, 2004). Thus, the specimens of *P. naumanni* supposedly younger than 23 ka are unreliably dated, suggesting *P. naumanni* was nearly extinct at that time, although it could have survived later in some regions.

The youngest AMS date for *S. yabei*, a faunal species in the Naumann's group, is about 31ka from Lake Nojiri, Nagano Prefecture. Also, since there are no reliable younger dates, *S. yabei*, like *P. naumanni*, possibly survived until around 23 ka.

In contrast, the youngest date for mammoths that were adapted to the colder climate in the Japanese Islands is 16,320±90RCYBP from the Yubari specimen, although its origin is unknown and it is even doubtful if it was from Japan (Takahashi *et al.*, 2006). The dates of 19 ka for the two specimens from Erimo are likely reliable. As we saw above, the landscape during the LGM in Hokkaido was covered with grassland and open spruce-larch-fir-oak taiga, very similar to the habitat of Siberian mammoths. After the LGM, the grassland rapidly decreased in Hokkaido, implying that mammoths started to decline. Therefore, the group with mammoths migrated into Hokkaido via Sakhalin in the colder period of the Late Pleistocene, and large to medium-sized mammals, such as mammoths and moose, disappeared at the end of the LGM. After 10 ka the Soya Strait formed as sea level rose in response to Holocene warming, and the present faunal communities in the Japanese Islands emerged. Our study, based on the dates and distributions of fossils, demonstrates that formation processes of modern faunal communities in the Japanese Islands are explained by climate fluctuations and associated environmental changes.

Human Impact on Megafauna Extinctions in the Japanese Archipelago

How did the emergence of behaviorally modern humans

related to the extinction events of large mammals in the Japanese Islands? It is certain that humans inhabited the Japanese Islands since the onset of the Early Upper Paleolithic 40-35ka (Izuho, 2010; Izuho and Sato, 2008). This suggests that mammal extinctions and the emergence of modern humans were not mutually related. Our model implies that *P. naumanni* was extinct at the onset of LGM in Honshu and *Mammuthus primigenius* was extinct after the LGM when the climate started to become warmer. The prime movers of the large mammal extinctions in the Japanese Islands might have been changes of ecosystems driven by climatic fluctuations. It is necessary to evaluate the causes of mammal extinction, however, by examining details of how humans impacted ecosystems and their hunting pressured large mammal populations.

As with the Japanese Islands, there was also a faunal group that included the genus *Palaeoloxodon* to the south of a group with mammoth in the eastern area of the Asian continent. Likewise, it is postulated that these two groups also migrated north and south in response to climate change as in Japan. The northernmost record of *Palaeoloxodon* in this area was around 41° N, and the definite southernmost woolly mammoth record is a specimen from Ji'nan (around 36° N), Shandong Province, China (Takahashi *et al.*, 2007). The AMS ^{14}C date of this latter specimen is 33,150±250 BP. It is known that the winter monsoon expanded in Asia 35-33 ka, and the age of the woolly mammoth specimen from Ji'nan corresponds to this period. The specimen suggests that this area became cold and dry at 33 ka, and grassland or open forest suitable habitat for woolly mammoth developed. This age is similar to that of the southernmost woolly mammoth in Europe, which supports a hypothesis that an important component of Chinese paleoclimate was linked to changes in the North Atlantic oceanic conditions. Recent studies of Eurasian fossil records conducted within the same framework as studies on the Japanese Islands have suggested that vegetation changes due to climatic changes were the dominant factors of mammal extinctions (e.g., Stuart *et al.*, 1999, 2002, 2004; Kuzmin and Orlova, 2004).

The direct evidence of fossil records and inferred animal behavior in Eurasia and the Japanese Islands implies that the large mammal extinctions were not caused by human factors such as mass killings. Rather, the fragmentary data examined in this paper are consistent with climatic change and subsequent vegetation shifts as the main causes.

Acknowledgments

We deeply thank Drs. Robin J. Smith (Lake Biwa Museum) and Yuichi Nakazawa (Zao Board of Education) for correcting the English.

References

Aiba, H., Baba, K., and Matsukawa, M., 2010, A new species of *Stegodon* (Mammalia, Proboscidea) from the Kazusa Group (Lower Pleistocene), Hachioji City, Tokyo, Japan and its evolutionary morphodynamics. *Palaeontology*, 53, pp.473-490.

Akagi, S., 1981, San'in oki kaitei no naumanzo kaseki [*Palaeoloxodon naumanni* (Makiyama) from the sea bottom off the coast of San'in]. *The Journal of the Faculty of Education, Tottori University, Natural Science*, 30, pp.57-69. (in Japanese)

Akiyama, M., Kamei, T. and Nakai, N., 1988, ^{14}C ages of Naumann's elephant *Palaeoloxodon naumanni* (Makiyama) from the sea bottom off the San'in district in the Sea of Japan by accelerator mass spectrometry: ^{14}C age of the Quaternary deposits in Japan (168). *Earth Science (Chikyu Kagaku)*, **42(1)**, pp.29-31. (in Japanese)

Akiyama, M., Kamei, T., and Nakai, N., 1989, Kaiteisan zo kaseki no ^{14}C nendai [^{14}C age of elephant fossil from sea bottom]. *Journal of Fossil Rearch*, **22**, pp.22-23. (in Japanese)

Akiyama, M., Nakamura, T., and Hoshimi, K., 1992, Kasokuki shitsuryo bunseki kei niyoru nihonkai san'in oki kaitei san no honyurui kaseki no ^{14}C nendai – nihon no daiyonki so no ^{14}C nendai - [^{14}C ages of three mammalian fossils from the sea bottom off the San'in district in the Japan Sea using an accelerator mass spectrometer: ^{14}C age of the Quaternary deposits in Japan (175)]. *Earth Science (Chikyu Kagaku)*, **46**, pp.241-242. (in Japanese)

Dodo, Y., Takigawa, W. and Sawada, J., 2003, *Kitakami sanchi ni nihon koshinsei jinrui kaseki wo saguru: iwateken Ohasama machi Abakuchi Kaza-ana dokutsu iseki no hakutsu [Search for Japanese Pleistocene Human Remains in the Kitakami Mountains: Excavation of the Abakuchi and Kaza-ana Cave Site in Ohasama, Iwate Prefecture*], 419p, Tohoku University Press, Sendai.

Fossil Elephant Research Group, 1973, Naganoken Chiisagata gun Maruko machi san no *Stegodon aurorae* [*Stegodon aurorae* from Maruko-machi, Chiisagata-gun, Nagano Prefecture]. *Journal of Faculty of Science, Shinshu University*, **8**, pp.65-79. (in Japanese)

Hanaizumi Site Excavation Research Group [Hanaizumi iseki hakkutsu chosadan], 1993, *Hanaizumi iseki. Hanaizumi machi kyoikuiinkai, [The report of Hanaizumi site excavations in Iwate Prefecture, northern Japan. Hanaizumi Boad of Education*], 161p. (in Japanese)

Hasegawa, Y., 1972, The Naumann's Elephant, *Paleoloxodon naumanni* (Makiyama) from the Late Pleistocene off Shakagahana, Shodoshima Is. in Inland Sea, Japan. *Bulletin of the National Science Museum*, **15-3**, pp.513-591, pls.1-22.

Hasegawa, Y., 1977, Sekitsui Dobutsu no Hensen to Bunpu [Vertebrates]. In Japan Association for Quaternary Research (ed.) *Nihon no Daiyonki Kenkyu: Sono Hatten to Genjyo*. pp.227-243. Tokyo University Press, Tokyo. (in Japanese)

Hasegawa, H., and Matsushima, Y., 1968, First Discovery of Fossil Elk Deer Antler from Japan. *Bulletin of the National Science Museum*, **11**, pp.77-84, pl.1.

Hoshimi, K., Morioka, H., 1987, San'in oki kaitei san naumanzo kaseki no ^{14}C nendai: Nihon no dai yonki so no ^{14}C nendai [^{14}C dating of *Palaeoloxodon naumanni* (Makiyama) from the sea bottom off the coast of San'in: ^{14}C dating of the Quaternary deposits on the Japanese Archipelago]. *Earth Science (Chikyu Kagaku)*, **41(4)**, pp.248-250. (in Japanese)

Igarashi, Y., Igarashi, T., Daimaru, H., Yamada, O., Miyagi, T., Matsushita, K., Hiramatsu, K., 1993. Hokkaido no kenbuchi bonchi to hurano bonchi niokeru 32,000 nenkan no shokusei hensen [Vegitation History of Kenbuchi Basin and Furano Basin in Hokkaido, North Japan, since 32,000 yrs BP]. *Quaternary Research*, **32**, pp.89–105. (in Japanese with English abstract).

Igarashi, Y., 1993, History of environmental changes in Hokkaido from the view point of palynological research. In: Higashi, O., Osawa, A., and Kanagawa, K. (eds.), *Biodiversity and ecology in the northernmost Japan*: 1-19, Hokkaido University. Press, Sapporo.

Igarashi, Y. 2008, Climate and Vegetation Changes since 40,000 Years BP in Hokkaido and Sakhalin. In Sato, H. (ed) *Human Ecosystem Changes in the Northern Circum Japan Sea Area (NCJSA) in Late Pleistocene*. pp.27-41. Research Institute for Humanity and Nature. Kyoto.

Inada, K., 1989, *Honyudobutsu kaseki no sanjo to kyusekkibunka* [*Mode of occurrence of mammalian fossils and palaeolithic culture in Japan*]. Publications of Okayama University, Faculty of Letters, **2**, 239p. (in Japanese)

Iwanai, A. and Hase, Y., 1986, Chu Hokubu Kyushu Koki Shinsedai no Shokusei to Kokankyo: Sono 2 Ajimu- Unnai Chiki (Jyobu Senshinto) [Late Cenozoic Vegetation and Paleoenvironment of Northern and Central Kyushu, Japan: Part 2 Ajimu-Innai area (Upper Pliocene)]. *Chishitsugaku Zasshi* , 92, pp.591-598. (in Japanese)

Izuho, M., 2010, Nihon Retto no Jyobu Kyusekki Jidai Zenhanki Kenkyu no Ichi Shiten: Gendaijinteki Kodo Tayosei to Hen'i no Hatsugen [A Note on the Upper Paleolithic Archaeology on Japanese Archipelago: Toward an Improvement Understanding]. *Jinbun Gakuho*, 38, pp.1-12. (in Japanese)

Izuho, M. and Sato, H., 2008, Landscape Evolution and Culture Changes in the Upper Paleolithic of Northern Japan. In A. P. Derevianko and Shunkov, M. V. (eds.) *The Current Issues of Paleolithic Studies in Asia; Proceedings of the International Symposium "The Current Issues of Paleolithic Studies in Asia and Contiguous Regions"*, pp.69-77, Publishing Department of the Institute of Archaeology and Ethnography SB RAS. Novosibirsk.

Kamei, T., 1967, Nihonkai nanbu no kaitei no naumanzo kaseki [Occurences of fossil Proboscidea, *Elephas naumanni* (Makiyama) from the sea bottom of southern Japan Sea]. *Tsukumo Chigaku*, **2**, pp.24-31. (in Japanese)

Kamei, T., 1990, Nihonkai to zo [The Japan Sea and elephant]. *The Quaternary Research*, **29(3)**, pp.163-172. (in Japanese with English abstract)

Kamei, T., Taruno, H., and Kawamura, Y. 1988a, Nihon Retto no Daiyonki Chishi heno Honyu Dobutsuso no Motsu Igi [Implication of Mammal Fauna for the Quaternary Geohistory of the Japanese Islands]. *Daiyonki Kenkyu* [*The Quaternary Research*], **26** (**3**), pp.293-303. (in Japanese with English abstract)

Kamei, T., Kawamura, Y. and Taruno, H., 1988b, Nihon no Daiyonkei no Honyu Dobutsu Kaseki niyoru Buntai [Mammalian stratigraphy of the Late Neogene and Quaternary in the Japanese Islands]. *Memoir of Geological Society of Japan*, 30. pp.181-204. (in Japanese)

Kawamura, Y., 1988, Quaternary rodent faunas in the Japanese Islands (Part 1). *Memoirs of the Faculty of Science, Kyoto University, Series of Geology and Mineralogy*, **53**(**1**,**2**), pp.31-348.

Kawamura, Y., Kamei, T. and Taruno, H., 1989, Nihon no Chu/ Koki Koshinsei no Honyu Dobutsuso [Middle and Late Pleistocene Mammalian Faunas in Japan]. *Daiyonki Kenkyu* [*The Quaternary Research*], 28, pp.317-326. (in Japanese with English abstract)

Kawamura, Y., 1989, Quaternary Rodent Faunas in the Japanese Islands (Part 2). *Memoirs of the Faculty of Science, Kyoto. University*, Series of Geology and Mineralogy, **54**, pp.1-235

Kawamura, Y., 1991, Naumanzo to Kyozon shita Honyurui [Mammalian species accompanied by *Palaeoloxodon Naumanni*]. In: Kamei, T. (ed.), *Nihon no Chobirui Kaseki*, pp.164-170. Tsukiji Shokan. Tokyo. (in Japanese)

Kawamura, Y., 2003a, Honyurui Kaseki Sojo [Mammalian Biozonation]. In Machida, H., Oba, T., Ono, A., Yamazaki, H., Kawamura, Y. and Momohara, A. (eds.) *Daiyonkigaku*, pp.245-248. Asakura Shoten, Tokyo. (in Japanese)

Kawamura, Y., 2003b, Kaza-ana douketsu no kanshinsei oyobi kokikoshinsei no honyurui itai [Holocene and Late Pleistocene Mammalian Remains in the Kaza-ana Cave]. In: Dodo, Y., Takigawa, W. and Sawada, J. (eds.), *Kitakamisanchi ni Nihon Kokikoshinsei Jinrui Kaseki wo Saguru*, pp.284-386. Tohoku University Press. Sendai. (in Japanese)

Kimura, M., 2004, Hokkaido naisan honyu dobutsu kaseki no kenkyushi [Research history of mammalian fossils from Hokkaido, Japan]. Kyodo to Shizen [*Nature and Science*], **116**, pp.34-56. (in Japanese)

Kirillova, I. V., 2003, Ostatki Pozvonochnyikh iz Grota Tronnyii (Tsentralinyii Sakhalin) [Mammal Fauna from Grotto Tronnyii (Sentral Sakhalin)]. *Kraevedcheskii Byulleteni*, **2**, pp.128-133. Sakhalinskii Gosudarstvennyii Universitet. Yujno-Sakhalinsk. (In Russian)

Kitagawa, H., Seto, K., Takahashi, K., 2006, Seto nai kai seibu syoto suido kaitei kara sansyutsu sita naumanzo kaseki to sono nendai [^{14}C dating of Naumann's elephant fossil from the channel of western islands in the Seto Inland Sea]. *Abstract with Programs, The 2006 Annual Meeting, The Palaeontological Society of Japan*, p.17. (in Japanese)

Kitamura, A. and Kimoto, K., 2004, 3.9Ma kara 1.0Ma no Nihonkai no Nanpo Kaikyo no Hensenshi [Reconstruction of the Southern Channel of the Japan Sea at 3.9- 1.0Ma]. *Daiyonki Kenkyu* [*The Quaternary*

Research], 43, pp.417-434. (In Japanese with English abstract)

Kitamura, A., and Kimoto, K., 2006, History of the Inflow of the Warm Tsushima Current into the Sea of Japan between 3.5 and 0.8 Ma. *Palaeogeography, Palaeoclimatology, Palaeoecology*, **236**, pp.355-366.

Kitamura, A., Takano, O., Takata, H., and Omote, H., 2001, Late Pliocene- Early Pleistocene Paleoceanographic Evolution of the Sea of Japan. *Palaeogeography, Palaeoclimatology, Palaeoecology*, **172**, pp.81-98.

Kobayashi, K., 1965, Naganoken Akashina cho Tochu shinyoju so no ^{14}C nendai – nihon no daiyonkiso no ^{14}C nendai XXV [Radiocarbon date of Tochu Confer Bed, Akashina-cho, Nagano Prefecture : ^{14}C age of the Quaternary deposits in Japan (25)]. *Earth Science (Chikyu Kagaku)*, **81**, pp.44-45. (in Japanese)

Koizumi, I., 1992, Biostratigraphy and Paleoceanography of the Japan Sea Based on Diatoms: ODP Leg 127. In: Tsuchi, R. and Ingle, J.C. (eds.), *Pacific Neogene: Environment, Evolution and Event*. pp.15-24. University of Tokyo Press.

Koizumi, A., 2003, Tokyoto Akishimashi Tamagawa no Senshin- Koshinto kara Sanshutu shita Nihon Hatsukiroku no Jun Nikushokusei Ookami Kaseki *Canis (Xenocyon) falconeri* [The First Record of the Plio-Pleistocene Hypercarnivorous Canid, *Canis (Xenocyon) falconeri* (Mammalia; Carnivora), from the Tama River, Akishima City, Western Tokyo, Japan]. *Daiyonki Kenkyu* [*The Quaternary Research*], **42**, pp.105-111. (in Japanese)

Kumon, F., Kawai, S., and Inouchi, Y., 2009, Nojiriko Taisekibutsu ni Motozuku Chubu Nihon no Kako 7.2 Mannenkan no Shosai na Kokiko Fukugen [High-resolution reconstruction of Paleoclimate during the Last 72 ka on th Basis of the Drilled Sediments from Lake Nojiri, Central Japan]. *Kyusekki Kenkyu* [*Palaeolithic Research*], **5**, pp.3-10. (In Japanese)

Kuzmin, Y. and Orlova, L., 2004, Radiocarbon Chronology and Environment of Woolly Mammoth (*Mammuthus primigenius* Blum.) in Northern Asia: Results and Perspectives. *Earth-Science Reviews*, **68**, pp.133-169.

Kuzmin, Y., Gorbunov, S., Orlova, L., Vasilevsky, A., Alekseeva, E., Tikhonov, A., Kirillova, I., and Burr, G., 2005, ^{14}C Dating of the Late-Pleistocene Faunal Remains from Sakhalin Island (Russian Far East). *Current Research in the Pleistocene*, **22**, pp.78-80.

Lister, A. and Burn, P., 2007, *Mammoths: Giant of the Ice Age (Revised Edition)*. 192p. University of California Press, Berkeley and Los Angeles, California.

Machida, H. and Arai, F., 2003, Shinpen Kazanbai Atlas: Nihon Retto to Sono Shuhen [Atlas of Tephra in and around Japan (Revised Edition)]. 336p. University of Tokyo Press. Tokyo. (in Japanese)

Makiyama, J., 1938, Japonic Proboscidea. *Memories of the College of Science, Kyoto Imperial University, B*, **14**, pp.1-59.

Matsuzawa, I. and Kosaka, T., 1987, On the Quaternary system in the Cape Erimo district - with special reference to the horizon of *Mammonteus primigenius* (M2). In: Prof. Masaru Matsui Memorial Volume Editorial Board (ed.), *Professor Masaru Matsui Memorial Volume*, pp.71-78. (in Japanese)

Minato, M., 1955, Zu den Mammonteusfauen Hokkaidos. *Jap. Journ. Geol. Geogr.*, **26**, pp.105-113.

Minato, M., 1967, On the age of mammoths in Japan and Siberia. *Earth Science (Chikyu Kagaku)*, **21**, pp.13-17.

Momohara, A., 1994, Floral and Paleoenvironmental History from the Late Pliocene to Middle Pleistocene in and around Central Japan. *Palaeogeography, Palaeoclimatology, Palaeoecology*, **108**, pp.281-293.

Morita, M., Che, K., Hibino, K., 1998, Chubu, Tokai Chiho no Shokuseishi [Vegetation History of the Chubu and Tokai Region]. In: Yasuda, K. and Miyoshi, N. (eds.), *Zusetsu Nihon Retto Shokuseishi*. pp.92-104. Asakura Shoten. Tokyo. (in Japanese)

Nakai, N., Arita, Y., Nakamura, T., Kamei, T., Akiyama, M., and Sawada, K., 1991, Naumanzo, otsunojika no AMS ^{14}C nendai to $^{13}C/^{12}C$ hi niyoru saisyu hyoki no kikohendo [AMS radiocarbon ages of mammal fossils from Lake Nojiriko, Nagano Pref. And environmental changes during the last glacial age]. *Summaries Researches Using AMS at Nagoya University*, **2**, pp.26-38. (in Japanese with English abstract)

Nakajima, R., Itoh, M., Kaneko, N., Taru, H., Toshimitsu, S., Nakazawa, T., and Isobe, I., 2004, Tsukuba shi hanamuro gawa no saijobu koshinto kara sansyutsu sita *Paleoloxodon naumanni* (Makiyama) (nauman zo) no kyushi kaseki [A fossil elephantoid molar of *Palaeoloxodon naumanni* (Makiyama) collected from the latest Pleistocene deposits of the Hanamurogawa River, Tsukuba City, Ibaraki, Japan]. *Daiyonki Kenkyu* [*The Quaternary Research*], **43**, pp.225-230. (in Japanese with English abstract)

Nakamura, T., Otsuka, Y., Okuno, M., Ota, T., 1996, Higashi shina kai no tairikudana oyobi ryukyuko no kaitei kara saisyu sareta honyurui kaseki no kasokuki shitsu ryo bunseki ho ni yoru ^{14}C nendai sokutei [Accelerator mass spectrometric ^{14}C dating of mammalian fossils collected from the sea bottom of continental shelves around the East China Sea and along the Ryukyu Island Arc]. *Journal of Geography*, **105(3)**, pp.306-316. (in Japanese with English abstract)

Nakaya, H., Akamatsu, M., and Yamada, G., 1992, Hokkaido yubari gun yuni cho kara sanshutsu shita manmosu oyobi otsunojika kaseki to sono igi [New late Pleistocene *Mammuthus primigenius* (Proboscidea, Mammalia) and *Shinomegaceros yabei* (Artiodactyla, Mammalia) from Yuni-cho, Hokkaido, Nortern Japan]. *Abstracts 1992 Annual Meeting the Palaeontological Soceity of Japan*, p.61. (in Japanese)

Nojo, A., Takahashi, Y., Tanaka, N., Takanashi, S., Hashimoto, C., and Iwamizawa Research Group, 2002, Hokkaido Chuo Bu Ishikari Teichi Tai Toen Chiiki no Jyobu Koshinto Sojyo to Kokankyo Hensen [Stratigraphy and Paleoenvironmental Changes of the Upper Pleistocene in the Eastern End Area of the Ishikari Low Land, Hokkaido, Japan]. *Earth Science (Chikyu Kagaku)*, **56**, pp.253-268. (in Japanese)

Okumura, K., Ishida, S., Kawamura, Y., Kumada, M. Nishida, S., and Ishida, K., 1978, Gifuken Kumaishi-do San no Herashika Kaseki nitsuite [Fossil Elk deer from Kumaishi-do Cave, Gifu Prefecture, Central Japan]. *Bulletin of the Osaka Museum of Natural History*, **31**. pp.13-24, pls.1-5. Osaka Museum of Natural History. Osaka. (in Japanese)

Okumura, K., Ishida, S., Kawamura, Y., Kumada, M., and Tamiya, S., 1982, Gifuken Kumaishi-do San Koki Kosekisei Honyu Dobutsu Gun to Sono ^{14}C Nendai no Igi [Latest Pleistocene Mammalian Assemblage of Kumaishi-do Cave, Gifu Prefecture and the Significance of its ^{14}C Age]. *Earth Science* (*Chikyu Kagaku*), **36**, pp.214-248. (in Japanese)

Ono, Y., 1991, Kita no Rikkyo [Northern Land Bridge] *Mongoloid*, **10**, pp.37-44. (in Japanese)

Rodgers, R., 2001, *Moose*. Voyageur Press. Vancouver. 72p.

Sawada, K., Arita, Y., Nakamura, T., Akiyama, M., Kamei, T., and Nakai, N., 1992, Kasokki Shitsuryo Bunseki Kei wo Mochiita ^{14}C Nendai Sokutei niyoru Nojiri-ko So no Hennen [^{14}C Dating of the Nojiri-ko Formation Using Accelerator Mass Spectrometry]. *Earth Science* (*Chikyu Kagaku*), **46**, pp.133-142. (in Japanese)

Shikama, T., 1962, Kaseki Honyurui Nado yori Mita Nihon Retto to Tairiku tono Rikuchi Setsuzoku [Quaternary Land Connections of Japanese Islands with Continent from the Viewpoints of Paleomammalogy]. *Daiyonki Kenkyu* [*The Quaternary Research*], **2**, pp.146-153. (in Japanese)

Sohma, K., 1986, Recent Progress of Palynology on Japan, with Special Reference to the Middle and Early Pleistocene. *Recent Progress of Natural Science in Japan 11: Quaternary Research*. Science Council of Japan. pp.61-70.

Stuart, A., 1999, Late Pleistocene Megafaunal Extinctions; a European Perspective. In MacPhee, R. (Ed.), *Extinctions in Near Time; Causes, Contexts and Consequences*. pp.257-269. Kluwer Academic/ Plenum Publishers, New York.

Stuart, A., Sulerzhitsky, L., Orlova, L. Kuzmin, Y., and Lister, A., 2002, The Latest Woolly Mammoths (*Mammuthus primigenius* Blumenbach) in Europe and Asia: a Review of the Current Evidence. *Quaternary Science Reviews*, **21**, pp.1559–1569.

Stuart, A., Kosintsev, P., Higham, T., and Lister, A., 2004, Pleistocene to Holocene Extinction Dynamics in Giant Deer and Woolly Mammoth. *Nature*, **431**, pp.684-689.

Suzuki, K., and Manabe, K., 1982, Pliocene-Pleistocene Chronology of the Yamato Group of Aizu Basin, Northeast Honshu, Japan. *3rd Report of Pliocene-Pleistocene Boundary in Japan*. National Working Group IGCP Project, pp.18-27.

Tada, R., 1994, Paleoceanographic Evolution of the Japan Sea. *Palaeogeography, Palaeoclimatology, Palaeoecology*, **108**, pp.487-508.

Tada, R., 1999, Late Quaternary Paleoceanography of the Japan Sea: An Update. *The Quaternary Research*, **38**, pp.216-222.

Tada, R. and Irino, T., 1994, Daiyonki Koki niokeru Nihonkai no Kaiyo Kankyo Henka [Sea Environment Change of the Japan Sea in Late Quaternary]. *Gekkan Chikyu*, **16**, pp.667-677. (in Japanese)

Takahashi, K., Izuho, M., and Sato, H., 2010, *Hokkaido Churui Naumanzo Sanshutu Chiten no Saichosa Hokoku [Reexamination Report of the Naumann's Elephant (Palaeoloxodon naumanni)-bearing site in Churui, Hokkaido]*. Kaseki Kenkyukai Kaishi, supplement No.4. 79p. Kaseki Kenkyukai, Kusatsu. (in Japanese)

Takahashi, K. and Kitabayashi, E., 2001, Ajimu Doubutsu Kaseki Gun [The Ajimu Fauna]. *Biwako Hakubutsukan Kenkyu Chousa Houkoku [Research Report of the Lake Biwa Museum]*. No. 18, 193p. (In Japanese)

Takahashi, K., Izuho, M., Soeda, Y., Chang, C., 2005, Nihonsan manmosu zo kaseki no nendai sokutei kekka kara wakatta sono seisoku nendai to ikutsukano shin chiken [The chronological record of the woolly mammoth (Mammuthus primigenius) in Japan, and its new findings]. *Journal of Fossil Research*, **38**(2), pp.116-125. (in Japanese with English abstract)

Takahashi, K., Shimaguchi, T., Kamiya, H., 2006b, Shimokita gun Higashi dori mura shikkari san no nauman zo kaseki to sono AMS ^{14}C nendai [*Palaeoloxodon naumanni* fossils from Shikkari, Higashi-doori, Shimokita-gun, Aomori Prefecture, Japan and its ^{14}C AMS dating]. *Journal of Fossil Research*, **39(1)**, pp.21-27. (in Japanese with English abstract)

Takahashi, K., Soeda, Y., Izuho, M., Aoki, K., Yamada, G., and Akamatsu, M., 2004, A New Specimen of *Paleoloxodon naumanni* from Hokkaido and Its Significance. *The Quaternary Research*, **43**, pp.169-180.

Takahashi, K., Soeda, Y., Izuho, M., Yamada, G., Akamatsu, M., and Chang, C., 2006a, The Chronological Record of the Woolly Mammoth (*Mammuthus primigenius*) in Japan, and Its Temporary Replacement by *Palaeoloxodon naumanni* during MIS 3 in Hokkaido (Northern Japan). *Palaeogeography, Palaeoclimatology, Palaeoecology*, **233**, pp.1-10.

Takahashi, K., Wei, G., Uno, H., Yoneda, M., Jin, C., Sun, C., Zhang, S., and Zhon, B., 2007, AMS ^{14}C Chronology of the World's Southernmost Woolly Mammoth (*Mammuthus primigenius* Blum.). *Quaternary Science Reviews*, **26**, pp.954–957.

Takakuwa, Y. and Fossil Mammal Research Group for Nojiriko Excavation, 2009, Nagonoken Shinanomachi no Joubu Koushintou niokeru Herajika Kaseki no Shinsansyutsu [New occurrence of fossil moose deer from the late Late Pleiocene Nojiri-ko Formation, Shinano-machi, Nagano, Japan]. *Nihon Koseibutu Gakkai 2009 Nen Nenkai Kouen Yokou Shu*, 24p. (in Japanese)

Taruno, H., 1991, Stegodonrui 3. Nihonsan Stegodon-ka Kaseki [Stegodons 3. Stegodont fossils from Japan]. In: Kamei, T. (ed.) *Nihon no Chobirui Kaseki*. Tsukiji Shokan, Tokyo, pp.82-99. (In Japanese)

Taruno, H., 1999, The stratigraphic positions of proboscidean fossil from Pliocene and lower to middle Pleistocene formations of Japanese Islands. *Chikyu Kagaku*, **53**, pp.258-264. (In Japanese)

Taruno, H. and Kamei, T., 1993, Kinki Chihou no Senshin Koushintou no Sekituidoubutu Kaseki [Mammalian fossils of the Pliocene ana Pleistocene in the Kinki District]. In: Ichihara, M. (ed.) *Osaka Sougun* [*The Osaka Group*]. Sougensha, Osaka, pp.216-231. (In Japanese)

Tomizawa, T., 1979, Nagano bonshi syuen no nauman zo kaseki sanshutsu so ni tsuite [On the stratigraphy of the beds with Palaeoloxodon naumanni (MAK.) around Nagono Basin, central Japan]. *The Journal of the Geological society of Japan*, **85**, pp.258-260. (in Japanese)

Tsuji, S., 2004, Chikyu Jidai no Kankyoshi [Environmental Historyp in Earth Era], In: Amuro, T. (ed.) *Kankyoshi Kenkyu no Kadai*. Yoshikawa Kobunkan. Tokyo, pp.40-70. (In Japanese)

Vasilevski, A., 2008a, Sakhalin niokeru Mammoth Dobutsugun to Jinrui no Tekio [Mammoth Fauna and Human Adaptation in Sakhalin]. In: Sato, H. (ed) *Human Ecosystem Changes in the Northern Circum Japan Sea Area (NCJSA) in Late Pleistocene*. pp.44-67. Research Institute for Humanity and Nature. Kyoto. (In Japanese and Russian)

Vasilevski, A., 2008b, Kamennuii Vek Ostrova Sakhalin [Stone Age of Sakhalin Island]. 409p. Sakhalinskoe Knijnoe Izdatelistovo. Yjno-Sakhalinsk. (In Russian)

Wu, R., Wu, X., Zhang, S., 1989, Zhong Guo Yuan Gu Ren Lei [Chinese Ancient Men]. Ke Xie Chu Ban She [Science Press], Beijing, 437p. (In Chinese)

Yamada, G., Akamatsu, M., Nakaya, H., and Kumasaki, N., 1996, Rausu oki kara hakken sareta manmosu zo kyushi kaseki no nendai ni tsuite [AMS ^{14}C data of Mammoth molar found on off the coast of Rausu, eastern part of Hokkaido]. *Bulletin of Historical Museum, Hokkaido*, **24**, pp.1-8. (in Japanese with English abstract)

Yamakawa, C., 2001, Oitaken Usagun Ajimumachi no Tsubusagawasousan Doubutsu Kaseki ni tomonau Ogata Shokubutsu Kaseki [Plant macrofossils obtained with animal fossils from the Tsubusagawa Formation in Ajimu, northern Kyushu, Japan]. In: Takahasi, K. and Kitabayashi, E. 2001 *Ajimu Doubutsu Kaseki Gun*. Biwako Hakubutsukan Kenkyu Chousa Houkoku. Kusatsu. pp.25-35. (in Japanese)

Yasui, K., Kusuhashi, N., Matsuoka, H., 2004. Kumaishido (Gifu ken Gujoshi) kara sansyutsu shita Naumanzo kaseki to sono nendai [The fossil of Naumann's elephant recorevered from Kumaishi-do Cave, Gujo City, Gifu Prefecture, Japan, and its ^{14}C dating]. *Abstract with Programs, The 2004 Annual Meeting, The Palaeontological Society of Japan*, p.48. (in Japanese)

Yasui, K., Matsuoka, H., 2002, Gifuken Yahata machi kumaishido kara sansyutu sita okami kaseki to sono igi [Wolf remains from Kumaisi-do Cave, Hachiman cho, Gifu Prefecture, and their significance]. *Abstract with Programs, the 151th Regular Meeting, the Palaeontological Society of Japan*. p.21. (in Japanese)

Some Issues on the Origin of Microblade Industries in Northeast Asia during OIS2

Anatoly Kuznetsov

Department of Social and Political Anthropology, School for Regional and International Studies, Far Eastern Federal University, Uborevicha st., 25, Vladivostok 690090 Russia.
Email: kuznetsov.2012@mail.ru

Abstract: *This paper discusses some issues on the origin of microblade industries, one of the most significant phenomena of North and East Asia in the Stone Age. According to current data, the oldest microblade industries were discovered at a number of sites in the southern area of Central Siberia, dating as early as the OIS3-OIS2 boundary or the beginning of OIS2. This period is very important archeologically because of the formation of microblade industries in North and East Asia and some parts of Eastern Europe. These industries were based on pressure flaking to remove microblades from cores. Such microblades were used very often as insets of composite tools for hunting, which had shafts made on antler, bone and in some cases ivory. Based on data from stratified sites in Siberia, it was concluded, despite contrary ideas of many scholars, that microblade industries were a stage of the Stone Age. Because of changes in the method of detachment, the earlier microblade industries were replaced by prismatic microblade industries in the beginning of the OIS1.*

Keywords: *Microblade industries, OIS2, Siberia, Northeast Asia*

Introduction

Before the OIS2-OIS1 boundary (Terminal Pleistocene - Early Holocene), a large number of sites with similar artifacts were reported in the vast territory of North Asia, part of East Asia, and northern northwest America (Fig. 1). The most distinctive artifact types of this huge archaeological complex are wedge-shaped microblade cores and the detached microblades (Fig. 3, 4) collectively referred to as the "microblade industry". Because wedge-shaped microblade cores are morphologically similar across the region, they were regarded in the beginning of the 20th century as very firm evidence for ancient human migrations across this vast territory including those related to the initial settlement of the New World. Some scholars, mainly archaeozoologists and geoarchaeologists, maintain the incorrect opinion that every assemblage, even those with butt-ended microblade cores, may be regarded as part of the microblade industry (Keates, 2007). Current archaeology does not recognize the morphological similarity of artifacts as a basis for drawing conclusions. Technological approaches and contextual data, for example, are more important for the interpretation of archaeological materials. Because of this, the microblade industry comprises archaeological assemblages unearthed from primary *in situ* geological contexts with wedge-shaped microblade cores, microblades, and frequently other implements. In turn, a distinctive attribute is reducing the microblade core by means of pressure flaking. If one studies collections of artifacts from redeposited sites, the material from them should be compared with assemblages from primary contexts with good stratigraphic control.

Aspects of distribution, assemblages, and the emergence of microblade industries

According to proposed criteria and current data, microblade industries are distributed from the Yenisei River basin in the west to Japan in the east, and from Alaska in the north to Kwan-he River in the South (Fig. 1). More than 200 sites with wedge-shaped cores and other microblade remains were discovered in Central and East Siberia; about 1,000 sites with microblades industries are known in Japan. At this time only several such sites have been excavated in north China, 17 in Korea, approximately 50 in the Russian Far East, and about 30 in Alaska and the Northwest Coast of North America. There is simultaneously a lack of data for some areas, but abundant evidence from others. Consequently, no one can examine all microblade assemblages in some areas. Yet, one can analyze some collections with microblades in a very cursory way because of data limitations. Despite these difficulties, the study of microblade industries is of great interest. In some regions, all the known sites of the Paleolithic period are represented by microblade industries (i.e., Primorye, Sakhalin, Kamchatka and other areas of the Russian Far East). It is true that microblade assemblages of some areas of continental northeast Asia, Japan, and Alaska are very similar. Resolving many questions concerning microblade industries depends on understanding their origin.

Siberian sites are the most important for the study of the origin problem because of their completeness. Geoarcheological data from stratified sites in the Yenisei, Angara, and Lena river basins can be used to prepare local

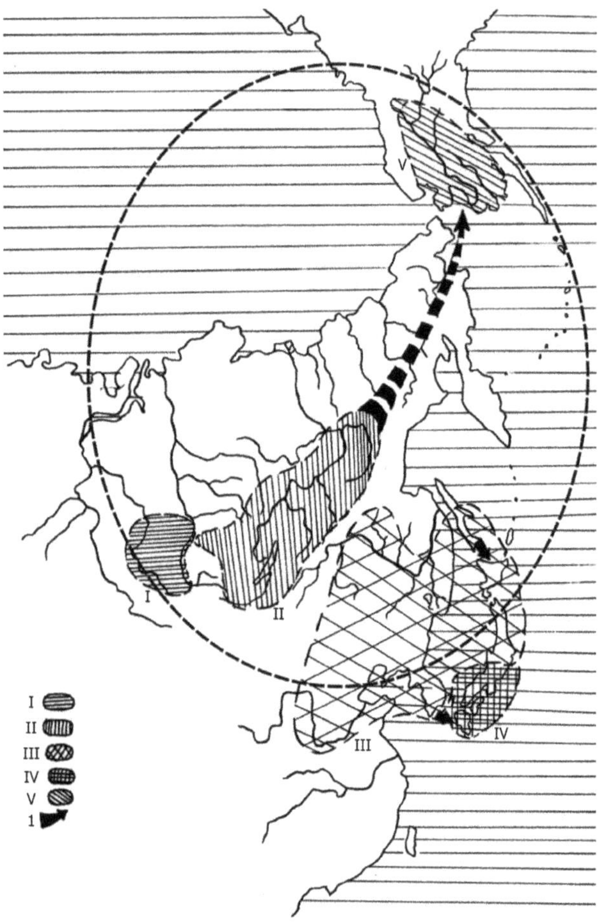

Figure 1 Distribution of microblade industries in North and East Asia and Northern North America
I: Yenisei River basin group of microblade industries, II: East Siberian group of microblade industries, III: South Far Eastern group of microblade industries, IV: small blade industries of Southern Japan, V: Alaskan microblade industry (Denali complex), 1: possible route of transmission of microblade industries from Asia to America

and regional stratigraphic sequences. These sequences very clearly display the chronological positions of key artifacts and their production techniques, and microblade industries as a whole (Fig. 2). Another reason the Siberian sites are of special value is the opportunity they provide to discover the association of different implements based on their positions in the same cluster of sites or on refitted artifacts from different clusters of the same site. Unfortunately, because of linguistic barriers and some other reasons, this important evidence is not yet accessible to many foreign scholars. It should be pointed out that stone and bone artifacts are not necessarily numerous at Siberian microblade industry sites (for example, Ust-Menza in the Transbaikal area and Kurla in the North Baikal area, Fig. 3). Consequently, it has been concluded that these were seasonal sites. In northern North America, microblade-bearing sites were also discovered with redeposited artifacts (but Onion Portage and Dry Creek are rare exceptions, Fig. 4). On the contrary, there are many sites with huge collections of stone artifacts (including tools) but lack stratigraphic control in the south of the Far East (Kamishirataki VIII with a million artifacts in Hokkaido is a good example). Despite the redeposited character of many assemblages, I can suppose that camp sites and workshop areas were combined at the same localities of this region.

It should be emphasized that the entire assemblage of microblade industry implements is not always represented at one site. For example, even artifacts such as end scrapers were not discovered at some Siberian sites with microblades. Antler and the rare ivory (mammoth tusk) shafts of composite tools were discovered at some Siberian sites (Kokorevo 1, Listvenka, Bolshoi Yakor 1, and so on) and at two Alaskan sites (Lime Hills 1 and Trail Creek 2). Grooved stones (shaft smoothers) were excavated at

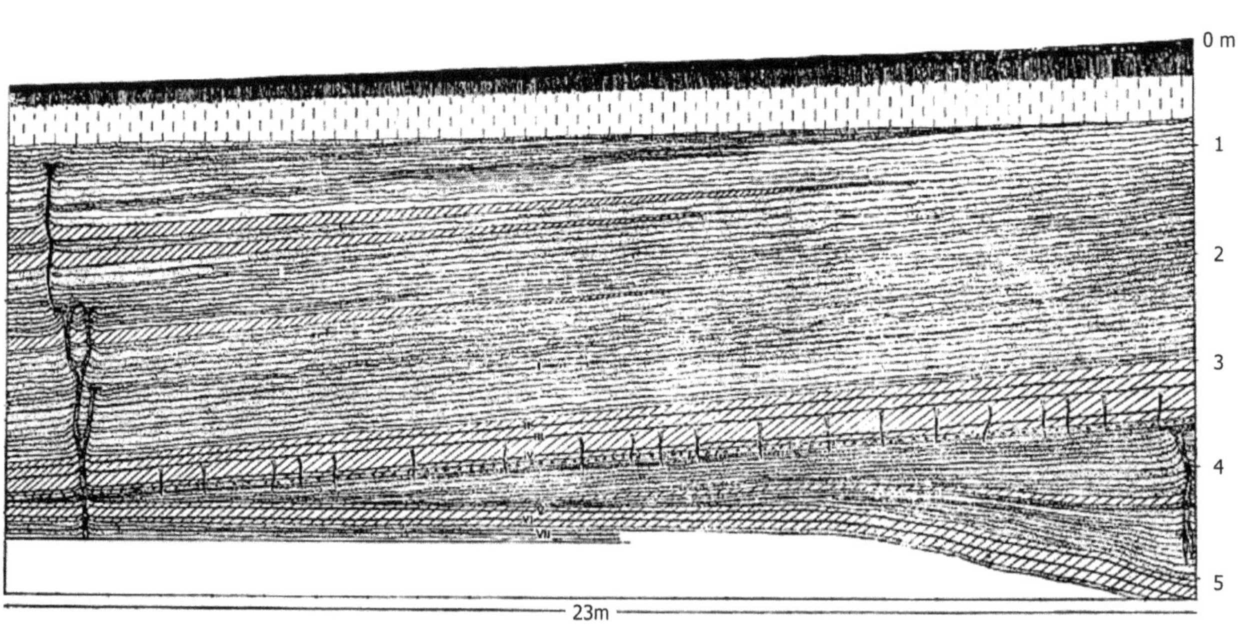

Figure 2 Stratigraphic cross-section of the multilayer site Krasnyi Yar (Angara River basin) with early (about 20, 000 RCYBP) microblade remains
There are ice-wedges on the left (after Medvedev 1966).

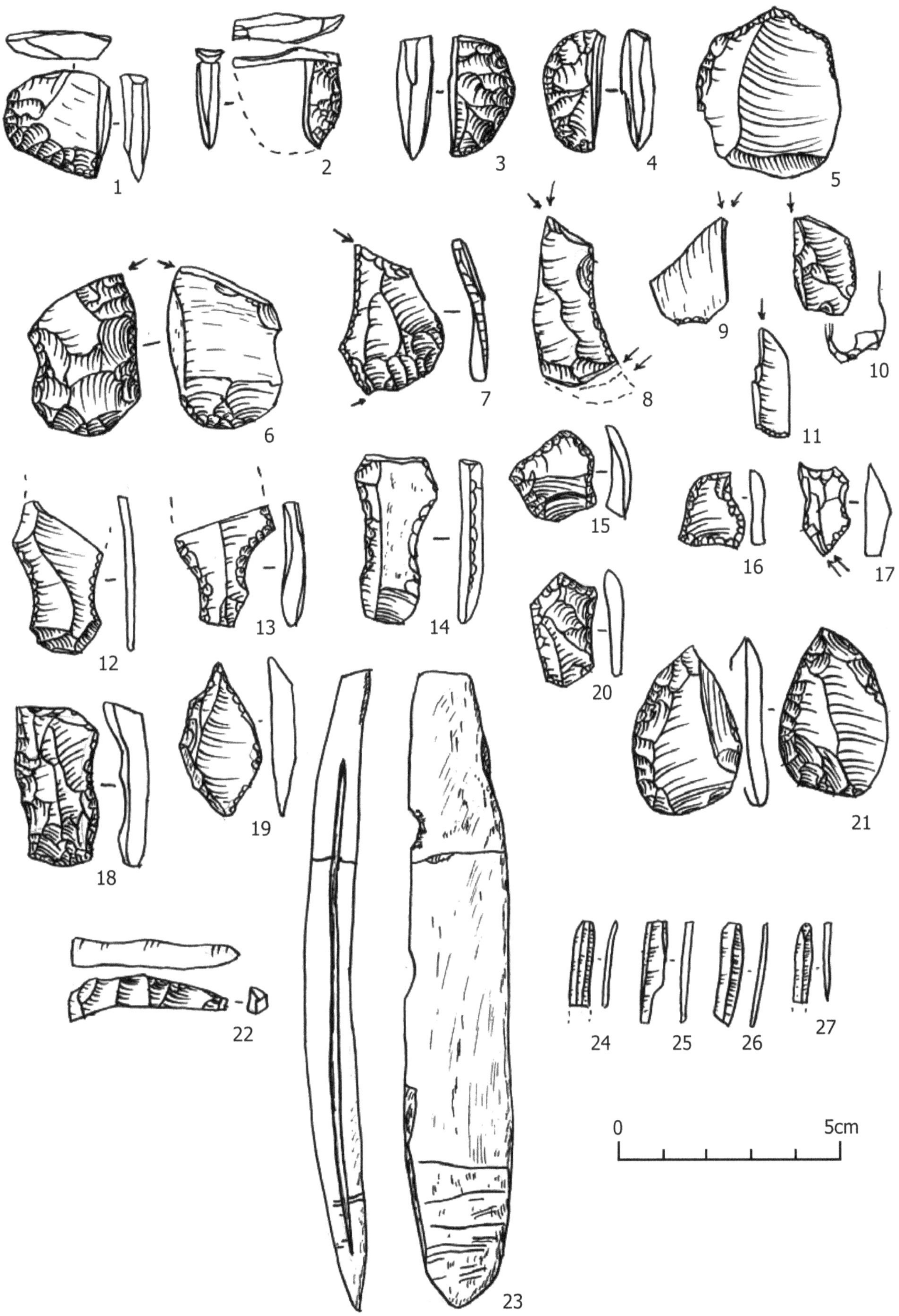

Figure 3 An early microblade assemblage from the second layer of the Kurla II-III site
1-4: small wedge-shaped cores, 6: transversal burin, 7-11: burins, 12-17: notched implements, 5,18-21: pointed implements, 22: spall of a wedge-shaped core platform preparation, 23: antler shaft of a composite tool, 24-27: microblades (after Shmygun and Philuppov 1982).

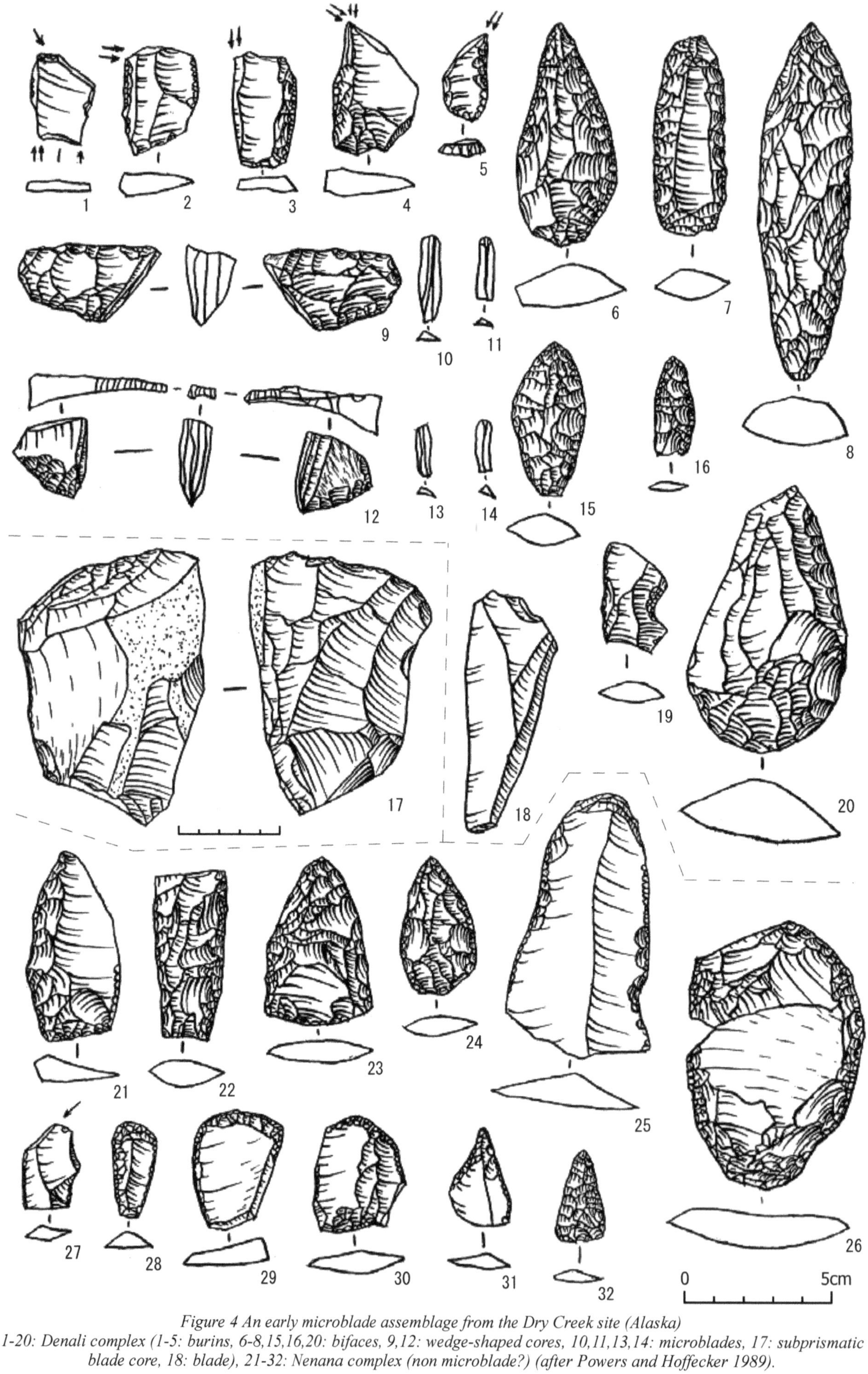

Figure 4 An early microblade assemblage from the Dry Creek site (Alaska)
1-20: Denali complex (1-5: burins, 6-8,15,16,20: bifaces, 9,12: wedge-shaped cores, 10,11,13,14: microblades, 17: subprismatic blade core, 18: blade), 21-32: Nenana complex (non microblade?) (after Powers and Hoffecker 1989).

three sites (Krasnyi Yar in Siberia, Ushki in the Far East, and Onion Portage in Alaska). One of the most rarely preserved artifact forms (except in the Yenisei River basin) is the composite tool for hunting (represented by antler spear heads and knives with microblade inserts). Despite differential representation, the real value of the artifact groups may be very distinctive. Because of the incompleteness of our initial data, we have to reconstruct the techno-typological context of microblade industries by comparing intra- and inter-site artifact clusters from the entire area of microblade distribution.

Despite some questions about the chronology of several important microblade industry sites (e.g., the lower level of Afontova Gora 2 in the Yenisei River Basin with dates between ca. 20,000 and 19,000 BP), it is clear the first evidence of pressure flaking for microblade production occurred at the beginning of OIS2. The formation of microblade industries in East Siberia may date to about 20,000 BP (the lower levels of Kurla II and III in the North Baikal area, levels 25-27 of Ust-Menza 2 in Transbaikal area) (Konstantinov, 2004:22, 28). The first microblade industries in the Yenisei River Basin are dated to about 16,000 BP (Abramova, 1979; Sinitsyn and Praslov, 1998; Astakhov, 1999). Archaeologically, microblade industries replaced blade industries in the early Upper Paleolithic and Mousterian periods in Siberia. From a paleoenvironmental perspective, microblade industries existed during the Last Glacial (Sartan in Siberia).

There is another opinion on the timing of the appearance of microblade industries based on materials from the Gornyi Altai area. Small cores (butt ended) and blades were excavated at Ust-Karakol 1 (layers 10 and 9); Anui 3 (layer 12 RTL date 54,000±13,000BP), Kara-Bom (levels 6 and 5 ^{14}C dates 43,200±1500 and 43,000±1600BP), Denisova cave (layers 11, 9 and 7). Data from these sites are regarded as important evidence concerning the origin of microblade industries during the the Middle to Upper Paleolithic transition in Siberia (Derevyanko *et al.*, 2002; Petrin and Chevalkov, 1992). This point of view was accepted by S. Keates (2007:128) who considered that Ust-karakol 1 and Denisova Cave represent the incipient phase of microblade technology, while Anui 2 represents true, fully formed microblade technology. I disagree with the approach of some authors in resolving the problem of the origin of microblade industries. It contradicts the techno-morphological character of microblade industries from the Altai and other parts of East Siberia and the Far East. As F. Ikawa-Smith (2007:191) concluded: "The Gornyi Altai area, then, is more likely to be part of the general area in Eurasia where blade-based technologies developed, rather than the direct ancestral homeland of the microblade industries of Northeast Asia and northern North America. In any event, few archeologists would object to describe these small cores in the Gorny Altai as 'proto-wedge-shaped'. Few would disagree, either, with a statement that what we have here in Altai Mountains area, dating back to at least c. 35,000BP, is probably a 'precursor' of the microblade industry..."

Transformation of microblade industries

Between the end of OIS2 and beginning of OIS1 (the Terminal Pleistocene-Early Holocene), traditional microblade industries were replaced by a number of new industries. In the East and Central Siberia, the northern area of the Far East, and northwest North America, industries appeared with cylindrical, conical, and semiconical microblade cores ('bullet-shaped', 'pencil-like'), small blades and microblades (including those produced through retouch), polyhedrical burins (prepared on these new microblade cores), as well as angle and side burins. Bifacial tools are represented in these assemblages by axe-like implements. The Sumnaginskaya 'culture' of Yakutia is a good example of such an industry. These new microblade cores are rather tall and microblades were removed from their entire surface but not from one butt end. So I propose to call these artifacts microprismatic cores. Another distinctive attribute of such microblade cores is a round, retouched platform (Fig. 7). It is clear that the microprismatic technique was also a kind of pressure flaking method, but it advanced to produce more small blades during the primary stages of use and then more microblades during the final stages. The importance of this technical change is the reason to separate microblade from microprismatic industries.

The relationship of microblade and microprismatic industries is now discussed. Some archaeologists consider these kinds of industries unrelated (Mochanov, 1977). Other scholars are using data from northern Chinese and Japanese sites to suppose that microprismatic industries appeared before industries with wedge-shaped microblade cores (Chen and Wang, 1989). The stratigraphy of Siberian multilayered sites indicates very clearly that microprismatic cores always overlie wedge-shaped microblade cores. According to geoarcharological data and radiocarbon dates from the sites of the Yenisei area, the earliest evidence of the microprismatic technique only dates to about 12,000 BP. In northern Japan, northern China, and Primorye wedge-shaped and microprismatic cores were discovered in the same assemblages. Yet, the stratigraphic position of these artifacts is not as clear as at the Siberian sites.

Another developmental trajectory of the archaeological industries is known in the southern Far East. After 12,000 BP, 'Neolithic' assemblages spread very quickly in southern Japan and sometime later around the coastal rim of the Sea of Japan. These assemblages are characterized by bifacially retouched and polished implements (arrowheads and axes, respectively). The Jomon of Japan is a good example of this early Neolithic culture. Microprismatic cores are known at some sites in this area during the Terminal Pleistocene-Early Holocene period. Yet real microprismatic industries did not develop here. In some assemblages in Japan, China, and the Amur River Basin, pottery was associated with

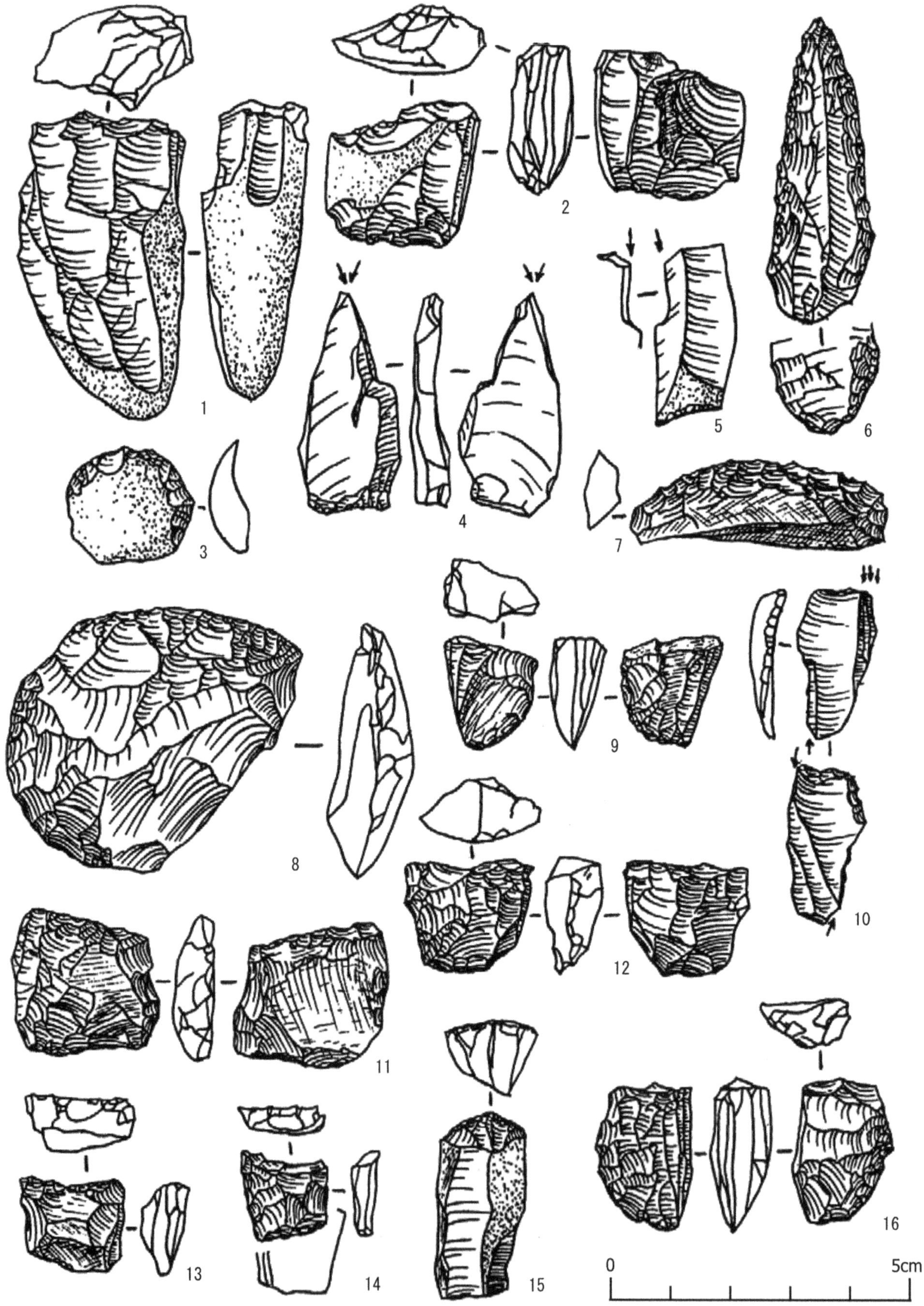

Figure 5 Microblade industry of Yenisei River basin

1: subprismatic core, 2,9,11,13,14,16: wedge-shaped cores, 3,15: end scrapers, 4,5,10: burins, 6: unifacial point, 7,8: scrapers (after Abramova 1979).

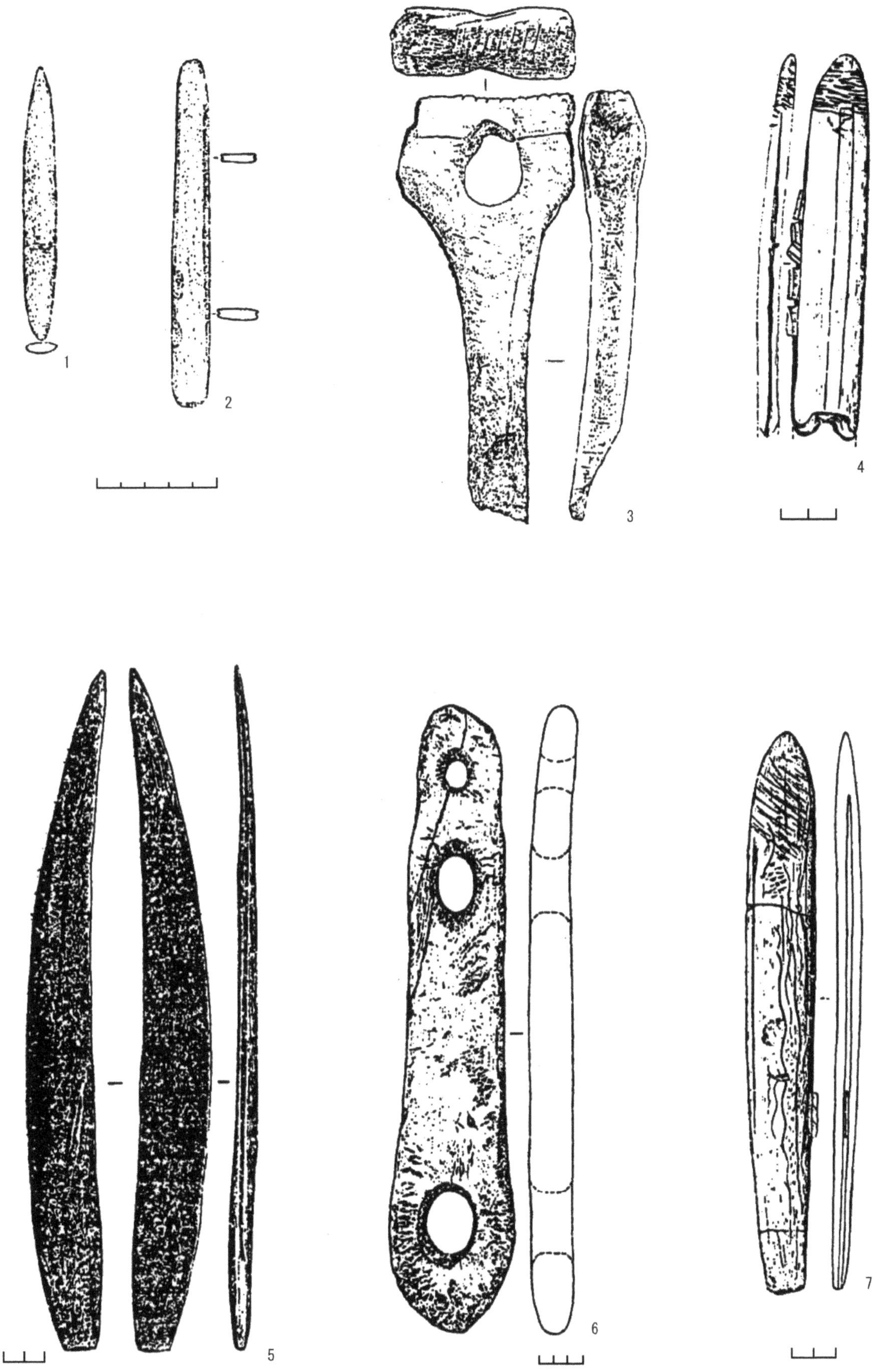

Figure 6 Bone implements of the Yenisei River basin sites of microblade industries
1-3: Bolshaya Slizneva site, 4-7: Listvenka site, 1,2, antler and 5 ivory shafts of composite tools, 4,7: composite tools with microblades, 3,6: shaft strightener (after Akimova *et al.*, 1999).

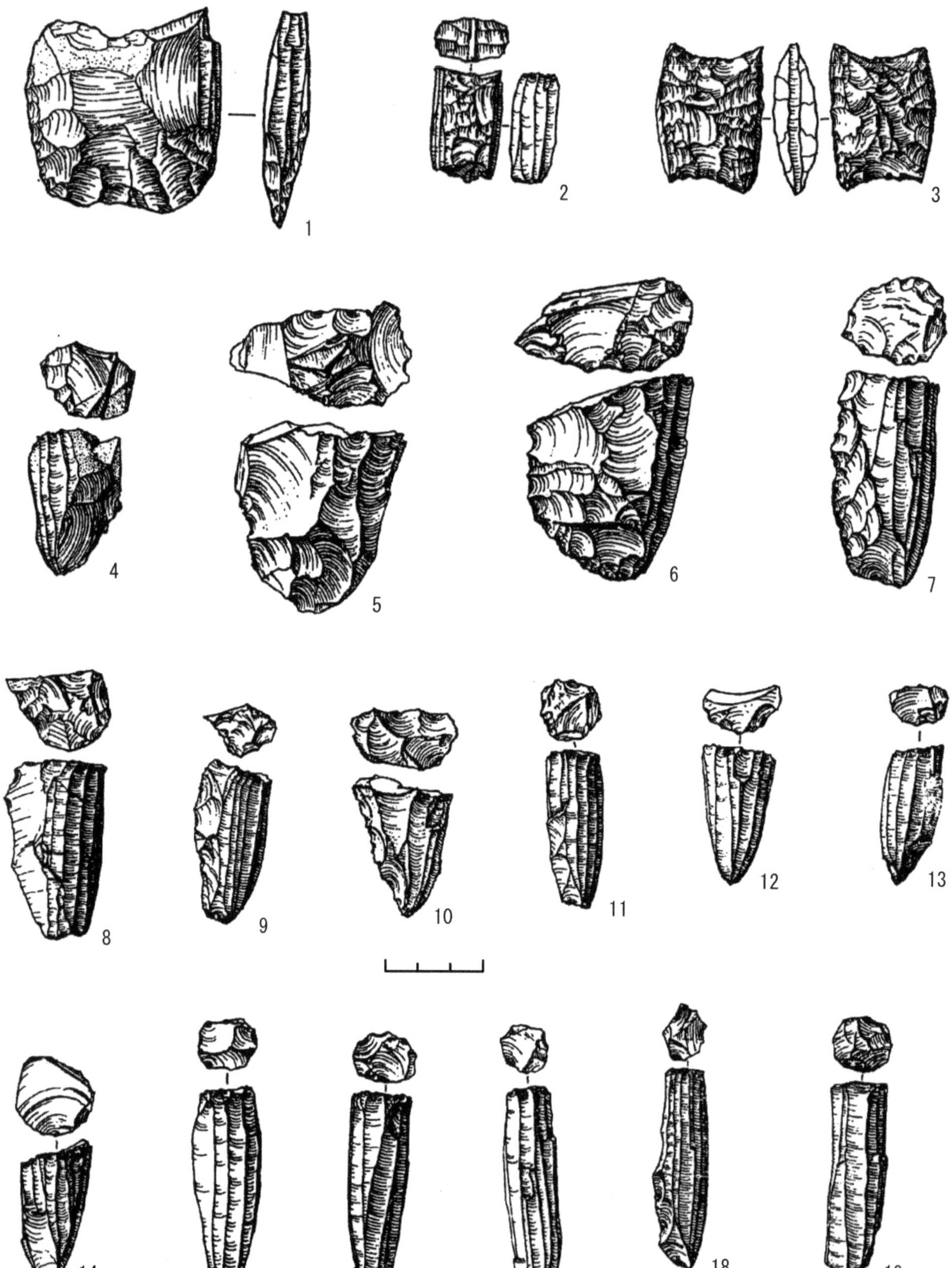

Figure 7 The Ust-Narym site
Microprismatic cores 8-19 and preforms 1-7. The Neolithic of Bashkortostan (after Korobkova, 1965).

microblade remains (e.g., Incipient Jomon of Japan) about 12,000 BP (e.g., at the Fukui cave in Kushu) (Sato and Tsutsumi, 2007).

Microblade industries and environments of OIS2

Despite contradictions about the timing of the origin of microblade industries, many archaeologists regard their appearance o as a kind of environmental adaptation. "These 'little things' called microblades made human adaptation to the temperate, subarctic and arctic environments of Siberia, East Asia, and northernmost North America very successful" (Kuzmin *et al.*, 2007:1). "From the distribution of microblade sites across Northeast Asia and northern North America, it has generally been assumed that microblades gave some advantages to the humans living in cold climates, where plant resources would be scarce and hunting would be the major subsistence activity" (Ikawa-Smith, 2007:197). I suppose that the temporal correlation of microblade industries and the last glacial maximum (LGM) is the reason for such conclusions. A more concrete explanation, that microblade industries were adaptations to fishing resources, was also declared (Sato and Tsutsumi, 2007:62). An alternative position was presented by R. Ackerman (2007:169-170): "Bow and arrow technology may serve as an explanation for the appearance of microblades beginning some 18,000-20,000BP in Siberia as it is for the popularity of microblades in Denali sites during the Late Pleistocene to early Holocene of Alaska".

Regarding the variety of explanations of the origin of microblade industries, I consider the adaptational approach as too narrow to embrace all factors influencing the process and their interrelationships. First, environments of the Sartan glacial were heterogeneous according to N. Kind (1974). There was a maximal phase between 23,000 to 22,000 BP (the Gydansky or Karaul) and again at 16,000 BP, with coldest temperatures around 20,000 to 18,000 BP. An early interstadial interrupted this initial cold phase around 15,000 BP, followed by another cold phase (the N'iapan) between 15,000 and 13,000 BP. The next warm period with double peaks (the Kokorevo and Taimyr) occurred around 13,000 to 11,400 BP. The final cold phase (the Noril'sk) was between 11,400-10,200 BP (Kind, 1974). The next reason for denying this explanation is the vast area of microblade distribution. Local differences in environments occurred during the Last Glacial period, which were responsible for several adaptations of microblade producers, such as cobble-lined dwellings at sites in Transbaikal area, seasonal sites in other parts of East Siberia and Alaska, and camp sites associated with workshop areas in the southern Far East (Vasil'ev, 2003b). To better understand microblade industries, we have to pay attention to data from the Upper Paleolithic of Eastern Europe. Wedge-shaped and butt-ended microblade cores and microblades were found at Kostenki I, the Severnyi Punkt of Kostenki II, and so on (Fig. 8). These are very important because they prove that the 'adaptation' of wedge-shaped and other microblade cores occurred independently in different environments (Lisitsyn, 1998). The Gornyi Altai material and some other small core and blade industries are also interesting because they represent the long trajectory of the minimization of artifacts size in the Paleolithic. Microblades were a part of this process.

Siberian data and the origin of microblade industries

To understand the origin of microblade industries, we have to examine Siberian data once again. As mentioned above, antler, bone, and rare ivory shafts with slots on one or both sides were discovered at a number of sites in Central Siberia (Fig. 6). The most impressive and well known of these artifacts is a bison scapula pierced by and antler spear point from Kokorevo 1 (cultural horizon 3) in the Yenisei River Basin. Another significant artifact from the same layer of this site is an inset point with nine unretouched microblade fragments in its slot. Layer 3 of Kokorevo 1 is dated by ^{14}C to between 13,300±50 RCYBP (GIN-911) and 14,450±150 RCYBP (Le-628) (Abramova, 1979:162). Several shafts were discovered at other sites in East Siberia as well (at Bol'shoi Yakor 1, Kurla 2, Studenoe 1, and so on). Unfortunately, at a large number of microblade-bearing sites neither shafts for composite tools nor fauna remains were discovered. These Siberian data are the key to resolve the question of microblade function. It is clear now that wedge-shaped microblade cores, despite their very elaborate preparation, were byproducts of microblade manufacture. In turn, unretouched microblades or their fragments were mainly used as insets for composite tools. Reindeer antler was the preferable raw material for the shafts of such tools. In some areas, for example the upper Yenisei River, reindeer were not represented in the Sartan Glacial fauna. Shafts from Maina and other sites of this area were made from antler and bones of other deer species (Vasil'ev, 1996).

The utility of the composite tool is in combining the flexibility of the antler or bone shaft with sharp cutting edge of stone microblades. It was a very successful technology used primarily for hunting weapons. Because of the efficiency of composite tools, some archaeologists regarded their appearance as a kind of technical revolution (Dikov, 1977). In any event, these tools with microblade insets were quickly distributed over a vast geographic range and they became an important part of assemblages with different techno-morphological characteristics (Fig. 3, 5). Regular microblades were produced primarily by pressure flaking during microblade core reduction. Sometime later the first variant of pressure flaking was improved in the microprismatic industries and even further in the Neolithic. The technique of wedge-shaped microblade core reduction was a stage in the evolution of stone knapping reached at the beginning of OIS2. It may be supposed that the origin of this technique was more complex than simply a direct adaptation to environmental change.

Maybe the spread of reindeer during the Sartan was related to the origin of microblade industries. For example, at Kokorevo 1 reindeer bones comprise 63% of the fauna as

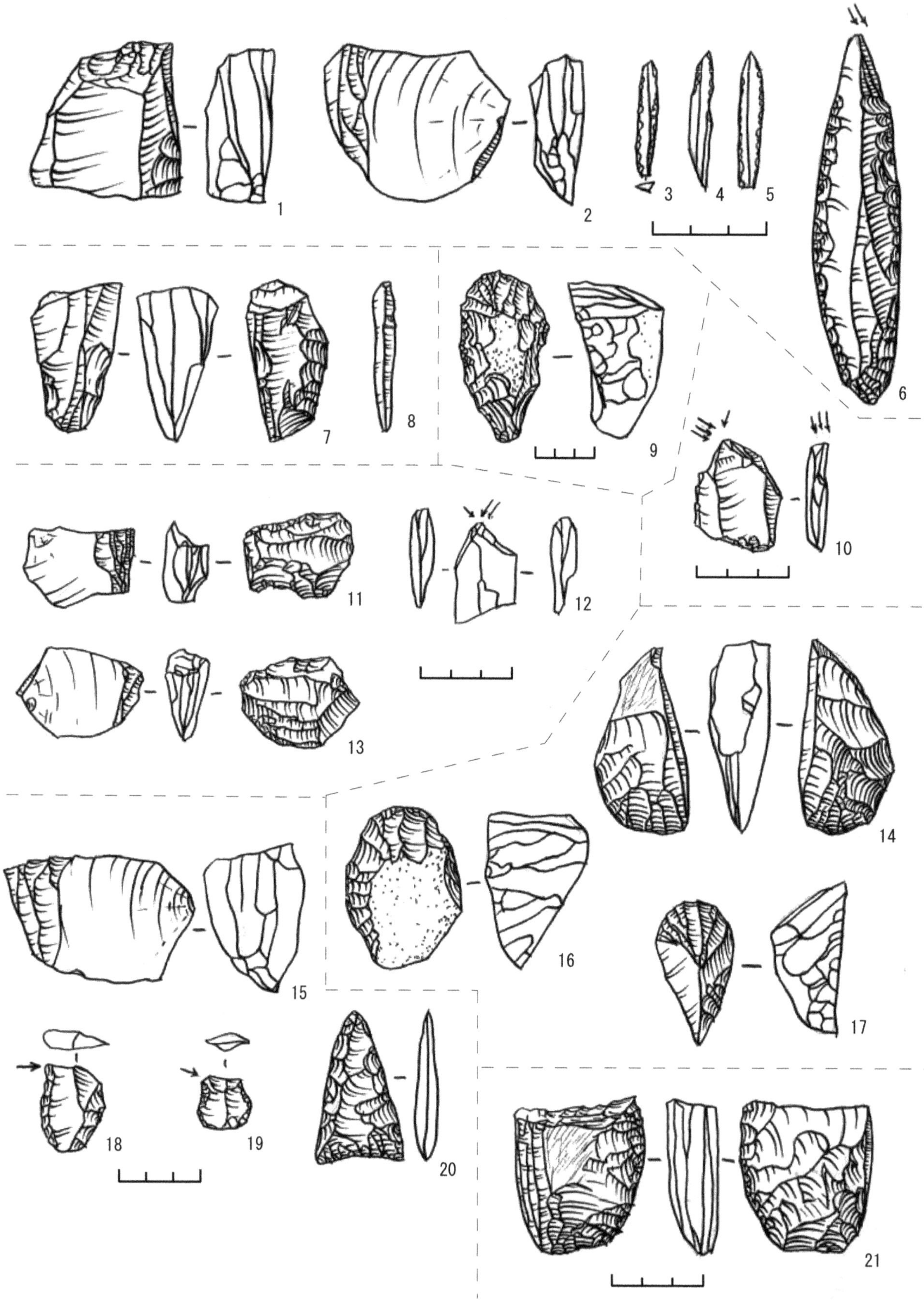

Figure 8 Microblade cores, microblades and artifacts from the Kostenki Area and Lipskaya group of sites of the Upper Palaeolithic of East Europe

1-4: Kostenki 4 (upper level), 7-8: Kostenki 11 (North point), 10: Kostenki 14 (level 4), 11-13: Kostenki 16, 15,18-20: Kostenki 1 (level 5), 14,16,17: Lipskaya culture (Ukraine) (after Rogachev and others).

a whole, 95% at Novoselovo VII (Abramova, 1979a:163). Reindeer also occurred, however, at Mousterian cave sites of the Yenisei River basin (Vasil'ev, 2003a). At the early Holocene Zhokhov Island site in the High Arctic, reindeer antler and mammoth ivory shafts for composite tools were discovered where microblades for insets were produced by a kind of microprismatic technique (Pituil'ko, 1998). There is no firm evidence of the correlation of knapping techniques with certain activities. The early farmers of the Central Asia and the Near East preserved microprismatic and blade industries similar to hunters-gatherers in the Neolithic of Siberia (Quintero and other, 1996).

Concluding remarks

In sum, the origin of microblade industries in north and east Asia and their appearance in northern North America were independent archaeological (technological) events. Environments of the Last (Sartan) Glacial in many ways influenced the people's existence in this area, but did not result in the appearance and distribution of the pressure technique of microblade production. This technique was a part of the whole techno-morphological context of microblade industries. According to current archaeological evidence, the oldest data on pressure flaking and composite tools are from southern East Siberia (about 20,000 BP). Because of the chronological gap and techno-morphological dissimilarity of microblade industries in the Yenisei River Basin and East Siberia, it may be proposed that the idea of composite tool production was transferred to the west across the Angara River Basin. From a technological point of view, the most elaborate techniques of microblades detachment from wedge-shaped cores are in the south of the Far East (Primorye, Korea, north China, and Hokkaido). Despite an absence of composite tools or their shafts, it may be the suggested that the core area of appearance of microblade industries was in this region. In any event, the previous conclusion that the first industry of such a kind was microprismatic should be denied (Inizan *et al.*, 1993). The microblade industry of Alaska (Denali complex), from the techno-typological point of view, is more similar according to our current level of understanding East Siberian assemblages.

References

Abramova, Z. A., 1979, *Paleolit Eniseya. Kokorevskaya kul'tura* [The Palaeolithic of Yenisei]. Izdatel'stvo Nauka, Novosibirsk.

Ackerman, R. E., 2007, 'The microblade complexes of Alaska and the Yukon: early interior and coastal adaptations' in Ya. Kuzmin, S. Keates and Chen Chen (eds.) *Origin and Spread of Microblade Technology in Northern Asia and North America*, Archaeology Press. Simon Fraser University (Burnaby, B.C), pp.147-170.

Astahov, S. N., 1999, *Paleolit Eniseya. Paleoliticheskie stoyanki na Afontovoi gore v g. Krasnoyarske* [The Palaeolithic of Yenisei. The palaeolithic sites of Afontova Gora locality in Krasnoyarsk]. Izdatel'stvo Evropriski Dom, Sanct-Petersburg.

Akimova, E. V., Vdovin, A. S. and Makarov, N. P., 1999, Pazovye orudiya Krasnoyarskogo arheologicheskogo raiona [Slotted tools of the Krasnoyarsk archaeological region], pp.62-82. In Drevnosti priEniseiskoi Sibiri, 1. Krasnoyarsk.

Chen, Chun and Wang, Xiang-Qian 1989 Upper palaeolithic microblade industries in North China and their relationships with Northweast Asia and America. *Arctic anthropology*, 26-2:127-156.

Derevyanko, A. P., Volkov, P. V. and Petrin, V. T., 2002, *Zarozdenie mikroplastinchatoj techniki rastsheplenia kamnya* [The Origin of Microblade Technique], Novosibirsk.

Dikov, N. N., 1977, K probleme mezolits na Kamchatke [On the problem of Mesolithic in Kamchatka]. *Kratkie Soobtsheniya Instituta Arheologii*, 149, pp.120-124.

Goebel, T., Powers, W. R. and Bigelow, N., 1991, 'The Nenana complex of Alaska and Clovis origin', in R.Bonnichsen and K.L.Turnmire (eds) *Clovis origin and adaptations*, (Corvallis).

Ikawa-Smith, F., 2007, Conclusion: in search of the origin of microblades and microblade technology. In: Ya. Kuzmin, S. Keates and Chen Chen (eds.), *Origin and Spread of Microblade Technology in Northern Asia and North America*, Archaeology Press, Simon Fraser University (Burnaby, B.C).

Inizan, M. L., Lechevallier, M. and Plument, P., 1993, A technological marker of the penetration into North America: pressure microblade debitage. It's origin in the paleolithic of North Asia and it's diffusion. In: H. Kimura (ed.) *The origin and dispersal of microblade industry in Northern Eurasia*, (Sapporo).

Keats, S., 2007, Microblade technology in Siberia and neighboring regions: an overview. In: Ya. Kuzmin, S. Keates and Chen Chen (eds.), *Origin and Spread of Microblade Technology in Northern Asia and North America*, Archaeology Press, Simon Fraser University (Burnaby, B.C.).

Kind, N. V., 1974, *Geohronologiya pozdnego antropogena po izotopnym dannym* [Geochronology of the Late Anthropogene on the isotopic data]. Izdatel'stvo Nauka, Moskwa.

Konstantinov, A.V. *Drevnie Zhilitsha Zabaikalia (paleolit, mezolit)*. Aftoreferat dissertatsii doktora istoricheskih nauk. Sankt-Peterburg. 2004:22, 28.

Kuzmin, Ya., Keats, S. and Chen Chen, 2007, Introduction: Microblades and beyond. In: Ya. Kuzmin, S. Keates and Chen Chen (eds.), *Origin and Spread of Microblade Technology in Northern Asia and North America,* Archaeology Press, Simon Fraser University (Burnaby, B.C.).

Lisitsyn, S. N., 1998, 'Mikroplastinchatyi inventar' verhnego sloya Kostenok 1 I nekotorye voprosy razvitia mikroorudij v verhnem paleolite Russkoi ravniny in Vostochnyi Gravette [Microblade implements of the upper layer of the Kostenki 1 site and some problems of microtools development in the Upper Palaeolithic of the Russian Plane in the Eastern Gravette]. In: H.A. Amirkhanov (ed.), *Vostochnyi Gravette*, pp.299-308. Nauchnyi Mir, Moskwa.

Mochanov, U. A., 1977, *Drevneishie etapy zaseleniya chelovekom Severo-Vostochnoi Azii* [The Early Stages of Settlement of North-East Asia]. Izdatel'stvo Nauka, Novosibirsk.

Petrin, V. T. and Chevalkov, L. M., 1992, O vozniknovenii tortsovoj techniki skalyvaniya na primere palolitícheskoj stoyanki Kara-Bom [On the Origin of Butt End Technique, the Case of Paleolithic Site Kara-Bom]. In: A. P. Derevyanko, N.I. Drozdov, E.V. Akimova, V.P. Chekha (eds.), *Paleoecologiya i rasselenie cheloveka v Severnoj Azii i Amerike,* pp.206-209. Krasnoyaesk.

Pitulko, V. V., 1998, *Zhohovskaya stoyanka* [The Zhokhov Site]. *Radiouglerodnaya chronologiya paleolita Vostochnoi Evropy I Severnoi Azii. Problemy I perspertivy 1997*, A. A. Sinitsyn, N. D. Praslov (eds.), Izdatel'srvo Nauka, Sankt-Peterburg.

Quintero, L. A., Wilke, F. J. and Wains, J. G., 1996, Pragmatic studies of near eastern Neolithic sickle blades. In: H. G. Gebel, G. O. Rollefson and Z. Kafafi (eds.), *Prehistory of Jordan ll: Perspectives from 1996. Studies in Early Near Eastern Production, Subsictence and Environment 4* (Berlin), pp.1-42.

Sato, H. and Tsutsumi, T., 2007, The Japanese microblade industries: technology, raw material procurement, and adaptations. In: Ya. Kuzmin, S. Keates and Chen Chen (eds.), *Origin and Spread of Microblade Technology in Northern Asia and North America*, Archaeology Press, pp.53-78, Simon Fraser University (Burnaby, B.C.)

Sinitsyn, A. A., and Praslov, N. D. (eds.) 1998 *Radiouglerodnaya chronologiya paleolita Vostochnoi Evropy I Severnoi Azii. Problemy I perspertivy 1997* [The Radiocarbon Chronology of the Palaeolithic of East Europe and North Asia, Problems and Perspectives 1997] , Izdatel'srvo Nauka, Sankt-Peterburg.

Vasil'ev, S. A., 1996, *Pozdnij Paleolit Verhnego Eniseya.* [The Late Palaeolithic of the Upper Yenisei]. Izdatel'stvo Peterburgskor Vostokovedenie, Sanct-Peterburg.

Vasil'ev, S. A., 2003a, Faunal explotation, subsistence practices and Pleistocene extinctions in Paleolithic Siberia. In: J. W. F Reumer., De Vos, J.and Mol, D (eds.), *Advances in Mammoth Research.* pp. 513-556.

Vasil'ev, S. A. 2003b, The Upper Paleolithic domestic structures in Siberia: a critical review of relevant evidence. In: S. A. Vasil'ev, O. Soffer, J. Kozlowski (eds.), *Perceived Landscapes and Built Environments. The cultural geography of Late Paleolithic Eurasia*, pp.155-160, BAR International Series 1122.

Vasil'ev, S. A., Kuzmin, Ya. V., Orlova, L. A., Dementiev, V. N., 2002, Radicarbon-based Chronology of the Paleolithic in Siberia and its Relevance to the Peopling of the New World. *Radicarbon*, 44:503-530.

Re-evaluation of the Chronology and Technology of Palaeolithic Assemblages in the Imjin-Hantan River Area, Korea

Yongwook Yoo

Department of Archaeology, Chungnam National University
220 Gung-Dong, Yuseon-Gu, Daejeon, 305-764, Korea
E-mail: palaeo@cnu.ac.kr

Abstract: *Recent excavations and research in the Imjin-Hantan River Area (the IHRA) furnished new data addressing a revised position on the age and technological characteristics of Palaeolithic industry. The earliest date of hominid occupation in this area is, if not earlier than, the later Middle Pleistocene (ca 0.2 Million Years Ago, hereafter MYA by the IRSL date of the Jangsanri site). The handaxes and associated small tools in the IHRA are predominantly bracketed into the OIS 3 (ca 60- 45 KYA by the luminescence/AMS dates of the Chongokni, Keumpari, and Kawolri). Contradicting the old notion that the age of the Chongokni site is to be the Middle Pleistocene (ca 0.30 MYA; e.g. Bae 2002; Danhara et al. 2002; Matsufuji et al. 2005; Norton et al. 2006), these new dates reveal that the handaxe-based assemblages are not chronologically equivalent to the Acheulian industry (or the Lower/Early Palaeolithic technocomplexes). Rather, the IHRA handaxe is believed not to be a direct output of either the hominid acculturation or technological transmission. The relatively simple and under-developed level of manufacturing technique suggests that this young handaxe might have been produced as a result of the provisional necessity in the demands of a reliable multi-purpose tool. This crude but instrumental tool-type had persisted until new high-quality raw materials (the obsidian and the porphyry; e.g. the Hwadaeri and the Janghungri sites) began to be heavily exploited and the small-tool-dominant Upper Palaeolithic technology finally emerged during the terminal Pleistocene in this area.*

***Keywords**: the IHRA, Handaxe, Chronostratigraphy of Chongokni, Raw material transition*

Introduction

In the Imjin-Hantan River Area at the midwestern region of the Korean Peninsula are distributed more than dozen of Palaeolithic localities. This region has been renowned for the discovery of Acheulian-like handaxes since the late 1970s (Kim and Chung 1979); the Chongokni and other neighboring localities have yielded as many handaxes as over 100 specimens and this led many international researchers to recognize this area as one of the most-handaxe-abundant Paleolithic regions in the East Asia. Several excavations and regional geoarchaeological research have been continuously proposing many important issues in the context of hominid's occupation history at the far eastern margin of Eurasian continent. Among these issues, the exact date of handaxe is one of the most debatable topics. In addition, the significance of the handaxe still remains questionable because this region is within the "chopper-chopping tool" area, the zone of Pleistocene cultural/technological "retardedness," suggested by Movius (1948).

Although much endeavor has been made to date this culturally unique lithic assemblage, some inherent difficulty in interpreting the geological sequence of the later Pleistocene (ca 0.7- 0.12 MYA) period hindered researchers from acquiring an indisputable chronology. It is a still matter of time that the Paleolithic research in the IHRA is enhanced by the aid of sound frameworks, not plagued by the plethora of unpersuasive proxy data. In an attempt to invigorate the current discussion with a broader perspective, this article is intended to evaluate the validity of the past results based on the currently available data. With new publications of chronometric dates and several unprecedented archaeological discoveries—Chongokni, Jangsanri, Juwolri, Kawolri, Keumpari, Hwadaeri, and Janghungri, for example—from 2000 to 2006 (Fig. 1), we can have a new perspective on the characteristics of the IHRA lithic assemblage. In short, some younger dates suggest that the formation of the IHRA assemblage was initiated no earlier than OIS 6, and that the variability of the IHRA assemblage witnessed a detectable temporal change, rather than a stagnant survival of the handaxe and other large crude tool types for a considerable long time.

Past Research: what was gained and what was overlooked?

The IHRA hosts numerous archaeological localities across the entire territory of the Imjin-Hantan channels. The initial research of the IHRA coincides with the first excavation of the Chongokni site in 1979. The age of its artifact horizon was early suggested to be around 0.21mya based on the controversial K-Ar dates of the basalt bedrock

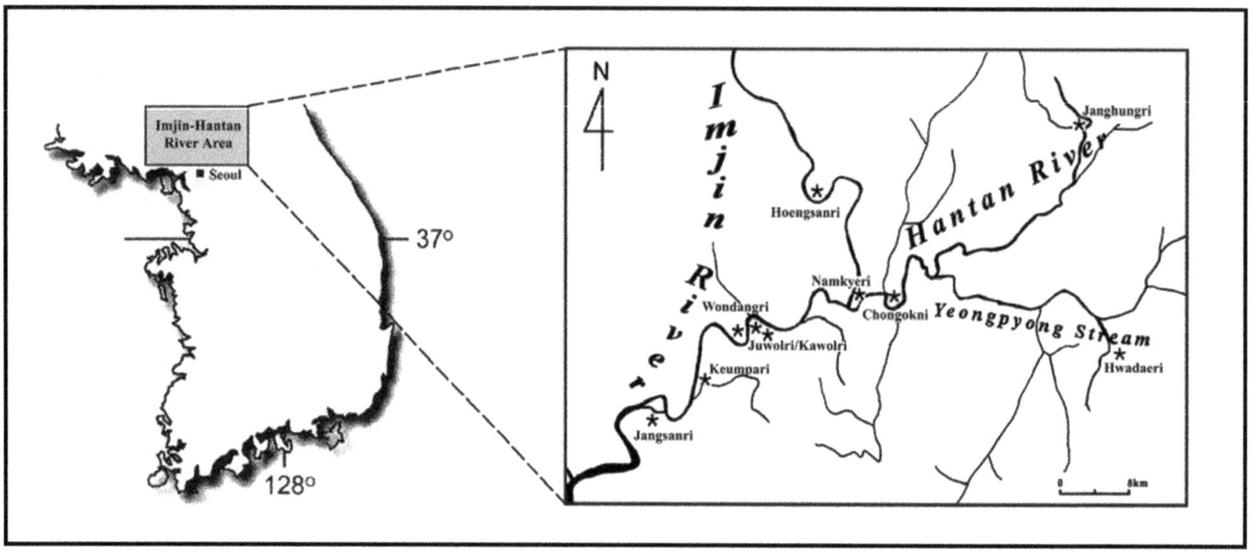

Fig. 1. Location of the IHRA and distribution of major archaeological sites

(Kojima 1983); some authors argued that the age of the IHRA assemblage were partly contemporaneous with the time range of the Lower/Early Palaeolithic sequence suggesting that the handaxe of the Chongokni site is the direct equivalent of the Acheulian industry in the western hemisphere, thereby contradicting the Movius' claim (Kim and Chung 1979; Bae 1988). However, several authors proposed a different opinion that, although Movius' groundless claim cannot be supported by current data, the Quaternary geomorphology of this area and several younger chronometric dates—e.g. TL and OSL dates presented by Yi (1989; 1996)—did not verify such a great "antiquity" of artifact horizon as dated to be Lower/Early Palaeolithic (Yi 1989, 1996, 2004).

Since then, several researchers have been vigorously engaged in the investigation of the IHRA assemblage, and recently made a claim that the Acheulian-like handaxe and associated lithic artifacts were estimated to be ca. 0.35 MYA, based on the new chronometric dates—K-Ar and fission track methods—of basalt bedrock and on the subsequent sedimentary rates above (Danhara *et al.* 2002; Matsufuji *et al.* 2005). Furthermore, others suggested that the IHRA handaxe, coeval to the Acheulian handaxe, was not technologically and morphologically parallel to the genuine Acheulian one supporting that the Movius' early claim is still *sensu lato* effective (e.g. Norton *et al.* 2006). Summing up these two-folded claims, the IHRA assemblage is arguably one of local East Asian Lower/Early Palaeolithic industries technologically adumbrating the Acheulian characteristics. This is how current argument is being made in the context of the chronology and technology of the IHRA assemblage.

Challenging this argument entails evaluating how the dates were acquired and elucidating to what extent these dates can contribute to the geochronology of the IHRA. As is known, the terrain of the IHRA reflects a sequential fluvial actions interrupted by multiple volcanic eruptions. Most archaeological horizons are distributed above the basalt bedrock formed as a result of cooled lava flows. In particular, the handaxes are discovered within the fine silty-clay layers positioned at the top of the entire stratigraphic column. Supposing that the age of the IHRA archaeological horizon is extrapolated by the time lapsed after the volcanic eruption, as indicated by Danhara *et al.* (2002), the formation of the handaxe assemblage is believed to be relatively young throughout the entire Pleistocene sequence of this area. Most chronometric dates are dependent on the K-Ar dates of the basalt bedrock as the chronological datum point. Subsequent dates of the fluvial layers above the basalt bedrock are dependent on these K-Ar dates as a maximum oldest limit. Nevertheless, the range of several K-Ar dates exceeds the tolerance of statistically significant value, from over 1 to 0.158 MYA (Yi 2005). The problematic distribution of the K-Ar dates leads to two fundamental questions about the nature of the basalt bedrock and the limitation of the K-Ar dating method: 1) why are the results almost always unreliable, and 2) to what extent is the K-Ar dating reliable in determining the age of such relatively recent geological events?

It is important to note that the lava flows directly and simultaneously covered the river channel including the river bed, the adjacent swamps, and the overbank area; these overlain zones have the potential to vaporize large quantities of minerals and of pyrolyzed organic materials that are critical to the ratio balance of atmospheric ^{40}Kr/^{40}Ar. As a result of thermal transformation occurring at the hydraulic zone, large amounts of miscellaneous isotopes are usually converted to either ^{39}Ar or ^{40}Ar. Consequently, the estimated age at which the lava was cooled and consolidated into basalt rock cannot but be biased. The relationship between the hydraulic condition of the sampling locations and the extent to which the dates are biased can neither be adequately addressed nor be correlated for these conditions are unpredictable and stochastic. Therefore, it is still beyond the current capacity of the K-Ar method to establish the reliable ages of lava flows.

On the contrary to the baffling situation of the lower limit, the upper limit, the possible youngest date of the IHRA archaeological horizon, is well-established. Yi (1996; see also Yi, Soda, and Arai 1998) firstly discovered the Aira-Tn (AT) tephras from the top crack of the silty-clay layer superseding the handaxe horizon; based on the solid dates of the AT (ca 27 to 24 KYA), he then postulated the handaxe assemblage of the IHRA might have been persisted until the global climatic deterioration of the OIS 2, when it was rapidly replaced by the new, high-quality-material-based Upper Palaeolithic technology (Yi 2004). What is clear from this position is that, regardless of the age of the initial handaxe production in the IHRA, the Acheulian-like handaxe assemblage does not chronologically coincide with the typical Acheulian of the western hemisphere. In the same vein, the temporal coevalness suggested by Norton *et al.* (2006) cannot but be contradicted based on the late survival of the handaxe manufacturing till the terminal Pleistocene. The remaining questions are, therefore, what "approximate dates" of the IHRA assemblage can be indicated by the current data, and how we can present a meaningful explanation on the nature of the IHRA handaxe. In an attempt to answer these questions, the next chapter will illustrate the general geological/archaeological features based on the new data.

New dates before and after the volcanic eruption

Although most IHRA assemblages are distributed above the basalt bedrock, some assemblages predate the volcanic eruption. An example is located at the downstream area of the Imjin River, far beyond the stretch of maximum lava flow range—the Jangsanri site (Yi 1996, 2004). Jangsanri site is the oldest Palaeolithic occupation in the IHRA and is positioned on top of the ancient river terrace formed by the fluctuation of the Imjin River. As the artifact horizon lies more than 20m above the current lava flow, its relative olderness to the basalt bedrock can be easily claimed. As indicated by the exposed outcrops at the volcanic area, the lava flow covers the fluvial channel bed formed after the downcutting of this deposit; there is no doubt that the terrace surface was formed much earlier than the formation of the basalt bedrock in the IHRA. In recognition of its importance in the Quaternary stratigraphic sequence of the IHRA, a formal name of the Jangsanri Terrace was given to this deposit (Yi *et al.* 2004; Fig. 4). While no indisputable handaxe was found in situ at Jangsanri, a large pointed tool indicates that a glimpse of handaxe manufacturing was on its way when hominids firstly resided in this area (1 of Fig 5). Notwithstanding that the real age is possibly older, the current available IRSL date of the Jangsanri Terrace is ca 0.23 MYA (Yi 2004). It would be, therefore, unpresumptuous to take this date as a provisional lower limit age of the IHRA assemblage.

In addition to the age of the Jangsanri Terrace, attempts to date the burnt gravels beneath the lava flow have been made. Danhara *et al.* (2002) obtained a single fission-track date of ca 500 KYA and assumed it to be the age of the basalt as it corresponds to their K-Ar dates. But it can be criticized on a technical ground because only the single date came from a single fission track etched on a single grain of mineral. Later on, new fission-track dates of ca 0.4 MYA from more and larger samples were published as the maximum age limit for the basalt flow (Yi *et al.* 2005; Yoo 2008). Additionally, several luminescence dates indicating the moment when the lava encroached in the basin were obtained by the burnt gravels beneath the basalt bed. Some of these dates— within the range of ca 200 to 60 KYA (Yi 2005; Choi, J. *et al.* 2004)—are apparently too young to be the real age for the basalt; another two samples from separate localities produced the same date of ca 148 KYA with slightly different error ranges (Choi, J. *et al.* 2004). These apparently young dates are not believed to be serendipitously acquired since there already exist several K-Ar dates of the basalt bedrock corresponding to these dates (Table 1). If we consider these dates valid for the age of volcanic eruption, then the real age of archaeological horizon above the basalt bedrock is necessarily to be younger than supposed before. Of course, these young K-Ar dates are not indisputable because, compared to the general dating range of the K-Ar dating method normally applied to the far older geological ages, the measured values are liable to be the result of insufficient amount of ^{40}Ar. More controlled sampling and innovative dating techniques should be substantiated in the future so that the K-Ar dates of the basalt bedrock are to be secured.

For the direct age of the archaeological horizons on top of the basalt, several scores of luminescence dates were newly obtained between the basalt bedrock and the artifact horizon from the 2004's excavation campaign of the Chongokni site (Fig. 2). Except for a few problematic results, the majority of the dates belong to the later part of the Upper Pleistocene—from the OIS 5e to the OIS 3—including those from the lowermost part (Yi 1995, 2005; Yi *et al.* 2005). This series of the OSL dates indicates that, after the termination of the volcanic activity, the fluvial sediments were continuously accumulated before the hominid's occupation was initiated at least after the OIS 4. Because the lower part of the fluvial layers is predominantly composed of clastic sediments transported by rapid and turbulent river flow, the assumption by several authors (e.g. Danhara *et al.* 2002; Matsufuji 2005) that the sedimentation of artifact layers was progressed in a steady rate cannot be supported.

In addition to these luminescence dates, several new ^{14}C (AMS) dates were acquired from organic samples directly associated with the artifact horizon (Table 2). At the Chongokni site, a biostratigraphic zone of fossil burrows distributed below the AT level were dated to be approximately 20- 30 KYA (Yi 2005). These burrows are believed to be dug by hibernating rodents and its hollow ducts were subsequently filled with organic sediments after

Table 1. Several "young" K-Ar dates from the basalt samples of the IHRA. Note that the estimate age extracted from these dates is no more than an approximation from late OIS 6 to early OIS 5. Past research did not give full consideration on these dates because of being too young to be comparable to the "believed age" of lithic assemblages.

Dates (KYA)	***Sources***	***Sampling location***	***Remarks***
108±158	Kojima (1983)	Chongokni	Too wide error range
138.4±5.7 and 136.5±5. 4	Park *et al.* (1996)	Chongokni	Double dating
160±25 and 18±2	Danhara *et al.* (2002)	Chongokni	Single sample with different methods
158	Yi (2005)	Shindapri	Peak comparison method

Table 2. AMS dates from the artifact horizons of the IHRA. Note that all four dates directly acquired from the same horizon as lithic assemblages are quite younger than supposed before. Two dates obtained from rodent burrows at Chongokni indicate that some post-depositional activities of hibernating animals seriously affected the in-situ condition of original assemblage horizons.

Dates (KYA)	***Sources***	***Archaeological Site***	***Remarks***
20.840±0.45	Lim *et al.* (2004)	Chongokni	Slightly below the top crack
30.490+4.76/-2.59	Lim *et al.* (2004)	Chongokni	Almost identical level to the artifact horizon
31.500±1.300	Bae *et al.* (2006)	Keumpari	Charcoal No. 2 from Grid W23N6-1
30.800±0.400	Bae *et al.* (2006)	Keumpari	Charcoal No. 3 from W23N6-1 Keumpari

Sequence of the OSL dates measured by two dating agencies

Sample Code	Age (KYA)	Dating Agency
CK8	N/A	IPCCPF
CK7	60±10	IPCCPF
CK6	65±10	IPCCPF
JG HT 1	57.8±3.1	KBSI
JG HT 3	48.8±3.8	KBSI
CK5	72±9	IPCCPF
JG HT 4	48.9±3.7	KBSI
CK4	83±13	IPCCPF
JG C 1	57.4±3.3	KBSI
JG C 2	38.2±3.2	KBSI
JG C 3	64.1±3.6	KBSI
CK2	92±8	IPCCPF
CK1	120±6	IPCCPF
JG A 1	60.9±5.1	KBSI
JG A 2	38.2±3.2	KBSI
JG A 3	56.7±3.2	KBSI
JG A 4	36.9±3.8	KBSI
JG A 5	43.6±4.7	KBSI
JG A 6	51.5±3.7	KBSI
JG A 7	58.0±3.5	KBSI
CHA	104±8	IPCCPF
CHB	127±6	IPCCPF
CHC	113±3	IPCCPF

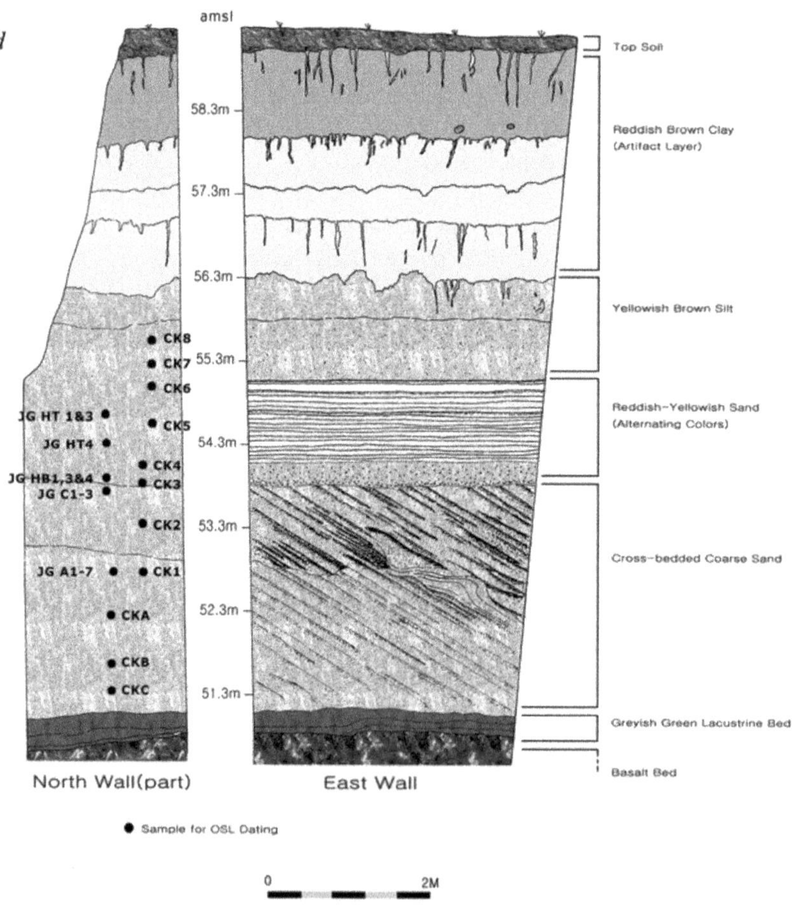

Fig. 2. Stratigraphic sequence of the Control Pit of the 2004's Chongokni Excavation Campaign (Dating Agency: KBSI-the Korean Basic Science Institute; IPCCPF-Institute of Paleoenvironment at the Chungcheong Cultural Property Foundation)

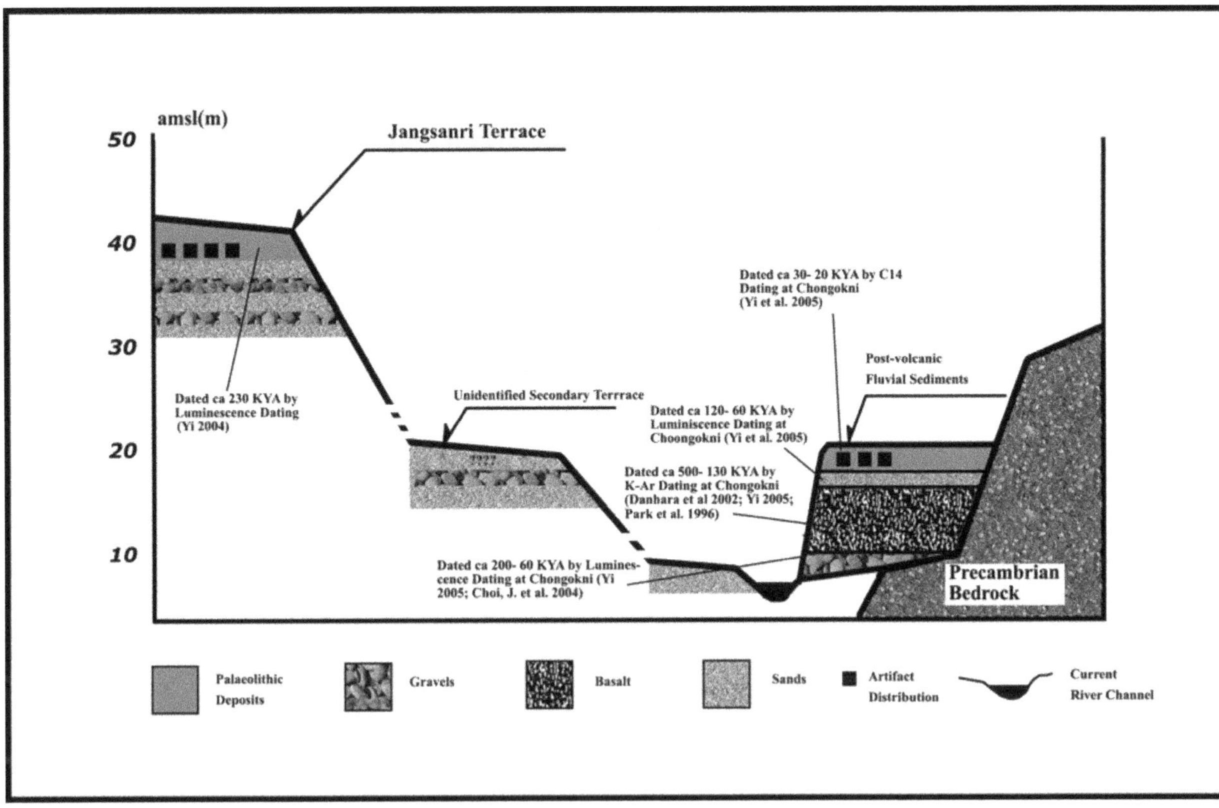

Fig. 3. Schematic crosssection of the IHRA archaeological/geological sequence. If the Jangsanri Terrace is the oldest fluvial deposit currently identified in the IHRA, most artifact-containing layers in the IHRA necessarily postdate to this stratum. The IRSL date of the Jangsanri Terrace is ca 0.21 MYA and it is highly probable that the actual age is quite older than the face value of the chronometric date

the rodents vertically penetrated and disturbed the original artifact horizon. Though its original age is believed to be somewhat older and needs to be verified, a handaxe was discovered from the same level as this burrow zone (2 of Fig. 5.). At the Keumpari site, two AMS dates are available from the charcoal samples in association with the artifact horizon; and their dates commonly suggest that the age of lithic assemblage is not older than the lower limit of the OIS 3. Taking these young AMS dates into consideration, the IHRA lithic assemblage was principally formed after the later phase of the OIS 4 and continuously survived until the terminal Pleistocene (Yi 1996, 2005). However, we are still far from understanding the exact provenience of handaxe in this area because most specimens are from the surface collection. Except for several pieces retrieved from the upper part of the silty-clay layer, the vertical position of "handaxe cultural layer" is yet to be clearly defined. In this regard, in spite of our knowledge that handaxe has survived until the final phase of the OIS 3, the maximum we can take is the bare possibility that the age of the handaxe in the IHRA is not so old as supposed by conventional approaches predominant among past researchers (e.g. Bae 1988; Chung 1985; Danhara *et al.* 2002; Kim and Bae 1983; Matsufuji *et al.* 2005; Norton *et al.* 2006).

The characteristics of the IHRA assemblage: some clarifications and hypotheses

Fig. 3 is an illustrated general scheme of archaeological horizons distributed in the IHRA. As can be seen, the exact correlation of the chronometric dates is hard to attain because the sequence of strata is different from location to location. The geographic range of site distribution reaches the upper region of the Imjin River beyond the DMZ, and is even distributed into the far downstream area in the western region (e.g. the Jangsanri site; see Fig. 1). The distribution of the basalt bedrock is concentrated in the upper and middle region of the IHRA, but it is not well-represented in the lower part of the Imjin River. Especially associated with the Keumpari, Juwolri, and Kawolri sites, the basalt layer principally exist as the remnants of stacked cobbles transported from the upper region. Yi and Lee (1993) also noticed that the Juwolri/Kawolri area was almost devoid of any evidence of basalt flows. In sum, the geological contexts in the entire IHRA are far from being the result of uniform and simultaneous formation processes, and the assumed "comparable" temporal ranges across the localities cannot be verified by current geological and archaeological evidence.

It is, therefore, clear that there might exist many variants of lithic assemblage in the IHRA with regard to the morphology and the manufacturing procedure of tool types. For example, some very refined handaxes, such as entirely bifacial shapes (e.g. 3 and 4 of Fig. 4), are included in the Juwolri/Kawolri assemblage and these have almost perfect symmetrical morphologies emulating the typical Acheulian specimens. Nevertheless, the frequency of these

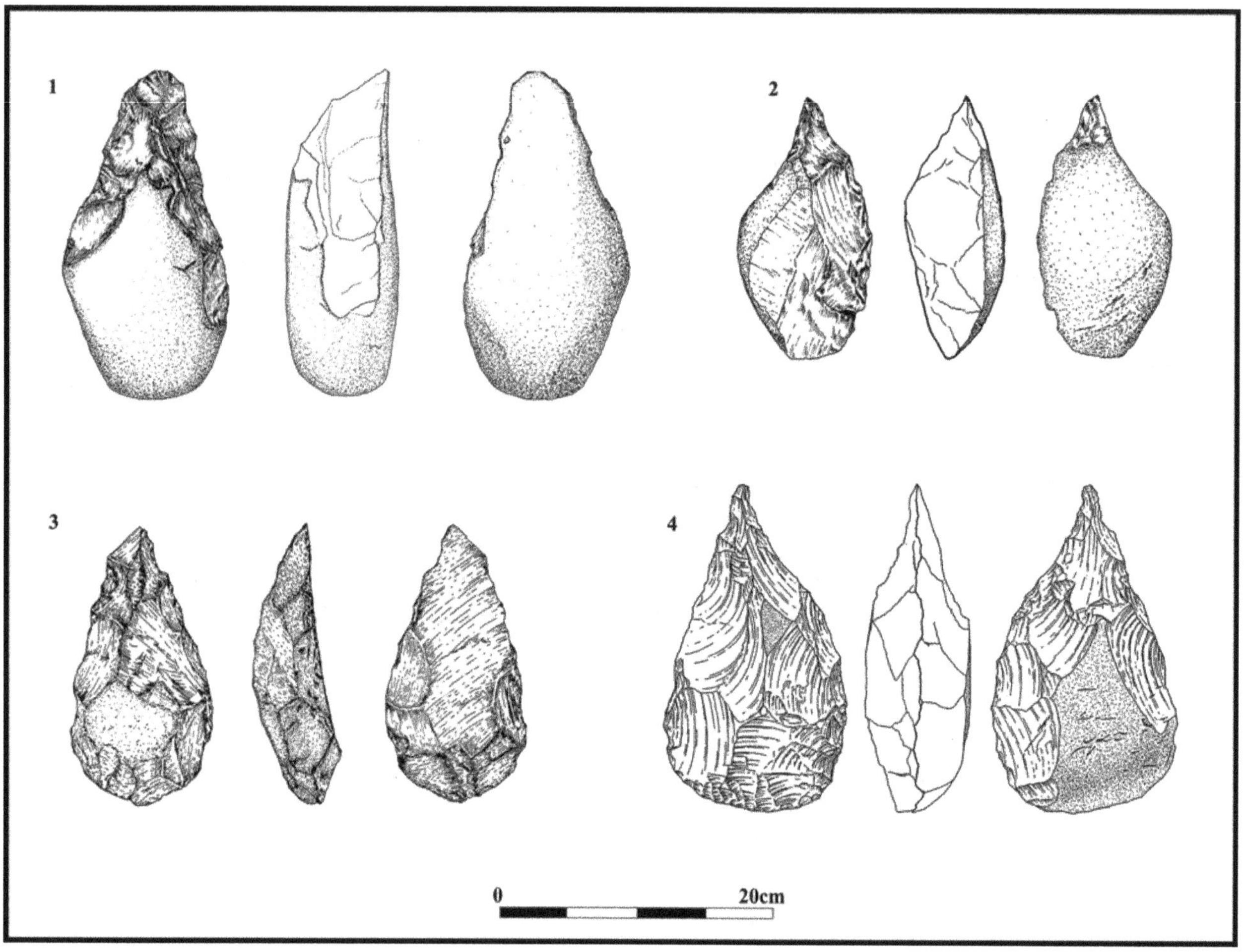

Fig. 4. Handaxes from the IHRA localities. 1: from Jangsanri lower layer; 2: from Chongokni upper layer; 3. from Juwolri disturbed top soil; 4. from Juwolri top soil (after Yi and Lee 1993; Yoo 2008)

refined handaxes in the total IHRA assemblage is very limited: a considerable number of the IHRA handaxes are crude, almost identical to the ficron type of biface under the Bordesian typological scheme (Débenath, and Dibble 1994: 144). Most handaxes distributed in the IHRA are expediently finished and their thickness is notably greater than that of typical Acheulian implements as Norton *et al.* (2006: 534) adequately presented. However, in contrast to what they claim (Norton *et al.* 2006: 533), the blanks of handaxe are equally composed of both large cobble-sized gravels and moderately reduced flakes. In either case, the reduction sequence is simple and the blanks are casually modified without being shaped to their perfection. At least three—unifacial, bifacial, and alternate shaping—manufacturing methods for producing handaxes were prevalent in the IHRA assemblages and each method is sufficiently effective to obtain workable lateral edges and a useful pointed tip at its distal end (Yoo 1997). If these two formal properties had been achieved, additional intensive retouch on the remaining part of the blank might have been redundant; its cortex was, therefore, mostly left intact, possibly for grasping. It is believed that the IHRA hominids had an underdeveloped technique or even lesser need to perform intensive reduction for the purpose of shaping: a so-called mental template of the "ideal" form a handaxe should display might have been minimally imposed during the manufacturing stages. In this regard, the raw material constraints are not solely responsible for the morphological crudeness of the IHRA handaxes (Yoo 2008).

In addition to the handaxe, newly excavated assemblages from Chongokni, Juwolri, Kawolri and Keumpari include a number of small tool types made of normal flake, broken shatter, and miscellaneous chips (1- 6 of Fig. 5). The distribution of these small tools is mostly concentrated at the level of OIS 3: for example, in Chongokni, Kawolri, and Keumpari, various small tools were found from just below the top crack where AT particles are usually discovered. The level of technological accomplishment shown in the small tool repertoires is not excellent; the position of retouch and formality of tool morphologies are rather arbitrary, comparable to those of handaxes. In this case, it can be addressed that the handaxe and various small tools were simultaneously made for a considerable long time before the OIS 2; and that both the productions of handaxe and of small tools were in the similar technological context. If the crudeness of general tool morphology can be attributed either to the poor quality of raw material (quartz/quartzite) or to the low level of manufacturing intensity, then the IHRA assemblage can be characterized as an

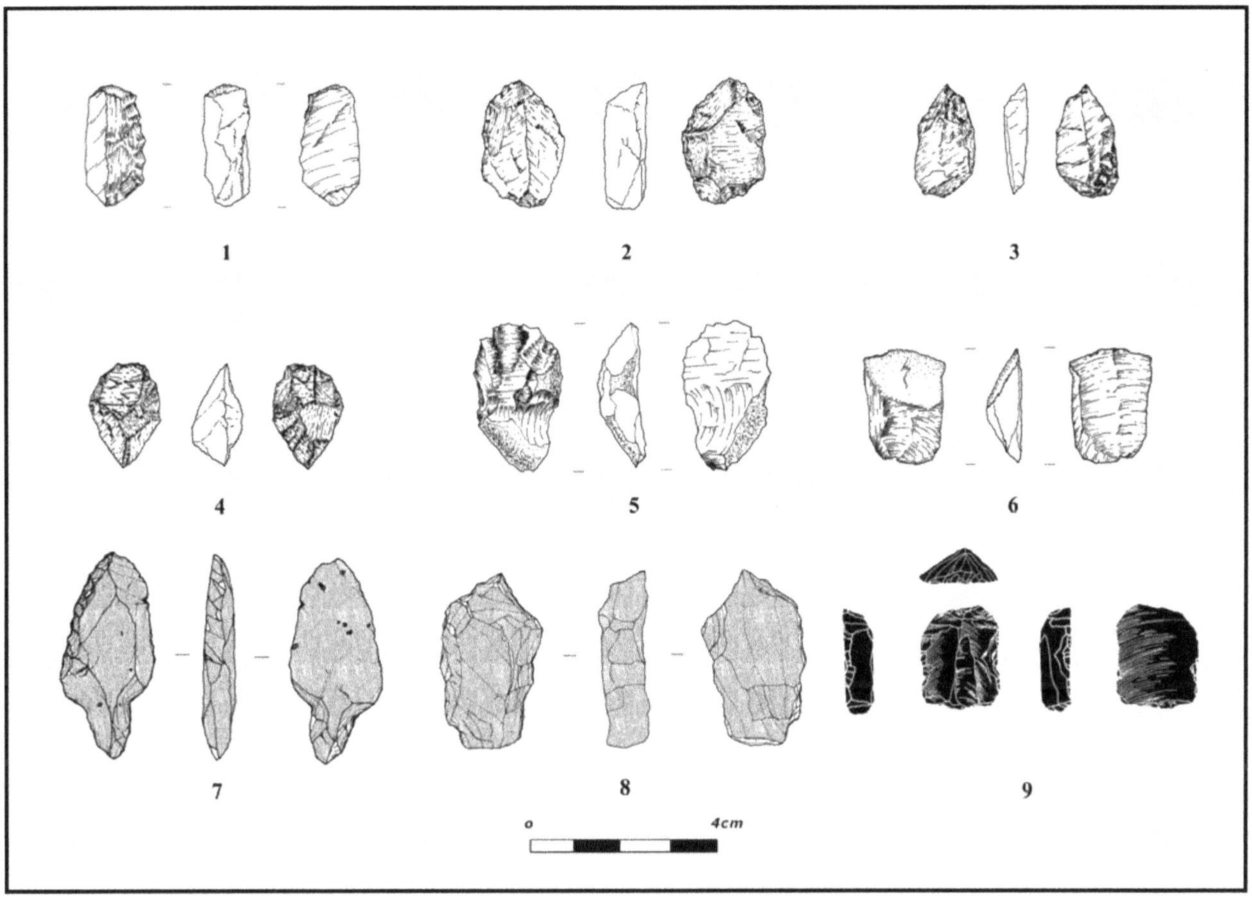

Fig. 5. Various small tools in the IHRA. In order to distinguish raw material types, porphyry specimens were shaded in grey and obsidian one in black. 1: side scraper from Chongokni; 2: sidescraper from Kawolri; 3: point from Kawolri; 4. endscraper from Kawolri; 5: endscraper from Chongokni; 6: tranchet from Chongokni; 7: porphyry stemmed point from Hwadaeri; 8: porphyry scraper from Hwadaeri; 9: obsidian endscraper from Janghungri (after Choi, B. 2001; Choi, B. and Yoo, H. 2005; Yoo 2008)

output of technological organization constrained by local raw material disadvantage and by low necessity to produce high-calibered specialized tools.

These apparently crude and expedient lithic assemblages show a relatively drastic change approaching the OIS 2 in the IHRA. The quartz and quartzite, principally exploited for both large and small tools, became less dominant and new high-quality materials such as porphyry and obsidian began to be heavily utilized (7- 9 of Fig. 5). Accordingly, the handaxe and other large tools became scarce and several small tools of low quality materials survived in less quantity. The Janghungri and Hwadaeri sites which have a good survival of younger layers above the top crack well-illustrate a transition from low to high-quality material assemblage. According to the Hwadaeri excavation result (Choi and Yoo 2005), the size of quartz tools are gradually reduced and new raw materials of rhyolite, porphyry and obsidian emerges around the terminal OIS 3 (39±1.4, 30±1.7 KYA by OSL dating from bottom to top before the AT crack; Choi B. and Yoo, H. 2005: 36). The Janghungri assemblage from the top cultural layer is composed of highly modified small tools and of such tiny blanks as heavily utilized chips and microblades.

Because most archaeological sites located at the midstream area of Imjin and Hantan River have been vulnerable to post-depositional transformation due to human disturbance and flood, it is extremely difficult to identify the general tendency of lithic changes in this area. In spite of this, while not found associated in situ, there are also several discoveries of high-quality materials at Chongokni (Yi, Yoo, and Kim 2006), Keumpari (Bae *et al.* 2006), and Heongsanri (Bae *et al.* 2006). Considering this, it is probable that the low-quality materials were finally replaced by exotic high-quality materials in the entire IHRA around the OIS 2; and that the manufacturing technique with these high-quality materials equals to the local Upper Palaeolithic tradition, both chronologically and technologically. The Middle/Upper Palaeolithic transition in the IHRA is not well-known, but, as far as the transitional phase shown at Hwadaeri and Janghungri is taken into consideration, the change into high-quality materials and the "extinction" of handaxe might have paralleled with the emergence of local Upper Palaeolithic technology in this area (Yi 1999; Yoo 2008). This transition could have been instigated either by the arrival of new hominid population at the far extreme of the Eurasian Continent or by the prompt technological adaptation of indigenous hominid in response to the rapid environmental deterioration around the onset of the OIS 2 (Yoo 2008).

Closing Remarks

During the last several decades, Paleolithic researchers have been opting for an explanation encompassing the materialistic appearance of various archaeological traits. Especially, in the field of lithic technology lies a general research trend concentrating on the relationship between the assemblage variability and the implications of hominid cultural/technological legacies. What is evident concerning this trend is that archaeologists try to extract meaningful dynamics from static archaeological records, and that they try to elucidate as many contexts as these records are formulated by. Notwithstanding the progress made in the research field, however, it seems still questionable to what extent the lithic data can be interpreted as a direct output of hominid's cultural/technological capacity in the course of their global colonization. Especially dealing with seemingly crude and notably "expedient" lithic technologies in East Asia, researchers are liable to succumb to the lack of comparable data; as a result, some feel for a near-obsolete tenet (e.g. resuscitating the Movius' claim of 1940s; Norton *et al.* 2006), or others tend to overestimate the nominal plausibility of published data in the hope of launching a general scheme applicable to a larger domain of data (e.g. extending the explicability of chronometric dates for ranging the artifact horizon; Danhara *et al.* 2002; Matsufuji *et al.* 2005). In a sense, the explanatory limitation of current East Asian Paleolithic data is arguably due to their constitutional dullness and uncharacteristic monotone over a long period. Therefore, it is an arduous task to conduct a region-based research with the reliable chronological schemes and the solid typological systematics that are compatible with those of Europe and Africa.

This article is not intended to set an agenda that the past research in the IHRA should be negated facing contradicting new data but to present a demand that the IHRA assemblage should be understood with a more flexible attitude. Taken from the current data, a new perspective that the handaxe of the IHRA is not in the same vein—chronologically, technologically, and contextually—with that of Acheulian industry can be proposed. Most IHRA assemblages are the output of hominid activity during the OIS 3 although some evidence older than OIS 5e also exists (e.g. the Jangsanri assemblage); the IHRA assemblage is not composed exclusively of large tools represented by handaxe but equally of small tools that were manufactured in a similar technological context with large tool types; the crude and underdeveloped technology dependent on the law-quality raw materials was abruptly replaced by a new technology based on the high-quality raw material such as porphyry, rhyolite, and obsidian from the terminal OIS 3 to OIS 2; this new technology is identical to the emergence of Upper Palaeolithic tradition in this area reflecting either the population replacement or technological innovation accomplished by local hominid.

Although still hypothetical, this perspective enables us to overcome an "idée fixe" that the Acheulian-like large tools discovered in the East Asia cannot but be interpreted in the context of global Lower/Early Palaeolithic technocomplexes. If we take the young age of the *in situ* Chongokni handaxe (2 of Fig. 4) presented above, the IHRA can be characterized as a domain of the East Asian Palaeolithic tradition in which the "non-Acheulian handaxe" was continuously manufactured throughout the Upper Pleistocene; and this should be the starting point from where future discussions will be progressed in investigating the significance of the IHRA lithic assemblage.

References

Bae, K., 1988, *The Significance of the Chongokni Stone Industry in the Tradition of the Palaeolithic Culture in East Asia.* Unpublished PhD Dissertation. University of California. Berkeley.

Bae, K., Yoon, G., Lee, H., Hwang, S. and Kim, Y., 2006, *Report of Excavation of Ten Keumpari Site.* Institute of Cultural Properties in Hanyang University, Ansan (in Korean)

Bae, K., Ahn, S., Lee, H., Chun, B. and Lee, C., 2006, *The excavation report of the Hoengsanri Site.* Institute of Cultural Properties in Hanyang University, Ansan (in Korean)

Choi, J., Murray, A. S., Cheong, C. S., Hong, D., and Chang, H., 2004, Age estimate of the Chongok Basalt with Luminescence dating: preliminary result. In *Proceedings of Annual Joint Conference,* pp. 7- 9, Petrological Society and Mineralogical Society of Korea, Cheongju.

Choi, B., 2001, *The Janghung-ri Palaeolithic Site.* Institute of Kangwon Archaeology. Chuncheon (in Korean)

Choi, B. and Yoo, H., 2006, The Hwadae-ri Shimteo Palaeolithic Site, pochen City, Korea. Institute of Kangwon Archaeology. Chuncheon (in Korean)

Danhara, T., Bae, K., Okada, T., Matsufuji, K. and Hwang, S., 2002, What is the real age of the Chongokni Palaeolithic site? In K. Bae (ed.) *Palaeolithic Archaeology in Northeast Asia,* pp. 77- 116, Institute of Cultural Properties in Hanyang University, Yeoncheon. (in English with Korean summary)

Débenath, A. and Dibble H. L., 1994, Handbook of Paleolithic Typology. University of Pennsylvania Press, Philadelphia.

Kojima, M., 1983, Dating of the basalt from the Chongokni Palaeolithic site. *Chongokni,* pp. 586- 588, National Institute of Cultural Properties of Korea, Seoul. (in Japanese)

Lim, Hyoun-Soo, Lee, Y., Yi, S., Kim, C., Lee, H. and Choi, J., 2004, A preliminary study of large burrows at Jeongok and Naju palaeolithic sites, Korea. *Journal of Geological Society of Korea* 40(4), pp. 559- 66.

Matsufuji, K., Bae, K., Danhara, T., Naruse, T., Hayashida, A., Yu, K., Inoue, N., and Hwang, S., 2005, New progress of studies at the Chongokni Palaeolithic Site, Korea: Korea-Japan Cooperative Project in 2001- 2004.

Palaeolithic Archaeology (Kyuseki Kogugaku) 66, pp. 1- 16. (in Japanese)

Movius, H. L. Jr., 1948, *The Lower Palaeolithic Cultures of Southern and Eastern Asia.* Transactions of the American Philosophical Society N. S. 38, pp. 329- 420.

Norton, C. J., Bae, K., Harris, J., Lee, H., 2006, Middle Pleistocene handaxes from the Korean Peninsula. *Journal of Human Evolution* 51, pp. 527-36.

Park, K., Kim, Y., Lee, I., Park, J., Choi, M., Lee, K., Cheong, C., Han, J., Lee, S. and Shin, H., 1996, *Study on the microcomposition of terrestrial and marine samples and it structural analysis. vol. 1*, Korean Basic Science Institute, Daejeon. (in Korean)

Yi, S., 1989, *The Study of Northeast Asian Palaeolithic.* Seoul National University Press, Seoul (in Korean)

Yi, S., 1996, Chronostratigraphy of Palaeolithic occurrences in the Imjin Basin. *Journal of the Korean Archaeological Society* 34, pp. 135- 60. (in Korean)

Yi, S., 1999, The temporal change of Korean Palaeolithic industry. *Proceedings of Symposium to Commemorate the 80s Birthday Celebrations of Professor Chosuke Serizawa- World Views on the Early and Middle Paleolithic in Japan*, pp. 115- 120, Sendai Fukushi University, Sendai.

Yi, S., 2004, *Jangsan-ri: A Lower Palaeolithic Site in Paju, Korea.* Seoul National University Museum, Seoul. (in Korean)

Yi, S., 2005, New Data on the formation of the basalt plain in the Imjin River Basin. *Journal of the Geomorphological Association of Korea* 12(3), pp. 21- 38. (in Korean)

Yi, S. and Lee, K., 1993, *Excavation Report of Juwolri/ Kawolri Palaeolithic Site.* Dept. of Archaeology and Art History, Seoul National University, Seoul (in Korean)

Yi, S., Soda, T. and Arai, F., 1998, New discovery of Aira-Tn ash (AT) in Korea, *Journal of the Korean Geographical Society* 33(3), pp. 447- 54.

Yi, S., Lee, Y. and Lim, H., 2005, *A Preliminary Study on the Geological Background of the Formation Process of Imjin Basin Palaeolithic Sites.* vol. I., Seoul National University Museum, Seoul.

Yi, S., Lee, Y. and Kim, J., 2004, The Jangsanri Terrace and lava deposit in the upstream area of the Imjin River. *Journal of the Geomorphological Association of Korea* 11(1), pp. 1- 14.

Yi, S., Yoo, Y. and Kim, D., 2006, *Excavation Report at Chongokni (aka Jeongok ACF) Site and its Vicinity.* Seoul National University Museum and Jeongok Agricultural Cooperative Federation, Seoul.

Yoo, Y., 1997, On the characteristics of Handaxes from Imjin-Hantan Riverine Area. *Journal of the Korean Archaeological Society* 36, pp. 147-80. (In Korean with English Abstract)

Yoo, Y., 2008, *Beyond the Movius Line: Hominin Occupation and Technological Evolution in thie Imjin-Hantan River Area, Korea.* British Archaeological Report International Series 1772. BAR Publishing. Oxford.

The Upper Paleolithic of Hokkaido: Current Evidence and Its Geochronological Framework

Masami Izuho

Archaeology Laboratory, Faculty of Social Sciences and Humanities
Tokyo Metropolitan University, 1-1, Minami Osawa, Hachioji City, Tokyo 192-0397, Japan.
Email: izuhom@tmu.ac.jp

Fumito Akai

Kagoshima-shi Board of Education
Shimofukumoto 3763-1, Kagoshima City, Kagoshima 891-0144, Japan.
Email: fumito-akai@nifty.com

Yuichi Nakazawa

Atsuma-cho Board of Education
Hongo 283-1, Atsuma Town, Hokkaido 059-1605, Japan.
Email: nkzw2011@live.jp

Akira Iwase

Archaeology Laboratory, Faculty of Social Sciences and Humanities,
Tokyo Metropolitan University, 1-1, Minami Osawa, Hachioji City, Tokyo 192-0397, Japan.
Email: yiu51057@nifty.com

Abstract: *This paper provides a critical examination of the lithic assemblages in Hokkaido. We present geochronologies based on tephras and AMS 14C dates, descriptions of stone tool assemblages, and core and tool reduction technologies. We then take a human-ecosystem approach to discuss changes in Upper Paleolithic industries in Hokkaido.*

The Small Flake Industry appeared at 30,000 B.P. and disappeared by 20,000 B.P. in western and southeastern Hokkaido. Flake, Blade, and Microblade Industries appeared between 22,000 and 20,000 RCYBP. The Microblade Industry appeared at the Last Glacial Maximum, and lasted until ca. 13,000 RCYBP.

The Small Flake Industry was contemporaneous with the faunal group of Palaeoloxodon naumanni and giant deer during the latter part of MIS 3. The Flake, Blade, and Microblade Industries were contemporaneous with patch grassland and open forest where the faunal community of Mammuthus primigenius lived. The post-LGM Microblade Industry was contemporaneous with the faunal community of Mammuthus primigenius that survived in gradually diminished patch environments of grassland and open forest. The temporal correspondence between lithic industries and floral/faunal groups imply that hunter-gatherer lifeway changed from small-scale hunting in the cold/temperate forests to a hunting strategy to target grouped mammals inhabiting cold grassland and open forest during the LGM. However, the change at the LGM as not just a replacement of the Small Flake Industry by microblades, but the divergence of Flakes, Blades, and Microblade industries.

***Keywords**: Hokkaido, Upper Paleolithic, Geochronology, Lithic Assemblage, Paleo-Sakhalin/Hokkaido/Kurile Peninsula*

Introduction

Over the last fifty years, archaeological excavations have produced an abundant Upper Paleolithic record in Hokkaido. The material has been categorized into four groups: the Small Flake Industry, the Flake Industry, the Blade Industry and the Microblade Industry (Sato, 2003; Izuho and Sato, 2008). Upper Paleolithic records in Hokkaido are technologically more diverse than in neighboring regions (Nakazawa *et al.*, 2005; Sato, 1992). The technological diversification in Hokkaido corresponded with changes in flora, fauna, and landscapes prompted by climatic conditions as recorded in the Marine Isotope Stages (MIS) (Izuho and Takahashi, 2005). In contrast,

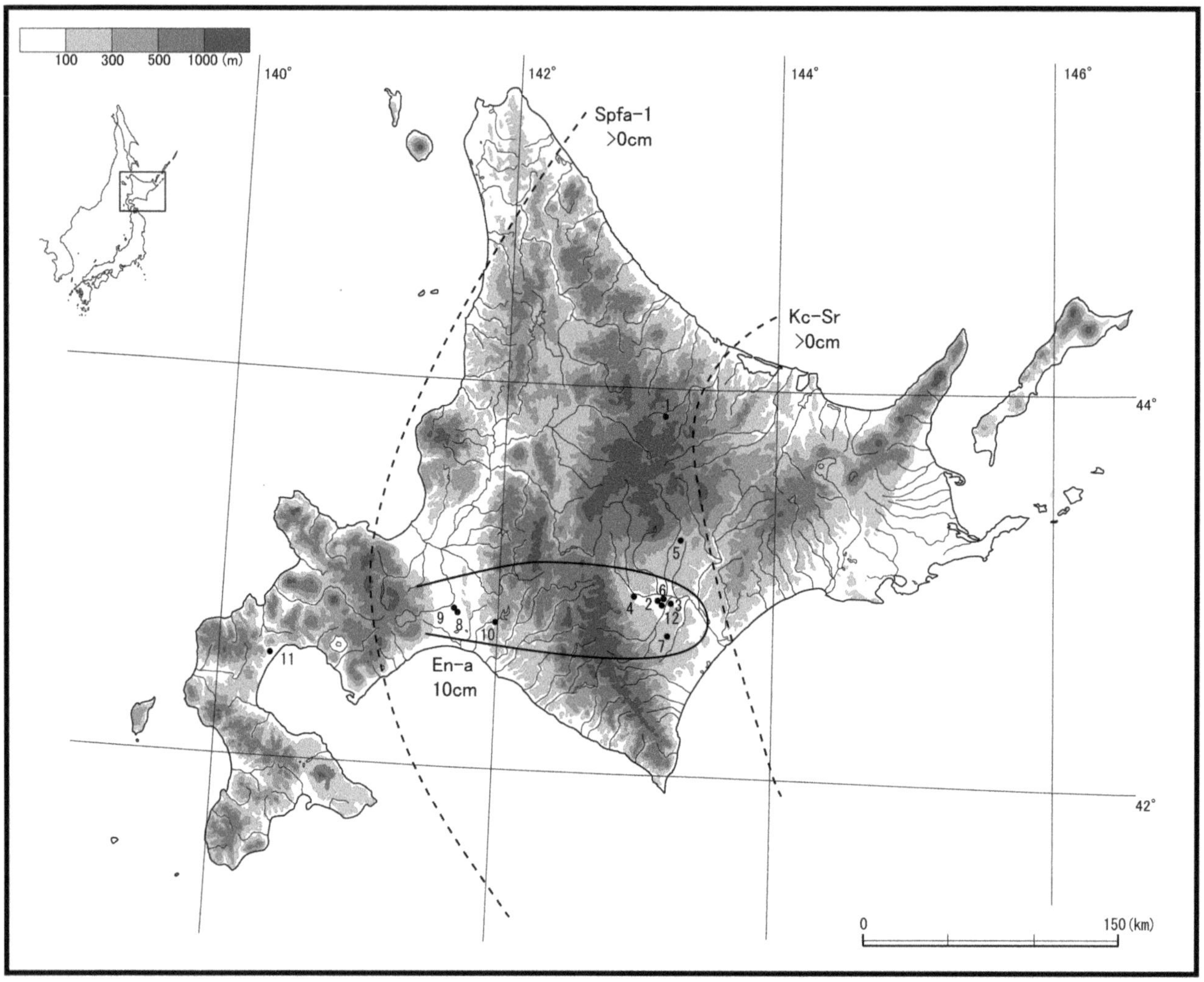

Figure 1. Locations of Hokkaido and sites mentioned in the text.
1: Okushiratakі 1, 2: Wakabanomori, 3: Kawanishi C, 4: Kyoei 3, 5: Shimaki, 6: Akatsuki, 7: Kukominami A, 8: Kashiwadai 1, 9: Shukubaisankakuyama, 10: Kamihoronai-Moi, 11: Obarubetsu 2, 12: Ozora. Numbers of sites correspond with the numbers in Table 1.

there has been little research that examines technological diversity across time and space. Upper Paleolithic sites in Hokkaido are primarily associated with eolian and colluvial depositional environments. Moreover, Hokkaido's humid climate and acid soils are not conducive to the preservation of organic material. These circumstances have had a great impact on how archaeological research is conducted in the area.

Until now, Upper Paleolithic research in Hokkaido has mainly focused on building chronological sequences based on detailed descriptions of lithic assemblages. Variability in stone tool assemblages and lithic technologies are so complex, however, that normative typological explanations based on *fossile directeur* are no longer applicable. It is now incumbent to provide explicit explanations of how behavioral (systemic) and natural processes influenced the perceived variability in lithic technologies and stone tool assemblages (Sato, 1992; Yamada, 2006; Nakazawa *et al.*, 2005). Here we provide a review of the dating and stone tool assemblages of 13 Upper Paleolithic sites in Hokkaido (Fig. 1) with focus on core and tool reduction strategies. Finally, we explain changes in Hokkaido's Upper Paleolithic record in terms of a human-ecosystem model.

Paleoenvironments in Hokkaido during MIS3 and MIS2

Hokkaido, one of biggest islands of the Japanese Archipelago, lies along the eastern margin of the Asian continent (Fig. 2). The cool and temperate modern climate is strongly affected by seasonal easterly winds from the Asian mainland during the winter and from the Pacific Ocean in the summer. Sea currents flowing north from the Pacific Ocean and from the Okhotsk Sea add variability to the local climate.

Climates of MIS3 and MIS2 in Hokkaido and its surrounding region are understood from terrestrial pollen analysis and sea-bottom and lake sediment cores (Ono and Igarashi, 1993; Fukusawa *et al.*, 2003; Igarashi and Oba, 2006; Kitamura and Kimoto, 2006). Although a

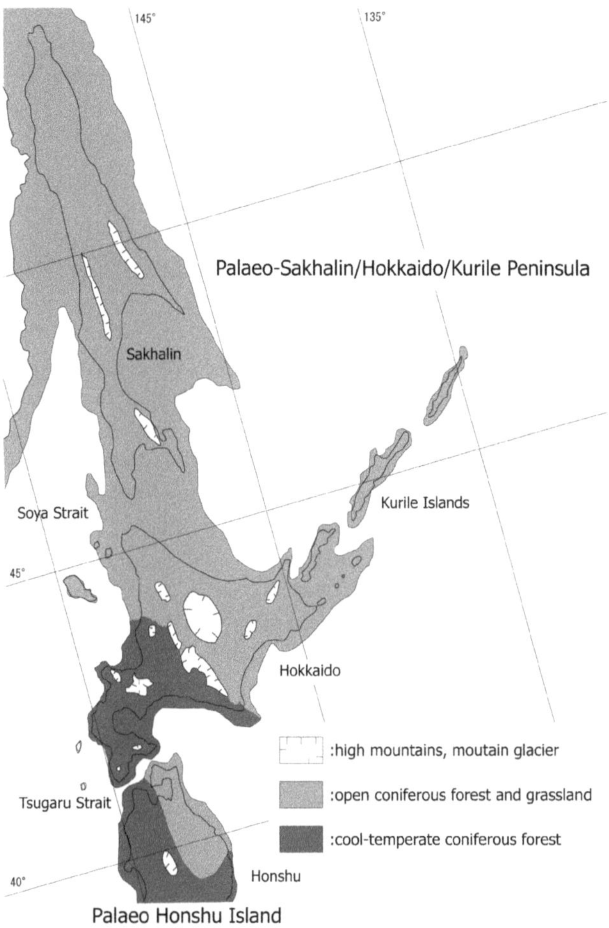

Figure 2. Map showing the paleogegraphy and paleovegetation of Paleo-SHK Peninsula during MIS3 to MIS2 (after Izuho and Takahashi, this volume).

regional climatic reconstruction for all Hokkaido has not yet been fully established, current evidence indicates that environments varied from cool-temperature to sub-arctic between MIS3 and MIS2. During the Last Glacial Maximum (hereafter, LGM), mean annual temperature is estimated to have been 7°C lower than today, and Hokkaido and the northern most part of Honshu were part of a widespread sub-arctic zone (Igarashi, 1990).

The landscape of late Pleistocene Hokkaido was dramatically changed from the last interglacial to the Holocene. As shown in Fig. 2, when sea levels were an estimated 105-130m lower than today, Hokkaido was connected with Sakhalin and the Asian continent at the mouth of Amur River (Ono, 1990; Machida *et al.*, 2004; Vasilevski, 2005). This geographic district is called Paleo-Sakhalin/Hokkaido/Kurile Peninsula (Paleo-SHK Peninsula). Below, we use "southern half of Paleo-SHK Peninsula" to designate Hokkaido (Sato, 2003; Vasilevski, 2008a). The ancient shoreline of Hokkaido during the LGM is estimated to have been about 10km farther out from the present coast. A land bridge existed in the Soya Strait connecting Hokkaido and Sakhalin, because the present depth is only about 60m (Ono, 1990). In contrast, a land bridge did not emerge in the Tsugaru Strait, because its present depth is about 130m, and Hokkaido likely was not connected with Honshu throughout LGM. Although the argument concerning a dry land connection across the Tsugaru Strait during the LGM is as yet inconclusive (Ono, 1990), it is at least presumable that the width of the strait was reduced to less than 2km. While the amount of exposed land of Paleo-SHK Peninsula expanded and contracted before and after the LGM, the peninsula was maintained during the Late Pleistocene.

LGM flora of the southern half of Paleo-SHK Peninsula has been reconstructed based on pollen and plant macro fossil records. A continuous reconstruction of this region between MIS3 and MIS2 has not yet been established. Flora during the latter half of MIS3 is characterized as open spruce-larch-fur-oak taiga based on the pollen record from the Kenbuchi Basin in central Hokkaido (Igarashi *et al.*, 1993). Post-LGM pollen records are available from long cores obtained at the Kenbuchi and Furano basins in central Hokkaido (Igarashi *et al.*, 1993; Igarashi, 1996). Vegetation of the LGM in this region are divided into two types; (1) a sub-arctic zone consisting of patches of open larch-pine-spruce taiga and grassland in central and southeastern Paleo-SHK peninsula, and (2) a sub-arctic zone consisting of open spruce-larch-fir-oak Taiga in southwestern Paleo-SHK peninsula (Ono and Igarashi, 1993). After the LGM, grassland gradually started to diminish as sparse coniferous forests spread (Igarashi, 2008; Izuho, 2008).

Faunal communities in Hokkaido during MIS3 and MIS2 were tied to the expansion and contraction of the foraging areas of the *Palaeoloxodon naumanni* and *Mammuthus primigenius* groups (Takahashi, 2007; Takahashi and Izuho, this volume). The basic components of the faunal groups in were established before the late Pleistocene. In MIS10 (336ka), a land bridge between Korea and southern Honshu appeared that allowed animals to migrate from northern China. It is believed that the formative processes, which produced the late-Glacial groups and modern endemic species, occurred when the last land bridge at the Tsushima Straits existed about 130ka (Takahashi, 2007).

After 50 ka, two kinds of fauna existed on the Paleo-SHK Peninsula, one that migrated from the west accompanied by *Palaeoloxodon naumanni*, *Panthera tigris*, *Canis lupus*, *Ulsus thibetanus*, *Sinomegaceros yabei*, *Sus scrofa*, and *Cervus nippon*, and another that migrated from the north accompanied by *Mammuthus primigenius*, *Bison* sp. (Takahashi, 2007), *Equus hydruntinus*, *Cervus elaphus*, *Megaroceros giganteus*, *Alopex tagopus*, *Raingifer tarandus*, *Pantera soelea*, *Lepus timidus*, *Ursus arctos*, and *Alces alces* (Kuzmin *et al.*, 2002). Both faunas have repeatedly migrated to the northern and southern areas in the Japanese Islands according to climatic changes (Takahashi *et al.*, 2006). It is estimated that elephants and large-sized deer went extinct by 23ka in Honshu and 16ka in Hokkaido, with the exception of a few relict populations (Takahashi, 2007).

The unique tectonic setting of the arc-trench systems in and around the Japanese Islands allowed human groups to procure various high-quality raw material from metamorphic, volcanic, and sedimentary sources. There are two major types of cryptocrystalline material in northern Japan—obsidian in the southeastern part of the Paleo-SHK Peninsula (Izuho and Sato, 2008), and siliceous shale widely distributed in the southwestern part of the Paleo-SHK Peninsula and the northern part of Paleo-Honshu Island. The volcanic tephras of Japan have facilitated the stratigraphic correlation of late Pleistocene archaeological layers. Particularly, the AT, Eniwa-a (En-a), and the Shikotsu Dai-1 (Spfa-1) tephras, dated at 24,500 RCYBP, 17,000 RCYBP, and 40,000 RCYBP, respectively, play important roles as geochronological marker horizons (Machida and Arai, 2003).

Lithic assemblages in the Paleo-SHK Peninsula

A total of 862 Paleolithic sites have been identified in Hokkaido, of which 146 were excavated (Japanese Paleolithic Research Association, 2010). University-led, academic excavations occurred from the 1950's to the 1970's, while salvage projects took the lead in between 1980 and 2000, but have since decreased (Tsurumaru, 2001). Below, we describe reduction strategies at some technologically and geochronologically important Upper Paleolithic sites. The assemblages are divided into (1) Small Flake, (2) Flake, (3) Blade, and (4) Microblade categories. In addition, lithic assemblages similar to those in Hokkaido have been recently reported in Sakhalin (Vasilevski, 2005, 2006, 2008b), but none from the Kurile Islands.

(1) Small Flake assemblages

1) Wakabanomori Site

The Wakabanomori site is located on the southern terraces in the southeastern Tokachi Plain, eastern Hokkaido (N42°53′, E143°10′) (B3-5), which are characterized by prominent, relatively flat-wavy tread surfaces (Koaze *et al.*, 2003). The site is situated on the edge of the Motomatsu Terrace along the Satsunai River at an altitude of 75m a.s.l. The artifact-bearing layer is in an eolian loam 0.5-1.0m below the surface positioned between primary units of the Spfa-1 and En-a tephras (Obihiro Board of Education, 2004; Izuho and Akai, 2005). Radiocarbon dated were reported between >27,000 and 24,000 RCYBP from three natural charcoal concentrations above the cultural layer (Table 1:1-6).

A total of 9,701 lithic specimens were recovered from four concentrations. Raw material consists of obsidian (n=9,660), "hard-shale" (n=3), agate (n=1), and other coarse igneous pebbles (n=37). Obsidian gravels, usually 4-7cm in diameter, were acquired from the bed of the Tokachi River near the site (Obihiro Board of Education, 2004). Small flakes removed from discoidal and multi-faced cores characterize primary reduction at the site. Secondary reduction strategies included marginal retouch, beaked retouch, as well as flat flaking on ventral surfaces. The lithic assemblage consists of trapezoids, beak-shaped tools, retouched flake, and utilized flakes, as well as small flake cores, hammer stones, pebble tools, and pebbles (Fig. 3-a).

2) Shukubaisankakuyama Site

The Shukubaisankakuyama site is located in the lowland of the central Ishikari Plain, Hokkaido (N42°49′, E141°41′) (C6-4) (Koaze *et al.*, 2003) situated in an ancient dune formed from reworked Spfa-l pumice along the Chitose River at an altitude of 25m a.s.l. The artifact-bearing layer is in the lower part of a thick eolian loam, 3.1m below the surface positioned between reworked Spfa-1 and En-a tephras (Chitose Board of Education, 1974; Izuho and Akai, 2005). Weak involution disturbed the cultural layer. A conventional radiocarbon date of 21,450±750 RCYBP (Gak-4346) on charcoal was reported from the loam with the cultural material, but this is inconsistent with the age of the tephras according to recent results of deep-ocean core drilling (Aoki and Arai, 2000) (Table 1:7).

A total of 211 lithic specimens were recovered from two discrete concentrations. Raw materials consist of obsidian (n=210) and andesite (n=1). The assemblage includes trapezoids, beak-shaped tools, retouched flakes, and utilized flake, as well as small flake cores, and hammer stones made on andesite (Fig. 3-b). Obsidian was transported from the Akaigawa source, 75km from the site, as well as from Shirataki, 170km from the site (Koshimizu, 1981; Naoe and Nagasaki, 2005). There is no evidence for the utilization of local raw material. Primary reduction was based on the production of small flakes. Marginal retouch, beaked retouch, and flat flaking on ventral surfaces were common secondary reduction strategies.

3) Okushirataki 1 Site

The Okushirataki 1 site is located in the middle part of the Kitami Mountains, northeastern Hokkaido (N43 ° 52′, E143 ° 07′), where river terraces develop within narrow valleys surrounded by mountains and hills (A1-2) (Koaze *et al.*, 2003). The site is situated at the edge of the Kamishirataki Terrace that formed during OIS5e along the Yubetsu River (Nakamura *et al.*, 1999) at an altitude of 450m a.s.l. The artifact-bearing layer is in an eolian loam superimposed on a soil with reworked Daisetsu Ohachidaira tephra (>30,000BP) 0.3m below the surface (Hokkaido Center for the Buried Cultural Property, 2002). Radiocarbon dates from 18,000 to 15,000 RCYBP were reported on 18 charcoal concentrations (Table 1:8-20). However, we rejected these dates because they were affected by severe post-depositional disturbance in a shallow buried context (Hokkaido Center for the Buried Cultural Property, 2002; Izuho and Akai, 2005).

A total of 84,105 lithic specimens were recovered from 44 concentrations that consisted of small flake industry and several groups of microblade industries. A total of 1,914

Table 1. List of radiocarbon dates from Upper Paleolithic sites mentioned in the text.

Region in Hokkaido	*Number in Figure 1*	*Site*	*Number in text*	*Geological Context*	*Archaeological Context*	*Material Dated*	*Laboratory Number.*	*Method*	*Measured Radiocarbon age*	*δ13c*	*Conventional Radiocarbon Age*	*Reference*
Northeast	1	Okushirataki1	8	abv Ds-Oh	M17, East margin of Sb-21, SHIRA-46	Charcoal	Beta-126157	AMS	18880±140	-24.3	18890±140	Hokkaido Center for Buried Cultural Property (2002)
			9	abv Ds-Oh	N16, North of Sb-20, Cb-6, SHIRA-45	Charcoal	Beta-126156	AMS	18350±140	-24.2	18360±140	
			10	abv Ds-Oh	M23, North margin of Sb-4, Cb-18, SHIRA-6	Charcoal	Beta-112878	AMS	18230±190	-23.5	18250±190	
			11	abv Ds-Oh	M23, North margin of Sb-4, Cb-18, SHIRA-5	Charcoal	Beta-112877	AMS	16040±130	-25.6	16030±130	
			12	abv Ds-Oh	J21・22, Sb-25, Cb-15(3), SHIRA-9	Charcoal	Beta-112881	AMS	15880±130	-27.0	15850±130	
			13	abv Ds-Oh	J21・22, Sb-25, Cb-15(2), SHIRA-8	Charcoal	Beta-112880	AMS	15850±150	-25.0	15850±150	
			14	abv Ds-Oh	M23, North margin of Sb-4, Cb-18, SHIRA-7	Charcoal	Beta-112879	AMS	15850±150	-25.0	15850 ±150	
			15	abv Ds-Oh	Q17, West of Sb-8, Cb-1(2), SHIRA-2	Charcoal	Beta-112874	AMS	15580±190	-25.6	15570±190	
			16	abv Ds-Oh	Q17, West of Sb-8, Cb-1(1), SHIRA-1	Charcoal	Beta-112873	AMS	15570±130	-25.1	15570±130	
			17	abv Ds-Oh	Q17, West of Sb-8, Cb-1(3), SHIRA-3	Charcoal	Beta-112875	AMS	15260±150	-24.1	15270±150	
			18	abv Ds-Oh	K18, North east margin of Sb-21, Cb-11, SHIRA-10	Charcoal	Beta-112882	AMS	15250±200	-23.6	15850±200	
			19	abv Ds-Oh	K18, North east margin of Sb-21, Cb-11, SHIRA-11	Charcoal	Beta-112883	AMS	15100±130	-24.3	15110±130	
			20	abv Ds-Oh	O5・6, South margin of Sb-11, Cb-4(1), SHIRA-4	Charcoal	Beta-112876	AMS	15010±120	-23.6	15030±120	
Southeast	2	Wakabanomori	1	blw En-a	Burned-soil in p-4, (21945) charcoal1	Charcoal	Beta-174960	AMS	27640±310	-24.8	27640±230	Obihiro Board of Education (2004)
			2	blw En-a	Burned-soil in p-4, (21946) charcoal2	Charcoal	Beta-174961	AMS	24680±230	-25.4	24670±230	
			3	blw En-a	Burned-soil in Sp-1, (19209)1	Charcoal	Beta-162682	AMS	24410±240	-25.3	24410±240	
			4	blw En-a	Burned-soil in North Sp-4, (21947) charcoal3	Charcoal	Beta-174962	AMS	24390±220	-23.9	24410±220	
			5	blw En-a	Sp-1, (129210)2	Charcoal	Beta-162683	AMS	23930±220	-25.0	23930±220	

Region in Hokkaido	*Number in Figure 1*	*Site*	*Number in text*	*Geological Context*	*Archaeological Context*	*Material Dated*	*Laboratory Number.*	*Method*	*Measured Radiocarbon age*	*δ13c*	*Conventional Radiocarbon Age*	*Reference*
			6	abv En-a	Burned-soil, (21948) charcoal4	Charcoal	Beta-174963	AMS	12280±080	-24.9	12280±080	Obihiro Board of Education (2004)
	3	Kawanishi C	57	blw En-a	17-15, Burned-soil in Sp-4, geo-5932	Charcoal	Beta-107731	AMS	21800±090	-26.3	21780±090	Obihiro Board of Education (1998)
			58		16-19, Burned-soil in Sp-3, geo-5805	Charcoal	Beta-106506	AMS	21420±190	-26.1	21420±190	
			59	abv En-a	21-21, Sp-17, (10325) No.3	Charcoal	Beta-126151	AMS	16940±050	-26.7	16920±050	Obihiro Board of Education (2000)
			60		21-21, Sp-17, (10324) No.2	Charcoal	Beta-126150	AMS	13070±040	-28.1	13020±040	
			61		23-22, Sp-16, (10913) No.1	Charcoal	Beta-127399	AMS	12920±050	-26.2	12900±050	
	4	Kyoei3	21	blw En-a	Unknown	Charcoal	KSU-2167	Radio.	>18000	-	-	Hokkaido Center for Buried Cultural Property (1992)
			22	abv En-a	Unknown	Charcoal	KSU-2168	Radio.	>18000	-	-	
	5	Shimaki	23	btwn Sipfa 1-2	Unknown	Charcoal	unknown	Radio.	23800	-	-	Kato and Yamada (1988)
			24	btwn Sipfa 1-2	Brown loam 40cm below cultural level	Charcoal	GaK-3262	Radio.	25,500±1200	-	-	Kosaka and Nogawa (1972)
	6	Akatsuki	65	abv En-a	Spot1, No data	Charcoal	KSU-717	Radio.	14700±250	-	-	Obihiro Board of Education (1985)
			66		Spot4, No data	Charcoal	KSU-718	Radio.	14700±130	-	-	
			67		Sp-8	Charcoal	KSU-889	Radio.	10900±500	-	-	Obihiro Board of Education (1986)
	7	KukouminamiA	62	blw En-a	Grid-F, Sample No.5	Charcoal	-	Radio.	23850+4480-2850	-	-	Tokachi Branch Office (1983)
			63		U-56	Charcoal	Gak-10746	Radio.	19420±1770	-	-	
			64	abv En-a	U-56	Charcoal	Gak-10580	Radio.	15280±340	-	-	Tokachi Branch Office (1982)
Central	8	Kashiwadai1	25	blw En-a	Area M-66, KD1-32	Charcoal	Beta-126179	AMS	37350±550	-24.9	37350±550	Hokkaido Center for Buried Cultural Property (1999)
			26		Area I-66, KD1-34	Charcoal	Beta-126181	AMS	33020±540	-24.6	33030±540	
			27		M-68, KD1-33	Charcoal	Beta-126180	AMS	32500±360	-25.8	32490±360	
			28		Area P-6, Sb-1,Hearth, KD1-35	Charcoal	Beta-126182	AMS	31350±330	-25.2	31350±330	

Region in Hokkaido	*Number in Figure 1*	*Site*	*Number in text*	*Geological Context*	*Archaeological Context*	*Material Dated*	*Laboratory Number.*	*Method*	*Measured Radiocarbon age*	*δ13c*	*Conventional Radiocarbon Age*	*Reference*
			29		Area O-7, KD1-6	Charcoal	Beta-112918	AMS	28200±480	-25.0	28200±480	
			30		Area F-59, Sb-9, Hearth, KD1-24	Charcoal	Beta-126171	AMS	22550±180	-24.7	22550±180	
			31		Area H-58, Sb-4a, Hearth, KD1-21	Charcoal	Beta-126168	AMS	22340±170	-24.7	22340±170	
			32		Area I-63, Sb-10, Hearth, KD1-26	Charcoal	Beta-126173	AMS	22330±200	-24.5	22340±200	
			33		Area D-58, Sb-7, Hearth, KD1-22	Charcoal	Beta-126169	AMS	22300±180	-24.7	22300±180	
			34		Area D-57, KD1-1	Charcoal	Beta-112913	AMS	22190±210	-23.9	22210±210	
			35		Area F-61, Sb-11, Hearth, KD1-36	Charcoal	Beta-126183	AMS	22180±170	-23.6	22200±170	
			36		Area F-61, Sb-11, Hearth, KD1-27	Charcoal	Beta-126174	AMS	21790±230	-24.7	21790±230	
			37		Area K-65, Sb-13, Hearth, KD1-14	Charcoal	Beta-120881	AMS	21000±100	-25.2	21000±100	
			38		Area P-69, Charcoal Concentration, KD1-31	Charcoal	Beta-126178	AMS	20910±190	-25.9	20900±190	
			39		Area F-64, Sb-12, Hearth, KD1-28	Charcoal	Beta-126175	AMS	20790±160	-24.9	20790±160	
			40		Area P-55, Sb-2, KD1-10	Charcoal	Beta-112922	AMS	20700±210	-26.6	20680±210	
			41		Area H-63, Sb-15, Hearth, KD1-29	Charcoal	Beta-126176	AMS	20700±150	-25.3	20700±150	
			42		Area F-64, Sb-12, Hearth, KD1-37	Charcoal	Beta-126184	AMS	20600±160	-24.2	20610±160	
			43		Area F-59, Sb-9, KD1-2	Charcoal	Beta-112914	AMS	20600±120	-26.8	20570±120	
			44		Area L-58, Sb-3, KD1-20	Charcoal	Beta-126167	AMS	20570±160	-24.5	20580±160	
			45		Area P-55, Sb-2, KD1-8	Charcoal	Beta-112920	AMS	20500±160	-24.4	20510±160	
			46		Area P-55, Sb-2, KD1-9	Charcoal	Beta-112921	AMS	20500±130	-25.6	20490±130	

Region in Hokkaido	*Number in Figure 1*	*Site*	*Number in text*	*Geological Context*	*Archaeological Context*	*Material Dated*	*Laboratory Number.*	*Method*	*Measured Radiocarbon age*	*δ13c*	*Conventional Radiocarbon Age*	*Reference*
			47		Area K-65, Sb-13, Hearth, KD1-13	Charcoal	Beta-120880	AMS	20410±070	-25.8	20390±070	
			48		Area N-63, Sb-14, Hearth, KD1-16	Charcoal	Beta-120883	AMS	20370±070	-25.3	20370±070	
			49		Area H-63, Sb-10, KD1-4	Charcoal	Beta-112916	AMS	20300±150	-23.7	20320±150	
			50		Area P-6, Sb-1, KD1-7	Charcoal	Beta-112919	AMS	20200±120	-26.1	20180±120	
			51		Area E-56, Sb-6, Hearth, KD1-23	Charcoal	Beta-126170	AMS	20140±150	-25.9	20130±150	
			52		Area N-63, Sb-14, Hearth, KD1-15	Charcoal	Beta-120882	AMS	19850±070	-25.6	19840±070	
			53		Area F-64, KD1-3	Charcoal	Beta-112915	AMS	19650±130	-24.4	19660±130	
			54		Area H-63, Sb-15, Hearth, KD1-30	Charcoal	Beta-126177	AMS	18840±150	-25.4	18830±150	
	9	Shukubai sankakuyama	7	blw En-a	Unknown, collected from outcrop	Charcoal	Gak-4346	Radio.	21450±750	-	-	Chitose Board of Education (1974)
	10	Kamihoronai-Moi	68	Abv En-a	Hearth (R-14 grid, IXCB-01: Water floatation No.1107)	Charcoal	PLD-5275	AMS	14570±50	-25.1	14565±50	Atsuma Board of Education (2006)
			69		Dence Charcoal with Hearth (R-14 grid, IXCB-01, G-block: Water floatation No.2489)	Charcoal	PLD-5276	AMS	14560±50	-25.1	14560±50	
			70		Dence Charcoal with Hearth (Q-14 grid, IXCB-01, A-brock)	Charcoal	IAAA-41577	AMS	14600±80	-22.0	14650±80	
			71		Dence Charcoal with Hearth (R-14 grid, IXCB-01, J-block, KM-01)	Charcoal	Beta-238725	AMS	14430±70	-24.0	14450±70	Izuho *et al.*, (2009)
			72		Hearth (R-14 grid, IXCB-01, KM-02)	Charcoal	Beta-238726	AMS	14440±60	-24.6	14450±60	
			73		Dence Charcoal with Hearth (R-14 grid, IXCB-01, F-block, KM-04)	Charcoal	Beta-240861	AMS	14770±70	-25.0	14770±70	
Southwest	11	Obarubetsu2	55	-	Spot-I, 24-12c, Sb 1 , No.203	Charcoal	Beta-156337	AMS	20490±200	-24.6	20500±200	Oshamanbe Board of Education (2002)

Region in Hokkaido	Number in Figure 1	Site	Number in text	Geological Context	Archaeological Context	Material Dated	Laboratory Number	Method	Measured Radiocarbon age	δ13c	Conventional Radiocarbon Age	Reference
			56		Spot-II, 35-16c, Near Sb-8	Charcoal	Beta-156336	AMS	7480±040	-27.6	7440±040	

・Abbreviation in Geological Context, abv : advance, blw : below, btwn : between

・Beta : Beta Analytic INC., NUTA : Nagoya University, Gak : Gakushuin University, KSU : Kyoto Sangyo University, N : Japan radioisotope association

・AMS: Accelerator Mass Spectrometry, Radio. : Radioactivity determination

specimens all on obsidian from six lithic concentrations, named Sb1 to Sb6, comprise the small flake industry. Refitted examples indicate that rounded or sub-angular cobbles were acquired from the bed of the Yubetsu River (Naoe, 2009) to produce trapezoids, beak-shaped tools, retouched flakes, and utilized flakes with micro-flaking, as well as small flake cores (Fig. 3-c). Primary reduction involved the removal of small flakes from discoidal and multi-faced cores. Trimming of distal ends of elongated flakes, marginal retouch, beaked retouch, as well as flat flaking on ventral surfaces were common secondary reduction strategies.

4) Kyoei 3 site

The Kyoei 3 site is located on the southeastern Tokachi Plain, eastern Hokkaido (N42°58′, E142°55′) (B3-5) (Koaze *et al.*, 2003). The site is situated at the edge of the Makubetsu Fan along the Tokachi River at an altitude of 158m a.s.l. Three archaeological layers were found 1.0m below the surface in an eolian loam. From top to bottom they were early Jomon component in the Black Humus soil above the Tarumae-D tephra (Ta-d, 7,000 RCYBP), a microblade assemblage below the Ta-d tephra but above the En-a tephra (17,000 RCYBP), as well as small flake industry below the En-a tephra and above the Spfa-1 tephra (40,000 BP) (Hokkaido Center for Buried Cultural Property, 1992; Izuho and Akai, 2005). Two conventional radiocarbon dates on charcoal which bracketed the En-a tephra, were >18,000 RCYBP (KSU-2168) and >18,000 RCYBP (KSU-2167), respectively (Table 1:21, 22). We rejected the dates due to inconsistency with the geologic context.

The archaeological component below En-a tephra included a total of 1,539 lithic specimens in eight lithic concentrations. Lithic raw materials consist of obsidian (n=1,480), "hard-shale" (n=2), mudstone (n=1), basalt (n=8), and other coarse igneous pebbles (n=48). Based on refits, it was concluded that 10-20cm-diameter obsidian gravels, acquired from the Tokachi River bed near the site, were the dominant raw materials (Hokkaido Center for Buried Cultural Property, 1992). The lithic assemblage consists of tools made on small flakes including trapezoids, beak-shaped tools, side scrapers, wedge-shaped tools, retouched flake, and utilized flakes with micro-flaking, as well as small flake cores, hammer stones, pebble tools, and pebbles (Fig. 3-d). Primary reduction involved small flaking technology, while secondary reduction included scraper edge retouch, marginal retouch, beaked retouch, as well as flat flaking on ventral surfaces.

5) Summary of the small flake assemblages

The archaeological levels in which the small flake assemblages are encompassed at Wakabanomori, Shukubaisankakuyama, and Kyoei 3 suggest that there is a distinctive archaeological component between the Spfa-1 and En-a tephras. Although samples of absolute dates are still small and inconsistent, the small flake assemblage at

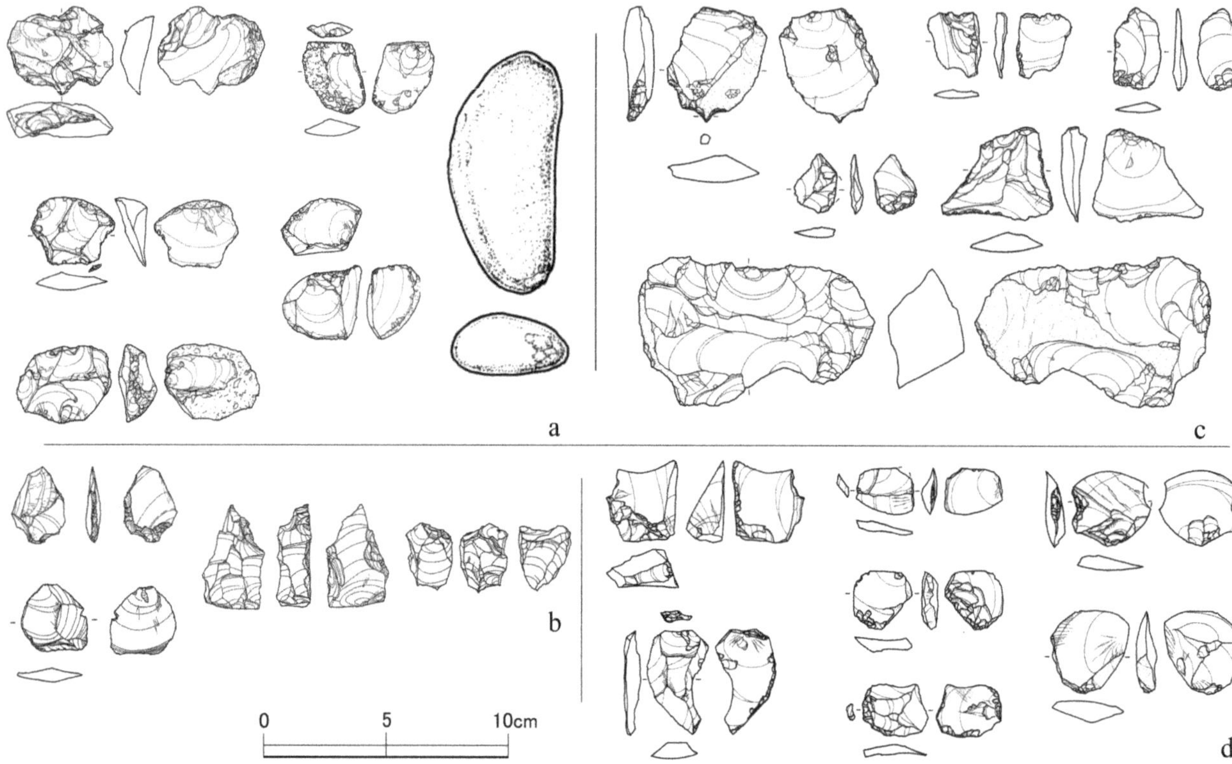

Figure 3. Stone tools of the Small Flake Assemblages in Hokkaido.
a: Wakabanomori (after Obihiro Board of Education, 2004), b: Shukubaisankakuyama (after Chitose Board of Education, 1974), c: Okushirataki 1 (after Hokkaido Center for Buried Cultural Property, 2001), d: Kyoei 3 (after Hokkaido Center for Buried Cultural Property, 1991).

Wakabanomori is dated to >24,000 B.P. Primary reduction of the lithic assemblages is characterized by the production of small flakes from discoidal and multi-faced cores. Tools include trapezoids (with partial retouch), scrapers on flakes, and beak-shaped tools. Because the technological and morphological characteristics of these small flake assemblages are similar to those of Honshu, the small flake assemblages are plausibly dated to 35,000-30,000 RCYBP (Sato, 2003; Izuho, 2010).

(2) Flake Assemblages

1) Shimaki site

The Shimaki site is located on the topographic division known as the northern terraces on the northeastern Tokachi Plain, eastern Hokkaido (N43°14′, E143°18′) in an area with prominent, relatively flat-wavy tread surfaces (Koaze *et al.*, 2003). The site is situated at the edge of the Kitaoribe II Terrace along the Otofuke River, a tributary of the Tokachi River at an altitude of 290m a.s.l. The artifact-bearing layer is in an eolian loam, 0.5-1.0m below the surface, and positioned between primary fall of the Shikaribetsu Dai 2 tephra (Sipfa2, <45,000BP) and Sipfa-1 (>20,000BP) tephras (Kato and Yamada, 1988; Izuho and Akai, 2005). Radiocarbon dates between >27,000 and 24,000 RCYBP were reported from three natural charcoal concentrations above the cultural layer (Table 1:1-6).

Previous work at the Shimaki site was conducted by Drs. Tsuji and Kato in 1967 and 1969, by the Kamishihoro Board of Education in 1984, as well as by the University of Tsukuba in 1986-1988 (Kato and Yamada, 1988). Tokyo Metropolitan University and Central Washington University reopened the site in 2010 (Buvit *et al.,* 2011). Several chronometric ages were obtained from the cultural layer including a fission track (21,700±1,800 BP) and obsidian hydration (BS.A1-26 19,000±800; A1-137 18,200±500; B1-43 18,200±700; B1-2896 17,200±800; B3-80 17,500±900) placing the layers within the LGM. A conventional radiocarbon age of 25,500±1,200 (GaK-3262) found 40 cm below the cultural level provided a lower age limit (Table 1: 23-24; Kosaka and Nogawa, 1974).

A total of around 10,000 lithic specimens were recovered from eight lithic concentrations. Raw material consists of obsidian, andesite, "hard-shale", agate, and other coarse igneous pebbles. Obsidian gravels, usually 10cm in diameter, were the dominant raw material acquired from the bed of the Tokachi River near where the site is located (Kato and Yamada, 1988). Primary reduction involved the production of flakes which were used to produce triangular and trapezoidals flakes from discoidal and conical cores, as well as bladelets from wedge shaped bladelet cores. Clear evidence of end scraper edge retouch, marginal retouch, beaked retouch, as well as flat flaking on ventral surfaces were common secondary reduction strategies. The tool

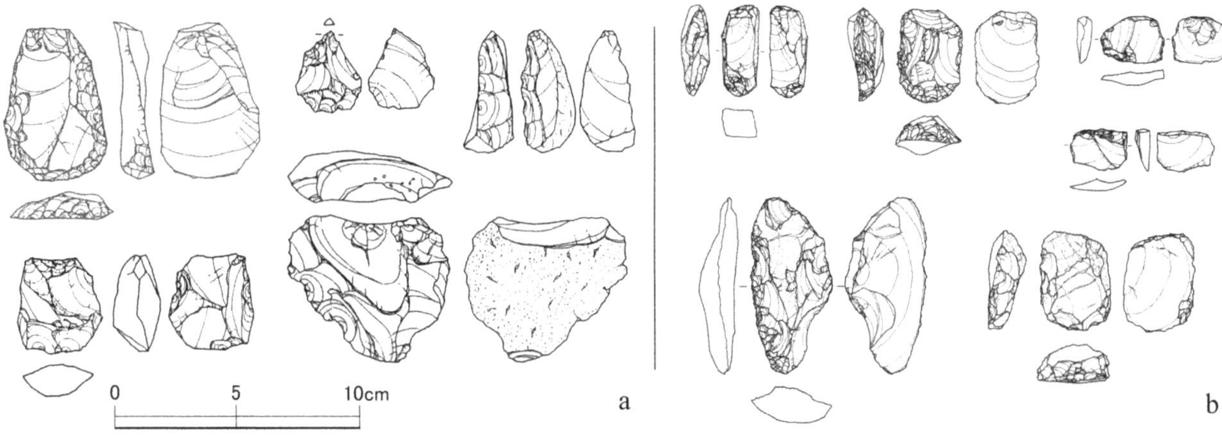

Figure 4. Stone tools of the Flake Assemblages in Hokkaido.
a) Shimaki (after Kato and Yamada, 1989), b) Kashiwadai 1 LC11 (after Hokkaido Center for Buried Cultural Property, 1999).

assemblage consists end scrapers, side scrapers, perforators, wedge shaped tool, retouched flakes, and utilized flakes with micro-flaking, as well as flake cores, ocher, hammer stones, pebble tools, and pebbles (Fig. 4-a).

2) Kashiwadai 1 site

The Kashiwadai 1 site is located in the central part of the Ishikari lowland, central Hokkaido (N42°48′, E141°41′) (Koaze *et al*., 2003) on a paleo-dune formed with reworked Spfa-1 material along the Chitose River at an altitude of 13m a.s.l. The artifact-bearing layer is in the lower part of a thick eolian loam 4m below the surface between reworked Spfa-1 (40,000 BP) and En-a (17,000 RCYBP) tephra (Hokkaido Center for Buried Cultural Property, 1999; Izuho and Akai, 2005). Paleolithic components were found in two block excavations; block excavation A and block excavation B. A lithic concentration associated with a hearth was found in block excavation A. Several lithic concentrations with 11 associated hearths and four charcoal concentrations were found in block excavation B. A total of 26 AMS radiocarbon dates were reported between 20,000 and 22,000 RCYBP (Table 1:25-54, Izuho and Akai, 2005).

A total of 32,822 lithic specimens were recovered from 14 concentrations in block excavation B (Fig. 4-b). As shown in Fig. 5, several lithic concentrations include microblade and flake assemblages. Of these concentrations, LC11 is classified as a tightly clustered flake assemblage consisting of 9,409 specimens. Lithic raw material in LC11 consists of obsidian (n=6,276; 2,416.9g), "hard-shale" (n=935; 2,435.9g), andesite (n=432; 5,429.3g), agate (n=298; 1,017.9g), chert (n=298; 1,303.27g), and other material (n=1,170; 2,682.1g). Obsidian and "hard-shale" were considered non-local, while the sources of agate, chert, and andsite are unknown (Hokkaido Center for Buried Cultural Property, 1999). Primary reduction includes bipolar and flaking technologies to produce triangular and trapezoid flakes from discoidal and conical cores. End scraper edge retouch and marginal retouch are common secondary reduction strategies. Tools made on flakes include end scrapers, side scrapers, wedge shaped tools, retouched flakes, and utilized flakes with micro-flaking, as well as flake cores, mobile art, ocher, hammer stones, pebble tools, and pebbles (Fig. 4-b).

3) Summery

The archaeological layers from which the flake assemblages were discovered at Kashiwadai 1 suggest that there is a distinctive archaeological component between the Spfa-1 and En-a tephras. The archaeological layer at Shimaki, on the other hand, is between the Sipfa-2 and Sipfa-1 tephras. While spatio-temporal correspondence between Sipfa-a tephra and marker tephras has not been established, since Sipfa-2 is bracketed by Spfa-1 (Machida and Arai 2003), Shimaki site is contemporaneous with the Kashiwadai 1 site. The reliable absolute dates for flake assemblages are 22,000-20,000 RCYBP from Kashiwadai 1. Primary reduction strategies associated with the flake assemblages include the removal of flakes from unifacial discoidal cores and multi-faced flake cores. Secondary reduction involves the production of end scrapers and side scrapers. Stone artifacts, pigments, and anvil stones are also part of these assemblages.

(3) Blade Assemblages

1) Obarubetsu 2 site

The Obarubetsu 2 site is located in the eastern Oshima lowland (D3-6) in the southeastern part of the Oshima Peninsula, southwestern Hokkaido (N42°31′, E140°22′) at an altitude of 9m a.s.l. (Koaze *et al*., 2003). The site is situated at the edge of the lower river terrace of the Oshamanbe River. The blade industry was recovered from block excavation II in an eolian loam below primary fall of the Nigorikawa Tephra (Ng, 12,000 RCYBP) 1.0m below the surface (Hokkaido Association of Cultural Resource Management, 2000; Izuho and Akai, 2005). Seven hearths

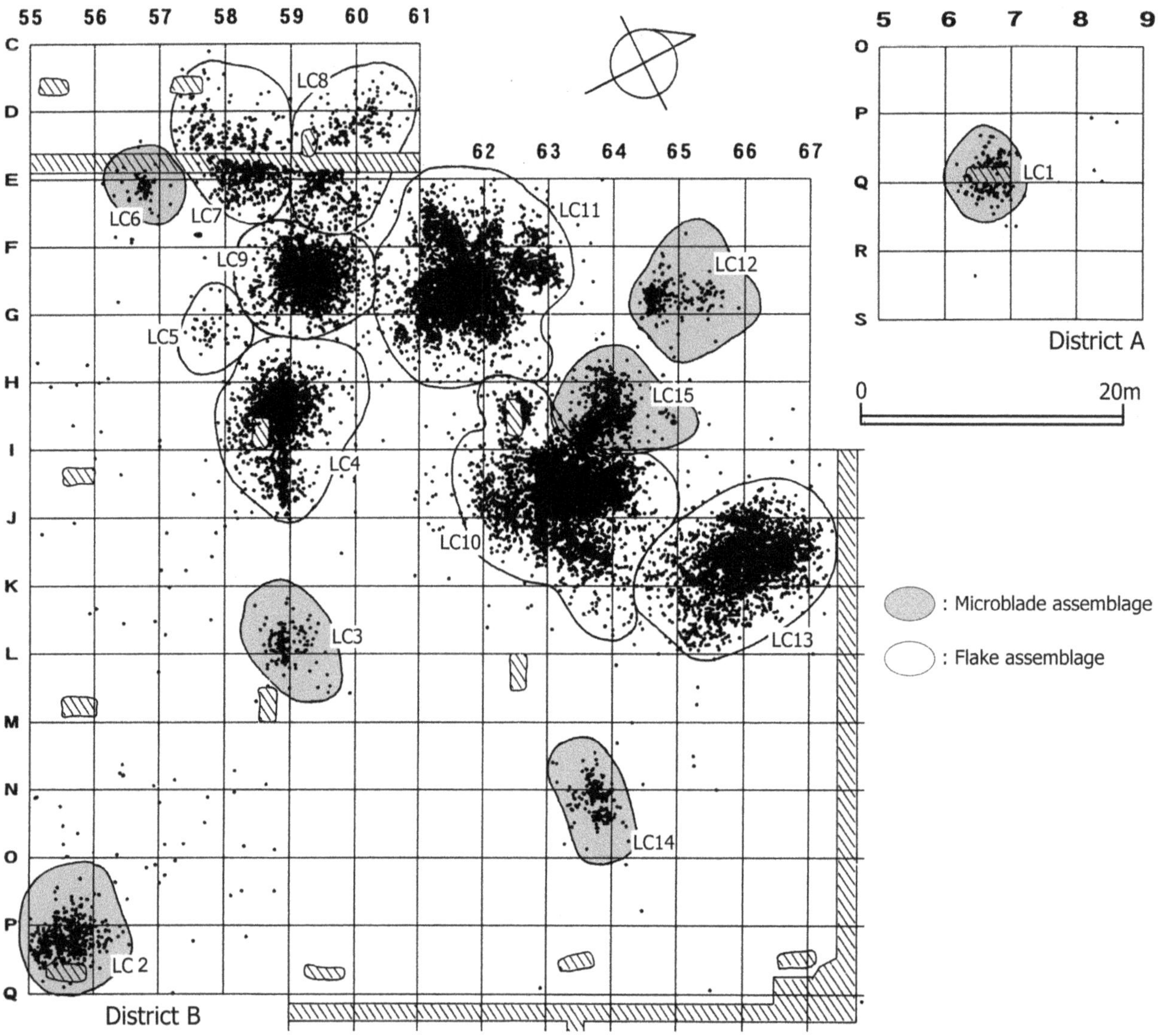

Figure 5. Distribution of the lithic concentrations at Kashiwadai 1 (after Hokkaido Center for Buried Cultural Property, 1999).

are reported from the site. Two dates from the cultural layer but not collected from the hearths are also reported (Table 1:55-56).

As for the Paleolithic assemblage, a total of 1,500 lithic specimens were recovered from five concentrations. LC 3 is classified as a backed blade assemblage. Lithic raw material consists of "hard-shale" (n=912), obsidian (n=6), agate (n=1), andesite (n=2), tuff (n=1), and rhyolite (n=1). "Hard-shale" gravels widely distributed in southwestern Hokkaido were acquired locally (Hokkaido Association of Cultural Resource Management, 2000). The lithic assemblage consists of tools made on blades including backed blades (n=8), perforators (n=1), retouched flakes (n=1), flakes, blades (n=66), and blade cores (n=2) (Fig. 6-a). Primary reduction was blade production. Backing was a common secondary reduction strategy. The backed blades strongly resemble those from the northeastern part of Honshu Island.

2) Kawanishi C site

The Kawanishi C site is located on the southern terraces (B3-5) in the southeastern Tokachi Plain, eastern Hokkaido (N42°52′, E143°11′) (Koaze *et al.*, 2003) at the edge of the Kamisatsunai I terrace of the Satsunai River at an altitude of 70m a.s.l. Three salvage excavations were done by the Obihiro Board of Education and a total 6,856m^2 were uncovered (Obihiro Board of Education, 1998, 2000a, 2000b). Three archaeological layers were found in eolian units classified from top to bottom as a Jomon component in the Black Humus soil above the Tarumae-D Tephra (Ta-d, 7,000 RCYBP), a microblade assemblage in an eolian loam between the Ta-d and En-a tephras, as well as a blade assemblage in an eolian loam between the En-a and Spfa-1 tephras 0.7m below the surface (Obihiro Board of Education, 1998; Izuho and Akai, 2005). Three AMS dates on charcoal from a hearth in the upper level of the Paleolithic component between the Ta-d and En-a

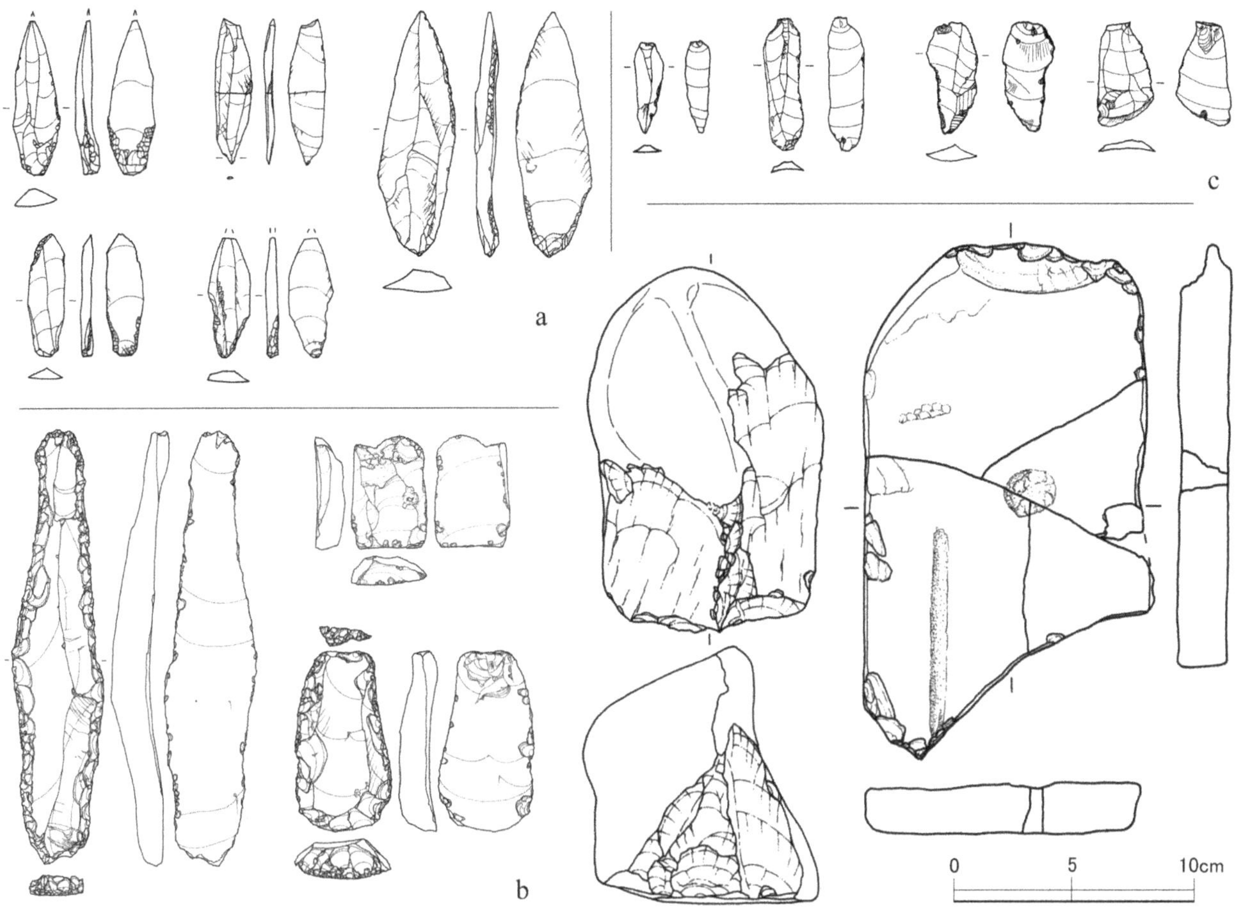

Figure 6. Stone tools of the Blade Assemblages in Hokkaido.
a: Obarubetsu 2 (after Hokkaido Association of Cultural Resource Management, 2000), b: Kawanishi C (after Obihiro Board of Education, 1998), c: Kukominami A (Tokachi Blanch Office, 1983).

tephras were reported at 16,940±050 RCYBP, 13,070±040 RCYBP, and 12,920±050 RCYBP (Table 1:59-61). In addition, four hearths were found in the lower part of the Paleolithic component between the En-a and Spfa-1 tephras where two AMS dates from hearth charcoal were reported at 21,800±090 RCYBP and 21,420±190 RCYBP (Table 1:57-58).

In the upper part of the Paleolithic component, a total of 2,000 lithic specimens were recovered from five concentrations. In contrast, a total of 19,000 lithic specimens were recovered from 12 concentrations in lower part of the Paleolithic component. Raw material mainly consists of obsidian, followed by a small amount of "hard-shale", agate, andesite, and coarse grained igneous cobbles and pebbles. The lithic assemblage consists of tools made on blades including side scrapers, end scrapers, burins, perforators, and wedge-shaped tools, as well as flakes, flake core, ocher, pebble tools, and hammer stones (Fig. 6-b). Primary reduction involved the production of blades. End scraper edge forming and burin technology were common secondary reduction strategies (Obihiro Board of Education, 1998, 2000a, 2000b).

2) Kukominami A site

The Kukominami A site is located on the southern terraces (B3-5) on the southeastern Tokachi Plain, eastern Hokkaido (N42°43′, E143°13′) (Koaze *et al.*, 2003) at the edge of the Kamisatsunai II terrace of the Satsunai River at an altitude of 150m a.s.l. The artifact-bearing layer was found in an eolian loam between the En-a and Spfa-1 tephras (40,000 BP) 0.5m below the surface (Tokachi Branch Office, 1983; Yamahara, 1996; Izuho and Akai, 2005). Two AMS dates on charcoal, 23,850±4480 RCYBP, 19,420±1770 RCYBP, were associated with the soil horizon below the En-a tephra, and one, 15,280±340 RCYBP, was associated with the soil horizon above the En-a tephra, but their archaeological context is not clear (Table 1:62-64).

A total of 30 lithic specimens were recovered from a concentration that was partially destroyed due to construction. Lithic raw material consists of obsidian for chipped stone tools and andesite for pebble tools. The lithic assemblage consists of retouched blades, blades, flakes and pebble tools (Fig. 6-c). Primary reduction involved bladelet technology. Not continuous retouches are appeared secondary reduction. Note for the existence of bladelets.

3) Summery

The archaeological layers in which the blade assemblages are encompassed at Kawanishi C and Kukominami A suggest that there is a distinctive archaeological component between the Spfa-1 and En-a tephras. Absolute dates obtained from the Kawanishi C site cluster around 21,000 RCYBP. Primary reduction at Obarubetsu 2 and Kawanishi C is characterized by the production of large blades, and consists of highly standardized burins, end scrapers, and side scrapers with intensive retouch, snaps, and burin facets. The Obarubetsu 2 assemblage has basally retouched pointed tools that are common among lithic assemblages in Honshu. In contrast, the Kukominami A assemblage is characterized by bladelets from small prismatic and conical cores.

(4) Microblade Assemblages

1) Kashiwadai 1 site

As given in brief introduction of the Kashiwadai site in (2-2), a microblade assemblage was recovered in a soil horizon situated in the lower part of a thick eolian loam, 4m below the surface between reworked Spfa-1 (40,000BP) and En-a (17,000 RCYBP) tephras (Hokkaido Center for Buried Cultural Property, 1999; Izuho and Akai, 2005). LC1 in block excavation A, and LC2, 3, 6, 12, 14, and 15 in block excavation B are the microblade assemblages (Fig. 5).

A total of 159 lithic specimens were recovered from a concentration in block excavation A. Raw material consists of non-local obsidian (n=2, 0.9g in total) and "hard-shale" (n=157, 267.8g in total) (Hokkaido Center for Buried Cultural Property, 1999). Primary reduction of this site involved production of blades and microblades from wedge shaped cores. End scraper edge retouch was a common secondary reduction strategy. The lithic assemblage consists of blades, microblades, platform- forming spalls, and flakes.

A total of 3,261 lithic specimens were recovered from six lithic concentrations in block excavation B. The Lithic assemblage consists of microblades, blades, burins, end scrapers, Rankoshi type microblade cores, flake cores, and flakes. Other material includes ocher, amber beads, hammer stones, and anvils (Fig. 7-a). Lithic raw material consists mainly of "hard-shale" followed by small amounts of obsidian, andesite, agate, and chert. Obsidian and "hard-shale" were considered non-local (Hokkaido Center for Buried Cultural Property, 1999). Primary reduction involved the production of flakes, blades and microblades. Continuous retouch, end scraper edging, and burin facet were common secondary reduction strategies.

2) Kamihoronai-Moi site

The Kamihoronai-Moi site is situated at the eastern edge of the southern Ishikari Lowland in central Hokkaido (42° 47' 15" N, 141° 59' 56" E; 65.6 m a.s.l.). The site is on the left bank of the Atsuma River, which flows down the west slope of the Yubari Mountains and runs into the present Pacific coast about 30 km to the southwest. Since 2004, archaeologists affiliated with the Atsuma Board of Education undertook full excavations after testing the site in 2002 and 2003. This was done under the mitigation of the construction of a water reservoir initiated for agriculture and flood control (Atsuma Board of Education, 2006:15-16). A total of 8,460 m^2 was excavated during 2004 and 2005 (Atsuma Board of Education, 2006; Nakazawa *et al.*, 2007).

Since the details of the topographic divisions of terraces around the site and the site stratigraphy are both reported in Izuho *et al.* (2009), here we only summarize the basic information on the Paleolithic layer. The Paleolithic occupation layer was found within a 295 m^2 area on T_3 (Atsuma Board of Education, 2006). The Paleolithic assemblage and an associated hearth (IXF-01) were discovered in a 10-cm thick loam (also called as "layer IXc") that was uniformly distributed horizontally across T_3 (260.2 m^2). AMS radiocarbon dates were obtained from charcoal samples contained in Unit VII. These dates consistently range between 14,700 and 14,500 RCYBP: 14,565±50 RCYBP (PLD-5275), 14,560±50 RCYBP (PLD-5276), and 14,650±80 RCYBP (IAAA-41577) (Table 1:68-73).

A total of 1,412 artifacts and fire-cracked rocks were recovered from Unit VII. These artifacts were horizontally scattered approximately 4 m east-west and 7 m north-south, while the vertical locations of artifacts are encompassed within 30 cm; 95% of the artifacts are distributed unimodally within 20-cm (95% of C.I. = 65.5–65.7 m a.s.l.). One hearth identified as a burnt patch with fragmented, fire-cracked rocks was located at the southeastern margin of the lithic scatter. Because no significant erosion and disturbance were observed in the sediments of Unit VII and all artifacts were plotted by the Electric Distance Meter, this site retains potential value to elucidate behavioral and physical processes that created the observed spatial pattern (Izuho et al., 2009; Nakazawa et al., 2009).

The chipped stone artifacts include 151 microblades (21 of them have edges with microflaking), 3 wedge-shaped microblade cores, 6 side scrapers, 2 end scrapers, a retouched flake, a perforator, a burin, and 10 burin spalls (Fig. 7-c). The wedge-shaped microblade cores are classified as Sakkotsu type, shaped by preparation known as the "Yubetsu method" (Bleed, 2001; Nakazawa *et al.*, 2005:281; Tsurumaru, 1979; Yoshizaki, 1961), which has been common among late-glacial sites of northeastern Asia. Microblades and microblade cores were solely made from obsidian, while burin and end scrapers were made on "hard-shale". The perforator was made from agate. All material used for the chipped-stone artifacts (i.e., obsidian, "hard shale", and agate) are "high-quality" and not locally available.

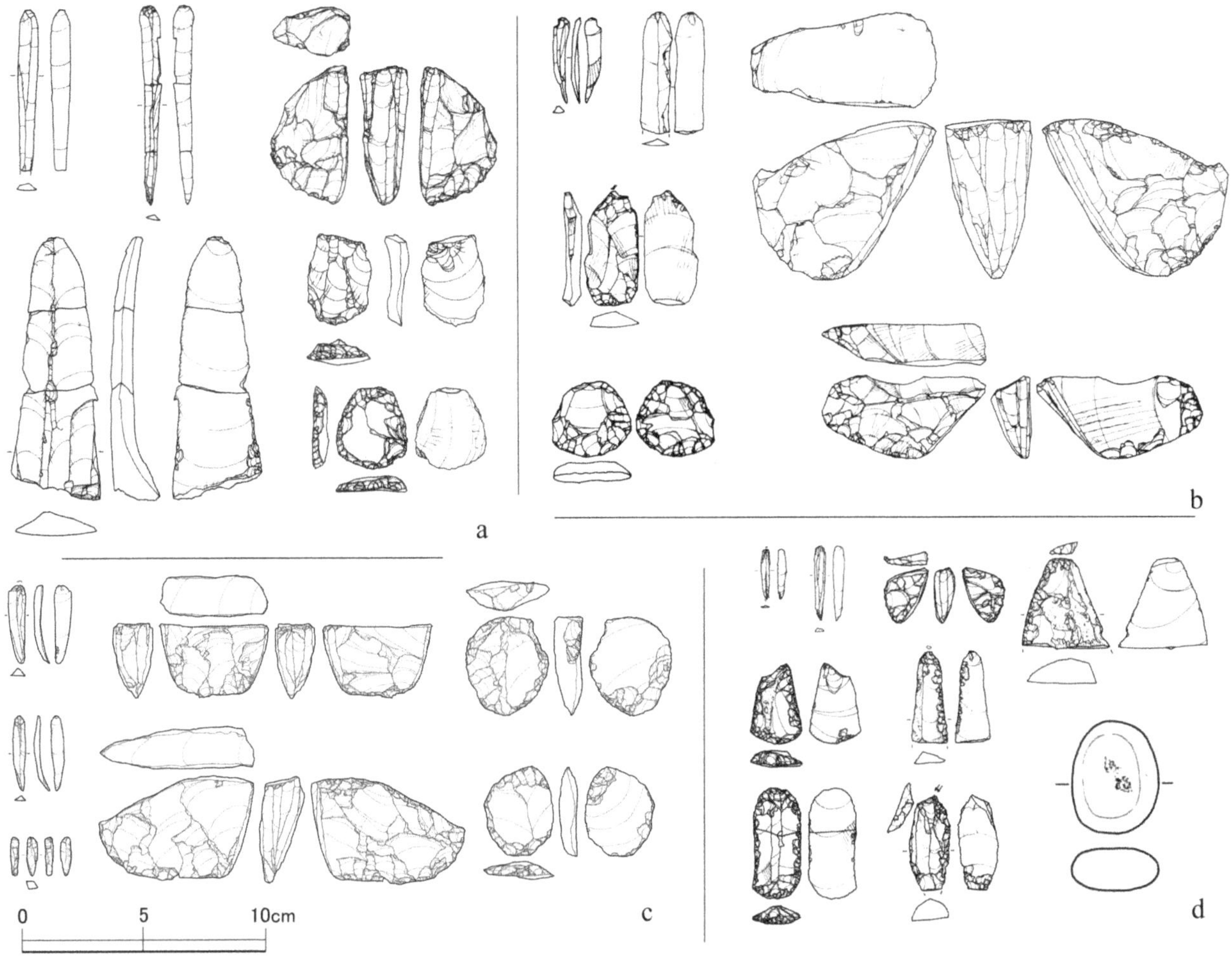

Figure 7. Stone tool components of Microblade assemblages in Hokkaido.
a: Kashiwadai 1 LC15 (after Hokkaido Center for Buried Cultural Property, 1999), b: Akatsuki (after Obihiro Board of Education, 1985, 1986), c: Kamihoronai Moi (after Atsuma Board of Education, 2006), d: Ozora (after Obihiro Board of Education, 1993)

3) Aktatsuki site

The Akatsuki site is located on the southern terraces (B3-5) of the southeastern Tokachi Plain, eastern Hokkaido (N42°54′, E143°11′) (Koaze *et al.*, 2003) at the edge of the Kamisatsunai IIb terrace of the Satsunai River at an altitude of 40m a.s.l. Because the site is situated in downtown Obihiro, a total of six salvage excavations were conducted by the Obihiro Board of Education that uncovered 4,500 m^2 (Tokachi Research Institute of Archaeology, 1989). Two archaeological layers were found in eolian units. From top to bottom they were a Jomon component in a Black Humus Soil above the Tarumae-D tephra (Ta-d, 7,000 RCYBP) and a microblade component between the Ta-d tephra and redeposited En-a tephra (17,000 RCYBP) 0.5m below the surface (Obihiro Board of Education, 1985, 1986). Two conventional dates on charcoal associated with the lithic concentration between the Ta-d tephra and redeposited En-a tephra were reported at 14,700±250 RCYBP and 14,700±130 RCYBP (Table 1:65-67).

Over 40,000 specimens were recovered from 22 lithic concentrations in the Paleolithic layers. The assemblage consists of microblades, Togeshita type microblade cores, Sakkotsu type microblade cores, end scrapers, burins, side scrapers, blades, blade cores, anvil stones, tools made on blades, including side scrapers, end scrapers, burins, perforator, and wedge-shaped tool, as well as flakes, flake cores, ocher, pebble tools, and hammer stones (Fig. 7-b). Primary reduction included microblade, blade, and flake productions. Continuous retouch of side scrapers on blades, end scraper edges and burins on flakes were common secondary reduction strategies. Lithic raw material mainly consisted of obsidian and "hard-shale". Obsidian was transported from the Shirataki and Tokachi-mitsumata sources for blade and microblade productioni, while "hard-shale" was transported from southwestern Hokkaido to produce flakes (Kimura, 1995; Obihiro Board of Education, 1991).

4) Ozora site

The Ozora site is located on the southern terraces (B3-5) of the southeastern Tokachi Plain, eastern Hokkaido (N42°38′, E143°10′) (Koaze *et al.*, 2003) on the Motomatsu terrace of the Satsunai River at an altitude of 83m a.s.l. The artifact-bearing layer was found in an eolian loam below the Ta-d tephra and above the En-a tephra 0.4m below the surface (Obihiro Board of Education, 1993; Izuho and Akai, 2005). No radiocarbon dates were reported.

A total of 8,674 lithic specimens were recovered from six lithic concentrations. The assemblage consists of microblades, Oshorokko type microblade cores, bifacial stemmed points, burins, end scrapers, side scrapers, hammer stones and anvils (Fig. 7-d). Lithic raw material mainly consists of obsidian (n=8,090; 1,716g), as well as "hard-shale" and agate (n=106; 452g). Primary reduction produced microblades and blades. End scraper edging, burin manufacture, and continuous retouch of side scrapers were common secondary reduction strategies.

5) Summery

The Paleolithic components from the Akatsuki, Kamihoronai-Moi, and Ozora sites are all from above the En-a tephra. In contrast, dates obtained from the Kashiwadai 1 are 22,000-20,000 RCYBP, consistent with tephro-stratigraphy. The tool components of the microblade assemblages from below the En-a tephra are similar to those from above the En-a tephra, implying that the earliest microblade assemblages are "pristine". A notable difference in microblade production between the two periods is that microblades in the earlier period were made from the same sequential removals of blades.

Among the microblade assemblages from the levels above the En-a tephra, there are principally seven groups including those with Sakkotsu-type microblade cores, with Oshorokko-type microblade cores, and with Shirataki-type microblade cores (Nakazawa *et al.*, 2005; Sato and Tsutsumi, 2007 ; Izuho and Sato, 2008). Primary reduction produced bifacially retouched blanks, various blade cores, and flake cores. A similarity in the assemblages from above and below the En-a tephra is high standardization of stone tools, namely the manufacture of microblades, burins, end scrapers, and side scrapers. Newly invented stone tools such as stemmed points and axe-shaped tools were found in the assemblages above the En-a tephra (Nakazawa *et al*, 2005; Yamada, 2006).

Lithic assemblages and environmental changes in the Paleo-SHK Peninsula

(1) Summary of temporal changes in lithic assemblages

Fig. 8 shows the geochronology of thirteen Upper Paleolithic sites in Hokkaido. The sites having clear correspondence between the tephras and absolute ages are represented by black letters, while those based on stratigraphic and techno-typological data from Honshu (Sato, 1992) are shown as gray letters. In Hokkaido, the Small Flake Industry appeared at least 30,000 B.P., and disappeared by 20,000 B.P. The sites are mainly distributed in central and southeastern Hokkaido. No sites of this industry have been found in northern Hokkaido. They are roughly dated to the latter part of MIS 3.

Following the Small Flake Industry, Flake, Blade, and Microblade industries appeared at least from 22,000 to 20,000 RCYBP. The emergence of Flake and Blade Industries would have been much earlier, considering the dates of similar assemblages in neighboring regions. The Flake Industry is widespread in central Hokkaido, while no sites have been found in the southern and northern parts. The Blade Industry is found in southwestern and southeastern Hokkaido. No sites were found in northern Hokkaido.

The Microblade Industry appeared at the LGM, and lasted until ca. 13,000 RCYBP. The microblade assemblages below the En-a tephra are widespread in southwestern and southeastern Hokkaido. On the other hand, a number of microblade assemblages found in layers above the En-a tephra are widely distributed across Hokkaido.

(2) A comparison of lithic assemblages and paleoenvironment

As we have suggested, the Small Flake Industry that emerged during the latter part of MIS 3 was contemporaneous with the faunal group of *Palaeoloxodon naumanni* and giant deer. The Flake, Blade, and Microblade Industries are contemporaneous with patch environments of grassland and open forest where the faunal community of *Mammuthus primigenius* lived. The Microblade Industry of post-LGM was contemporaneous with the faunal community of *Mammuthus primigenius* that survived in gradually diminishing patch environments of grassland and open forest (Izuho and Takahashi, 2005; Izuho and Sato, 2008).

The temporal correspondences between lithic industries and floral/faunal groups imply that hunter-gatherer lifeways changed from small-scale hunting in cold/temperate forests to a hunting strategy that targeted group mammals that inhabited cold grassland and open forests during the LGM. However, the change at the LGM was not just a replacement of the small flake Industry by microblade, but three the divergence Flake, Blade, and Microblade Industries diverged. We would like to continue investigating questions about when, where, how, and why these changes occurred through detailed geochronology of the lithic industries, their distribution, and lithic analyses employing behavioral perspective.

Acknowledgements

We deeply thank Dr. Ian Buvit (Tokyo Metropolitan University) for correcting the English.

Reference Cited

Aoki, K., and Arai, F. 2000 Sanriku Oki Kaitei Core

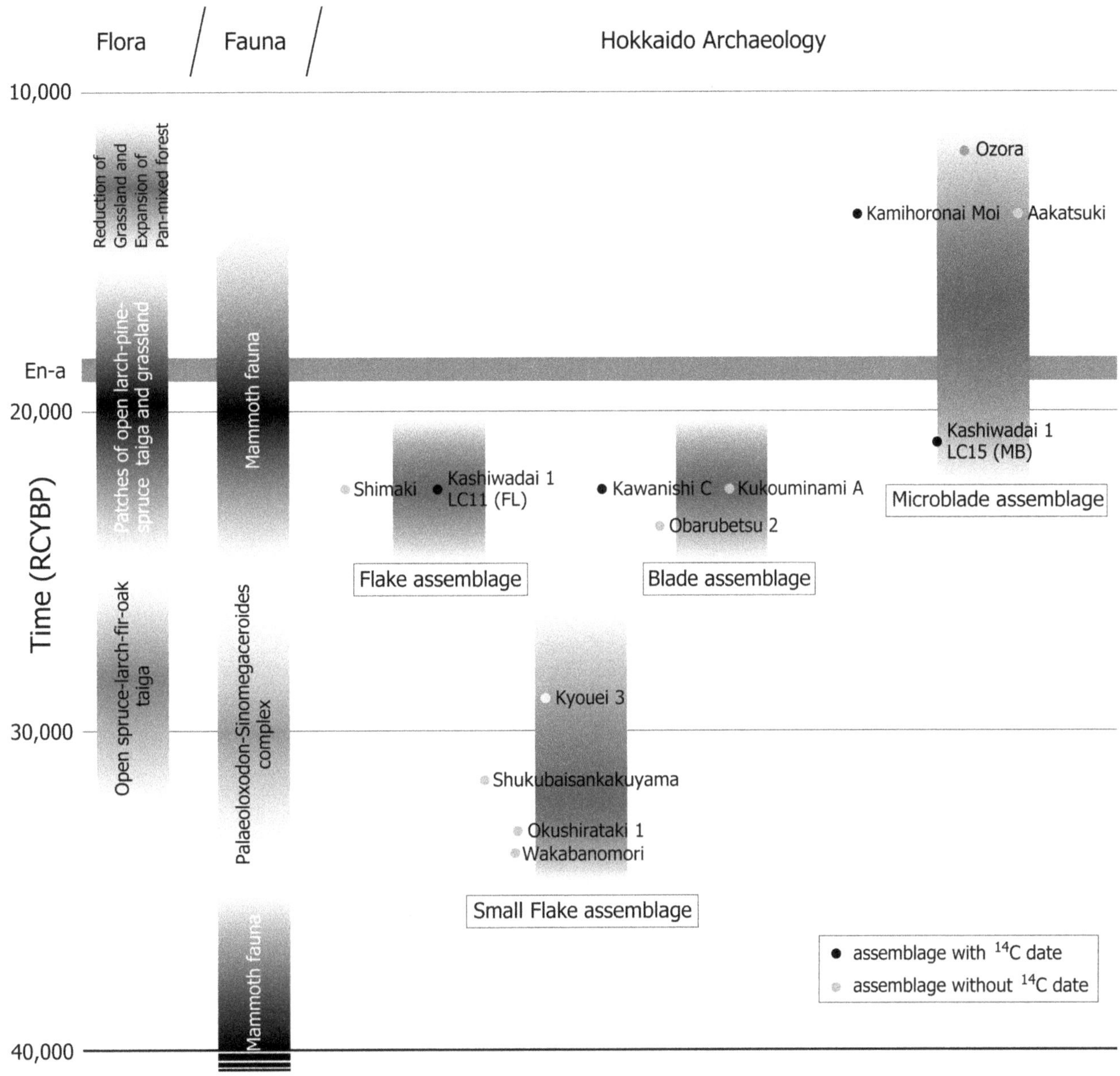

Figure 8. Geochronology of the Upper Paleolithic of Hokkaido.

KH94-3, LM-8 no Koki Koshinsei TephraSojyo [Late Quaternary Tephrostratigraphy of Marine Core KH94-3, LM-8 off Sanriku, Japan]. *Daiyonki Kenkyu* 39:107-120. (In Japanese)

Atsuma Board of Education 2006 *Kamihornai-Moi Iseki (1)* [Kamihoronai-Moi site (1)]. Atsuma-cho Kyoiku Iinkai. (In Japanese)

Bleed, P. 2001 Trees or Chains, Links or Branches: Conceptual Alternatives for Consideration of Stone Tool Production and Other Sequential Activities. *Journal of Archaeological Method and Theory,* 8(1):101-127.

Buvit, I., Terry, K., Izuho, M., and Hamaguchi, K. 2011 Recent Excavations at the Shimaki Paleolithic Site, Hokkaido, Japan. *Current Research in the Pleistocene*, in press.

Chitose Board of Education 1974 *Shukubaisankakuyama Chiten* [Shukubaisankakuyama Site]. Chitose-shi Kyoiku Iinkai. Chitose. (In Japanese)

Fukusawa, H., Saito, K., and Fujiwara, O. 2003 Nihon Retto niokeru Koshinsei Koki Iko no Kiko Hendo no Trigger wa Nanika? Tibet Kogen to West Pacific Warm Water Pool no Yakuwari [Climatic Change Since the Late Pleistocene in the Japanese Islands: The Role of the Tibetian Plateau and West Pacific Warm Water Pool]. *Daiyonki Kenkyu*, 42: 165-180. (In Japanese)

Hokkaido Association of Cultural Resource Management 2000 *Oshamanbe-cho Obarubetsu 2 Iseki* [Obarubetsu 2 Site, Oshamanbe Town]. Hokkaido Bunkazai Hogo Kyokai. Sapporo. (In Japanese)

Hokkaido Center for Buried Cultural Property 1992 *Shimizu-cho Kamishimizu 2 Iseki, Kyoei 3 Iseki(2), Higashimatsuzawa 2 Iseki, Memuro-cho Hokumei 1 Iseki* [Kamishimizu 2 Site, Kyoei 3 Site(2), and Higashimatsuzawa 2 Site in Shimizu, Hokumei 1 Site in Memuro]. Hokkaido Maizo Bunkazai Center. Sapporo. (In Japanese)

Hokkaido Center for Buried Cultural Property 1999 *Kashiwadai 1 Iseki* [Kashiwadai 1 Site]. Hokkaido Maizo Bunkazai Center. Sapporo. (In Japanese)

Hokkaido Center for Buried Cultural Property 2002 *Shirataki Isekigun III* [The Shirataki Site Group III]. Hokkaido Maizo Bunkazai Center. Sapporo. (In Japanese)

Igarashi, Y. 1990 Kafun Kaseki kara Saguru Shinrin no Rekishi: Hokkaido no 3 Man'nenkan [Pollen Analysis for the Forest History of Past 30,000 years in Hokkaido]. *Nihon Ringakkai Hokkaido Shibu Ronbunshu*, 38:1-9. (In Japanese)

Igarashi, Y. 1996 A Late Glacial Climatic Reversion in Hokkaido, Northeast Asia, Inferred from the *Larix* Pollen Record. *Quaternary Science Reviews*, 15:989-995.

Igarashi, Y. 2008 Climate and Vegetation Changes since 40,000 Years BP in Hokkaido and Sakhalin. In Sato, H. (ed.) *Human Ecosystem Changes in the Northern Circum Japan Sea Area (NCJSA) in Late Pleistocene*, pp.27-41. Research Institute for Humanity and Nature. Kyoto.

Igarashi, Y., Oba, T. 2006 Fluctuations in the East Asian monsoon over the last 144ka in the northwest Pacific based on a high-resolution pollen analysis of IMAGES core MD01-2421. *Quaternary Science Reviews*, 25:1447-1459.

Igarashi, Y., Igarashi, T., Daimaru, H., Yamada, O., Miyagi, T., Matsushita, K., Hiramatsu, K. 1993 Hokkaido no kenbuchi bonchi to hurano bonchi niokeru 32,000 nenkan no shokusei hensen [Vegitation History of Kenbuchi Basin and Furano Basin in Hokkaido, North Japan, since 32,000 yrs BP]. *Daiyonki Kenkyu*, 32:89-105. (In Japanese)

Izuho, M. 2008 Hokkaido Yufutsu-gun Atsuma-cho Kamihoronai-Moi Iseki Kyusekki Chiten niokeru Kafun Bunseki to Shokubutsuso [A Report of Pollen Analysis at the Paleolithic Locality of the Kamihoronai-Moi Site, Hokkaido (Japan)]. *Ronshu Oshorokko* II, pp.33-39.

Izuho, M. 2010 Nihon Retto no Jyobu Kyusekki Jidai Zenhanki Kenkyu no Ichi Shiten: Gendaijinteki Kodo Tayosei to Hen'i no Hatsugen [A Note on the Upper Paleolithic Archaeology on Japanese Archipelago: Toward an Improvement Understanding]. *Jinbun Gakuho* 38, pp.1-12. (in Japanese)

Izuho, M., and F. Akai. 2005 Hokkaido no Kyusekki Hennen: Iseki Keisei Kateiron no Tekiyo [Geochronology of Palaeolithic Sites on Hokkaido]. *Kyusekki Kenkyu* 1:39-55. (In Japanese)

Izuho, M., Nakazawa, Y., Akai, F., Soda, T., and Oda, H. 2009 Geoarchaeological Investigations at the Upper Paleolithic Site of Kamihoronai-Moi, Hokkaido, Japan. *Geoarchaeology* 24, pp.492–517.

Izuho, M. and Sato, H. 2008 Landscape Evolution and Culture Changes in the Upper Paleolithic of Northern Japan. In A. P. Derevianko and Shunkov, M. V. (eds.) *The Current Issues of Paleolithic Studies in Asia; Proceedings of the International Symposium "The Current Issues of Paleolithic Studies in Asia and Contiguous Regions"*. pp.69-77. Publishing Department of the Institute of Archaeology and Ethnography SB RAS. Novosibirsk.

Izuho, M. and Takahashi, K. 2005 Correlation of Paleolithic Industries and Paleoenvironmental Change in Hokkaido (Japan). *Current Research in the Pleistocene*, 22:19-21.

Japanese Paleolithic Reseach Association 2010 *Nihon Retto no Kyusekki Jidai Iseki* [Paleolithic Sites in the Japanese Islands: A Database]. 377p. Nihon Kyusekki Gakkai. Tokyo. (In Japanese)

Kato, S., and M. Yamada 1988 Hokkaido Kato-gun Kamishihoro-cho Shimaki Isekino Sekki Bunka [A Paleolithic Culture at Shimaki- The Oldest Palaeolithic Components in Hokkaido]. *Rekishi Jinrui* 16, pp.252-340. (In Japanese)

Kimura, H. 1995 Kokuyouseki, Hito, Gijutsu [Obsidian, Humans, and Technology]. *Hokkaido Kokogaku*, 31:3-63. (In Japanese)

Kitamura, A., and Kimoto, K. 2006 History of the Inflow of the Warm Tsushima Current into the Sea of Japan between 3.5 and 0.8 Ma. *Palaeogeography, Palaeoclimatology, Palaeoecology*, 236:355-366.

Koaze H., Nogami, M., Ono, Y., and Hirakawa, K. 2003 *Nihon no Chikei 2 Hokkaido* [Regional Geomorphology of the Japanese Islands vol. 2: Geomorphology of Hokkaido]. 359p. The University of Tokyo Press. Tokyo. (In Japanese)

Kosaka, T., and K. Nogawa 1974 Hokkaido Tokachi Kamishihoro-cho niokeru Kyusekki Hoganso no ^{14}C Nendai [14C date of Paleolithic layer in Kamishihoro town, Tokachi, Hokkaido]. *Chikyu Kagaku*, 26:137-138. (In Japanese)

Koshimizu, S. 1981 Ishikari Teichitai ni Shutsudo suru Kokuyosekihen no Gensanchi [Source Areas of Obsidian Found in the Prehistoric Sites in the Ishikari-Tomakomai Low-land Area in Hokkaido, Japan]. *Chikyu Kagaku*, 35:267–273. (In Japanese)

Kuzmin, Y., Glascock, M., and Sato, H. 2002 Sources of Archaeological Obsidian on Sakhalin Island (Russian Far East). *Journal of Archaeological Science* 29:741-749.

Machida, H. and Arai, F. 2003 *Shinpen Kazanbai Atlas: Nihon Retto to Sono Shuhen* [Atlas of Tephra in and around Japan (Revised Edition)]. 336p. University of Tokyo Press. Tokyo. (in Japanese)

Machida, H., Oba, T., Ono, A., Yamazaki, H., Kawamura, Y., and Momohara, A. 2004 *Daiyonkigaku* [Quaternary]. 325p. Asakura Shoten. Tokyo. (In Japanese)

Nakamura, Y., Hirakawa, K., and Naganuma, T. 1999 Hokkaido Shirataki Iseki to Shuhen no Tephra[Tephras at and araound a Paleolithic Site in the Shirataki Basin, Eastern Hokkaido, Japan]. Chigaku Zasshi, 108:616-628. (In Japanese)

Nakazawa, Y., Izuho, M., Takakura, J., and Yamada, S. 2005 Toward an Understanding of Technological Variability in Microblade Assemblages in Hokkaido, Japan. *Asian Perspectives* 44, pp.276-292.

Nakazawa, Y., Izuho, M., and Akai, F. 2007 The late-glacial microblade assemblage from the Kamihoronai- Moi site in Hokkaido, northern Japan: A newly discovered

Yubetsu site. *Current Research in the Pleistocene*, 25:42-44.

Nakazawa, Y., Izuho, M., Akai, F. 2009 Between the Two Hearths: Site Formation Processes and Spatial Organization at the Upper Paleolithic Open-air Site of Kamihoronai-Moi, Hokkaido (Japan). The Quaternary Reserch (Daiyonki Kenkyu) 48-2, pp.85-96."

Naoe, Y. 2009 Shirataki San Kokuyoseki no Kakutoku to Sono Hirogari [Procurement of Obsidian in Shirataki Region and Its Distribution]. *Kyusekki Kenkyu*, 5:11-22. (In Japanese)

Naoe, Y., and Nagasaki, J. 2005 Hokkaido Koki Kyusekki Jidai Zenhan Ki no Sekizai Shohi Senryaku; Shirataki Ia Gun to Shukubai Iseki Sankakuyama Chiten [Raw Material Consumption Strategy at Late Upper Palaeolithic on Hokkaido]. *Hokkaido Kyusekki Bunka Kenkyu* 10, pp.45-58.

Obihiro Board of Education 1985 *Obihiro Akatsuki Iseki* [Akatsuki Site, Obihiro]. Obihiro-shi Kyoiku Iinkai. Obihiro. (In Japanese)

Obihiro Board of Education 1986 *Obihiro Akatsuki Iseki 2* [Akatsuki Site, Obihiro volume 2]. Obihiro-shi Kyoiku Iinkai. Obihiro. (In Japanese)

Obihiro Board of Education 1991 *Obihiro Akatsuki Iseki 4* [Akatsuki Site, Obihiro volume 4]. Obihiro-shi Kyoiku Iinkai. Obihiro. (In Japanese)

Obihiro Board of Education 1993 *Obihiro Ozora Iseki* [Ozora Site, Obihiro]. Obihiro-shi Kyoiku Iinkai. Obihiro. (In Japanese)

Obihiro Board of Education 1998 *Obihiro Kawanishi C Iseki* [Kawanishi C Site, Obihiro]. Obihiro-shi Kyoiku Iinkai. Obihiro. (In Japanese)

Obihiro Board of Education 2000a *Obihiro Kawanishi C Iseki 2* [Kawanishi C Site, Obihiro volume 2]. Obihiro-shi Kyoiku Iinkai. Obihiro. (In Japanese)

Obihiro Board of Education 2000b *Obihiro Kawanishi C Iseki 3* [Kawanishi C Site, Obihiro volume 3]. Obihiro-shi Kyoiku Iinkai. Obihiro. (In Japanese)

Obihiro Board of Education 2004 *Obihiro, Wakabano Mori Iseki* [Wakabano Mori Site, Obihiro]. Obihiro-shi Kyoiku Iinkai, Obihiro. (In Japanese)

Ono, Y. 1990 Kita no Rikkyo [Northern Land Bridge]. *Daiyonki Kenkyu*, 29:183-192. (In Japanese)

Ono, Y. and Igarashi, Y. 1993 *Hokkaido no Shizenshi: Hyoki no Shinrin wo Tabi suru* [Natural History in Hokkaido: A Journy of Ice Age Forest]. 219p. Hokkaido University Press. Sapporo. (In Japanese)

Sato, H. 1992 *Nihon Kyusekki Bunka no Kozo to Shinka* [Evolution and Structure of Paleolithic Culture in Japan]. 362p. Kashiwa Shobo, Tokyo. (In Japanese)

Sato, H. 2003 Hokkaido no Koki Kyusekki Jidai Zenhanki no Yoso [The Early Upper Paleolithic in Hokkaido]. *Kodai Bunka*, 55:181-194. (In Japanese)

Sato, H. and Tsutsumi, T. 2007 The Japanese Microblade Industries: Technology, Raw Material Procurement, and Adaptations. In Y. Kuzmin, S. Keates and C. Shen (eds.) *Origin and Spread of Microblade Technology in Noerthern Asia and North America*. pp.53-78. Archaeology Press, Burnaby.

Takahashi, K. 2007 Nihon Retto no Senshin- Koshin Sei niokeru Rikusei Honyu Dobutsuso no Keisei Katei [The Formative History of the Terrestrial Mammalian Fauna of the Japanese Islands during the Plio- Pleistocene]. *Daiyonki Kenkyu*, 3:5-14. (In Japanese)

Takahashi, K., Soeda, Y., Izuho, M., Yamada, G., Akamatsu, M., and Chang, C. 2006 The Chronological Record of the Woolly Mammoth (*Mammuthus primigenius*) in Japan, and Its Temporary Replacement by *Palaeoloxodon naumanni* during MIS 3 in Hokkaido (Northern Japan). *Palaeogeography, Palaeoclimatology, Palaeoecology*, 233:1-10.

Tsurumaru, T. 1979 Hokkaido Chihono Saisekijin Bunka [Microblade Culture in Hokkaido District]. *Sundai Shigaku* 47, pp.23-50. Sundai Shigakkai, Tokyo. (In Japanese)

Tsurumaru, T. 2001 Hokkaido Kyusekki Kokogaku no Ronten: Kon-nichiteki Shoten to Tenbo [Contemporary Issues and Perspectives of Palaeolithic Archaeology on Hokkaido]. *Hokkaido Kokogaku*, 37: 3-22. (In Japanese)

Tokachi Blanch Office 1982 Obihiro Kukominami A Iseki Miyamoto Iseki [Kukominami A Site and Miyamoto Site]. Tokachi Shicho. Obihiro.

Tokachi Blanch Office 1983 Kukominami A Iseki [Kukominami A Site]. Tokachi Shicho. Obihiro.

Tokachi Research Institute of Archaeology, 1989 Obihiro-shi Akatsuki Iseki no Hakkutsu Chosa: Dai 5 Ji Chosa Hokokusho [A Report on the 5th Excavation at the Akatsuki Site, Obihiro City]. Tokachi Kokogaku Kenkyujo, Obihiro. (In Japanese)

Vasilevski, A. A. 2005 Periodization of the Upper Paleolithic of Sakhalin and Hokkaido in the Light of Research Conducted at the Ogonki-5 Site. In A. P. Derevianko (ed.) The Middle to Upper Paleolithic Transition in Eurasia: Hypothesis and Facts. Institute of Archaeology and Ethnography Press, Novosibirsk.

Vasilevski, A. A. 2006 The Upper Paleolithic of Sakhalin Island. In S. M. Nelson, A. P. Derevianko, Y. V. Kuzmin and L. B. Richard (eds.) *Archaeology of the Russian Far East: Essays in Stone Age Prehistory*, pp.75-100, BAR International Series 1540, BAR Publishing, Oxford.

Vasilevski, A. 2008a *Kamennuii Vek Ostrova Sakhalin* [Stone Age of Sakhalin Island]. 408p. Sakhalinskoe Knijhoe Izdatel'stvo. Yujino-Sakhalinsk

Vasilevski, A. 2008b Sakhalin niokeru Mammoth Dobutsugun to Jinrui no Tekio [Mammoth Fauna and Human Adaptation in Sakhalin]. In H. Sato (ed.) *Human Ecosystem Changes in the Northern Circum Japan Sea Area (NCJSA) in Late Pleistocene*, pp.44-67. Research Institute for Humanity and Nature. Kyoto. (In Japanese and Russian)

Yamada, S. 2006 *Hokkaido niokeru Saisekijin Sekkigun no Kenkyu* [A Study of Microblade Assemblages in Hokkaido, Japan] 244p. Rokuichi Shobo. Tokyo. (In Japanese)

Yamahara, T. 1996 Hokkaido niokeru Saisekijin Bunka Izen no Sekkigun nituite [On the Lithic Industry before Microblade Culture on Hokkaido]. *Obihiro Hyakunen Kinenkan Kiyo* 14, pp.1-28. (In Japanese)

Yoshizaki, M. 1961 Shirataki to Hokkaido no Mudoki Bunka [The Shirataki Site and Pre Ceramic Culture on Hokkaido]. *Minzokugaku Kenkyu*, 26:13-23. (In Japanese)

Pioneer Phase of Obsidian Use in the Upper Palaeolithic and the Emergence of Modern Human Behavior in the Japanese Islands

Kazutaka Shimada

Meiji University Museum, 1-1, Kanda-surugadai, Chiyoda Ward, 101-8301, Tokyo, Japan E-mail: shimameiji@hotmail.co.jp

Abstract: *This paper focuses on the beginnings and subsequent changes of obsidian use during the Eary Upper Palaeolithic (ca. 28-33ka 14C BP) in the central Japanese Islands, as viewed from the emergence of modern human behavior. The discovery of obsidian sources in the initial stage of Upper Palaeolithic both in high-altitude mountainous ranges and on the ocean islet is discussed, as well as on the development of obsidian conservation systems through time in Central Japan. Representative circular settlement structures of the pioneer phase of obsidian use have been intensively discussed with emphasis on obsidian procurement and exchange networks. It is suggested that the pioneer phase of obsidian use, reflecting the discovery and development of obsidian outcrops, is the first archaeological manifestation of modern human behavior. The appearance of modern human behavior varies from region to region associated with different palaeoenvironmental conditions. As such, the early acquisition of obsidian in the middle of OIS3 of the Japanese islands appears to be one of the characteristic features of successful regional adaptation of modern humans.*

***Keywords**: obsidian, modern human behavior, initial Upper Palaeolithic, Japanese islands*

Introduction

In archaeological studies of Oxygen Isotope Stage (OIS) 3 in Japan, one of the most important issues concerns the timing and identity of the first colonizers of the Japanese Islands. The process of migration and dispersal of hominids in the Japanese Islands is also of great interest. The exposure of "the Japanese Early and Middle Paleolithic forgeries" (Japanese Archaeological Association 2003) in the year 2000 set the study of the Lower and Middle Palaeolithic in Japan back by over twenty years almost to its starting point. Since then, Japanese researchers have re-initiated the pursuit of earlier sites belonging to the Lower or Middle Palaeolithic. Contrary to the rarity of sites prior to ca. 40 ka, over 14,000 Upper Palaeolithic sites are known throughout Japan except on the Ryukyu Islands at southernmost part of the Japanese Islands (Japanese Palaeolithic Research Association 2010). At present, the initial Upper Palaeolithic is tentatively defined by the first appearance of human records in the Kanto Region, Central Japan and the Kyushu Region in the southern part of Japan dating 30-33ka ^{14}C BP, along with a rapid increase in the number of sites and the beginning of long distance transportation of lithic raw materials.

A trait list for the technological, ecological, economical, and cognitive abilities of modern humans in the European Upper Palaeolithic and the African Middle Stone Age seems to be well established by the archaeological records and make a clear contrast to previous hominid populations (Mellars 1991, McBrearty and Brooks 2000). Interestingly, in Australia, which is the southernmost periphery for modern human dispersal and one of the endpoints in the last glacial period, a part of the typical trait list, especially relating to lithic technology and subsistence economy, cannot be taadopted directly (Mellars 2006, Habgood and Franklin 2008, Stern 2009). This may also be the case concerning the emergence of modern human behavior in the Japanese Islands, which should be considered as a particular set of socioeconomic activities (Stern 2009) within a given regional context. In this paper, a hypothesis is put forth that the global movement of *Homo sapiens* which resulted in the settlement of the Japanese Islands triggered the beginning of obsidian use in Central Japan and the emergence of modern human behavior.

Issues on the Middle Palaeolithic prior to 40 ka

In recent years, some archaeological sites with dates preceding 40 ka have been reported in the Kyushu Region situated close to the Korean Peninsula. The lithic assemblage of Layer 3b of the Iriguchi C 713 Site (Shiotsuka 2004) in Nagasaki Prefecture has been dated later than 90±11 ka by the OSL method (Nagatomo and Shitaoka 2004). The Sozudai Site in Oita Prefecture was excavated in 1953, 1963, 1964, and a re-excavation was conducted in 2001 (Fig. 1). Yanagida *et al.* (2007) concluded that a large quantity of the findings, in addition to a likewise large quantity of andesite gravels, found in Layer 5 fell within the time period between 70-80 ka based on tephra analysis and the typology of crude forms of the findings. The OSL date of Layer 5, however, indicates an age of 27±8

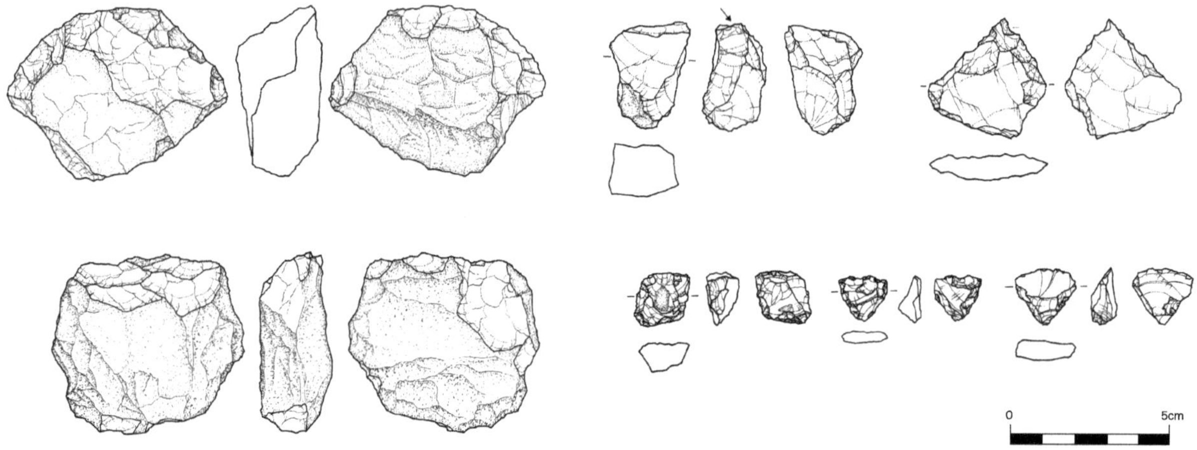

The Sozudai Site, Layer 5 in Oita Prefecture, the Kyushu Region (after Yanagida and Ono 2007)

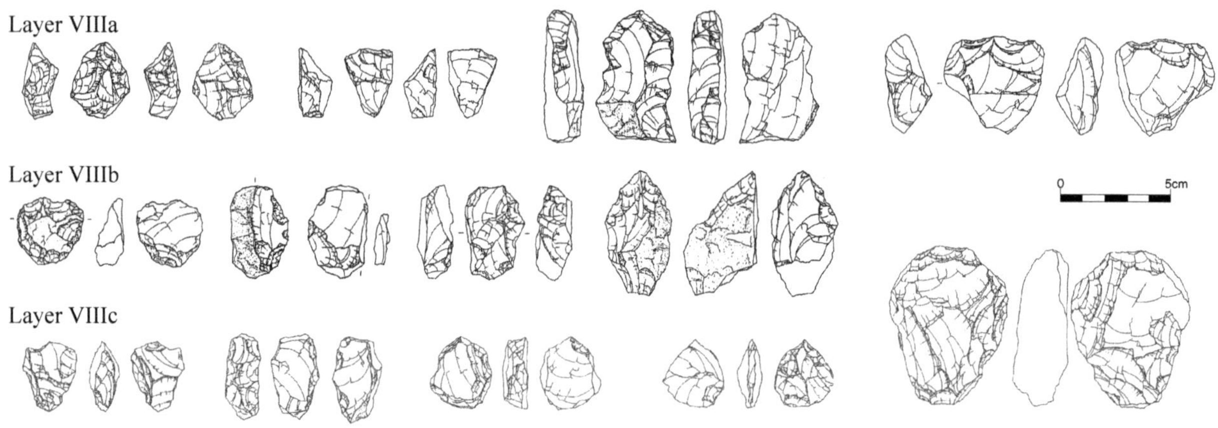

The Ono D Site in Kumamoto Prefecture, the Kyushu Region (after Kitamori 2003)

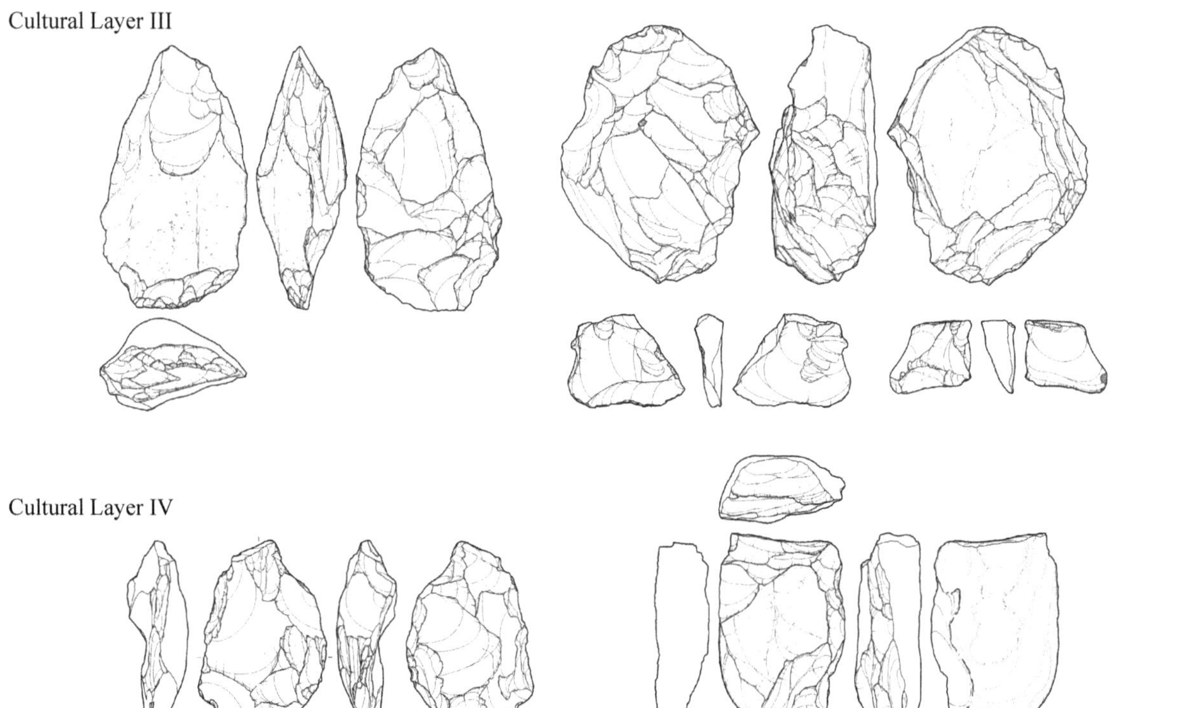

The Kanedori Site in Miyagi Prefecture, the Tohoku Region (after Kuroda *et al.* 2005)

Fig. 1 The Middle Palaeolithic assemblages in Japan, excavated or re-evaluated after the year of 2000.

ka (Nagatomo and Shitaoka 2007). In the Usiromuta Site, Miyazaki Prefecture, the Layer below the Kr-Iw (Kirishima-Iwaokoshi) tephra yielded Cultural Layer IV and Cultural Layer V, of which the dates were estimated to be ca. 40 ka and between 40-90 ka, respectively (Tachibana *et al.* eds. 2002). In the Ono D Sites in Kumamoto Prefecture, 2,585 artifacts along with 7,758 natural stones were discovered throughout six layers VIIa, VIIb, VIIIa, VIIIb, VIIIc and VIIIe (Fig. 1). In the upper part of the artifact-bearing layers, trapezoids were discovered, which are diagnostic of the early Upper Palaeolithic. The excavators concluded that the upper limit of the date for the findings yielded from the lower part of the cultural layers was ca. 60 ka based on comparison with layers dated by OSL in Locality E of the Ono Site near Locality D (Kitamori 2003). Some researchers recognize these sites in Kyushu as archaeological belonging to the Middle Palaeolithic, and a few chronological sequences concerning the Middle/Upper Palaeolithic transition have been proposed (Sato 2002, Hagiwara 2004, Yanagida *et al.* ibd.). In the Kanto Region (Central Japan), the Tsurugaya-higashi Site in Gunma Prefecture was excavated by Tohoku University in 2003. Serizawa *et al.* (2006) argued in the preliminary report that based on the tephra analysis, the findings were between 40-80 ka. These recent studies on sites older than 40 ka represent an effort to discover a Middle Palaeolithic record and to reconstruct its archaeological framework within Japan. Presently, it is generally considered, however, that more discussions are required of the accuracy of dates and whether the findings are true lithic artifacts (Okamura 2008, Yamada 2008).

The Kanedori Site in Iwate Prefecture, located in the northern part of Japan, was excavated in 1985 (Kikuchi *et al.* 2002), and again between 2003 and 2004 (Kuroda *et al.* 2005) (Fig. 1). Based on tephra analysis and OSL and radiocarbon dating, the excavators concluded that the date of Cultural Layer III falls within the time range of 35-50 ka, while Cultural Layer IV is estimated to date between 50-90 ka (Kuroda 2005). It seems to be generally accepted that the findings of Cultural Layer III and IV of Kanedori are true lithic artifacts as these assemblages included refits. As the author indicated in the report, however, these date estimations should be treated with caution because although the Aso-4 tephra (ca. 85-90 ka), which was discovered in the lower part of Layer 4a, was used for estimating the upper limit of the time range of Cultural Layer IV, it is represented by only one grain (Yagi 2005). In addition, the radiocarbon dates are inconsistent with the stratigraphic order relating to Cultural Layer III and overlying Cultural Layer II (Kuroda ibd.). In 2009, a new radiocarbon date of 46,480±710 ^{14}C BP (IAAA-81798) was obtained on charcoal from Layer IIIc which yielded Cultural Layer III in 1985 (Institute of Accelerator Analysis 2009).

In the Upper Palaeolithic, sites in the same horizon ordinarily show a pattern of distribution in which they are mapped along a river basin or comprise site clusters (Fig. 3). By contrast, each controversial site prior to 40ka have been discovered as isolates within the area and demonstrate unique characteristics independently in terms of lithic raw materials, conditions of the location, lithic artifact morphology, and patterns of lithic distribution in a site. Also regional variability is uncertain. The understanding of the framework of the Middle Palaeolithic still remains a subject of substantial debate in Japan.

Initial Stage of the Upper Palaeolithic in Japan

Based on tephrochronology and oxygen isotope chronology, Machida (2005) estimated a date of ca. 40 ka for Layer X and its lithic industry at the bottom of the Tachikawa loam layers. Lithic industries of the last glacial period excavated between Layer III to Layer X on the Musashino Upland have been arranged in chronological order. For example, Suwama (1988) divided the sequence of lithic industries on the Sagamino Upland, Kanagawa Prefecture, into twelve archaeological stages. Based on these chronologies, a regional change of lithic technologies and of residential patterning has been studied in detail. Based on a database assembled by the Japanese Palaeolithic Research Association, the total number of Palaeolithic sites, including a few sites dating before 40 ka, is at least 14,542 as of 2010 (Japanese Palaeolithic Research Association 2010).

The lithic industry from Layer X was characterized by flake tools such as trapezoids, blades with retouched bases, scrapers, heavy-duty tools such as bifacial stone axes with polished edges, choppers and chopping-tools (Figs. 2, 6). Some techniques of core reduction for blades or elongated flakes appeared in the prehistoric stage represented by Layer X and were further developed in the stage represented by Layer IX (Kunitake 2004). In addition, circular settlements, in which lithic clusters are arranged in circular layouts of ca. 10 to 50 m in diameter (Figs. 8, 9) emerged at the end of the Layer X Stage or in the beginning of the Layer IX Stage (Hashimoto 2003).

Archaeological records parallel to the Layer X Stage have also been discovered in other regions of Japan. In the Kyushu Region, the following sites can be tentatively assigned to this earliest stage of the Upper Palaeolithic; Shizume (Kisaki 2002), the lithic assemblage from Layer VIa of Ishinomoto, Loc. 8 (Ikeda 1999: 24-121), Cultural Layer 1 of Mimikiri C (Murasaki 1999: 150-158), the lithic assemblage from a lower part of Layer VI of Uwaba (Iwasaki *et al.* 2007: 152-167) in Kumamoto Prefecture; Cultural Layer I of Yamada (Akasaki 2007: 13-37), Cultural Layer III of Ushiromuta (Tachibana *et al.* ed. 2002) in Miyazaki Prefecture; and the lithic assemblage from Layer XIII below the Tane-4 tephra (ca. 30 ka) of Tachikiri (Tabira 1999: 45-100) and Cultural Layer I of Maeyama (Samukawa *et al.* 2007: 23-50) in Kagoshima Prefecture. The artifact-bearing layer of Ishinomoto provided dates of 32,740±1060 ^{14}C BP (Beta-84289), 33,720±430 ^{14}C BP (Beta-84290), 33,140±550 ^{14}C BP (Beta-84291), and 31,460±270 ^{14}C BP (Beta-84292) (Palaeoenvironmental Research Institute 1999c). Also in the Tohoku Region,

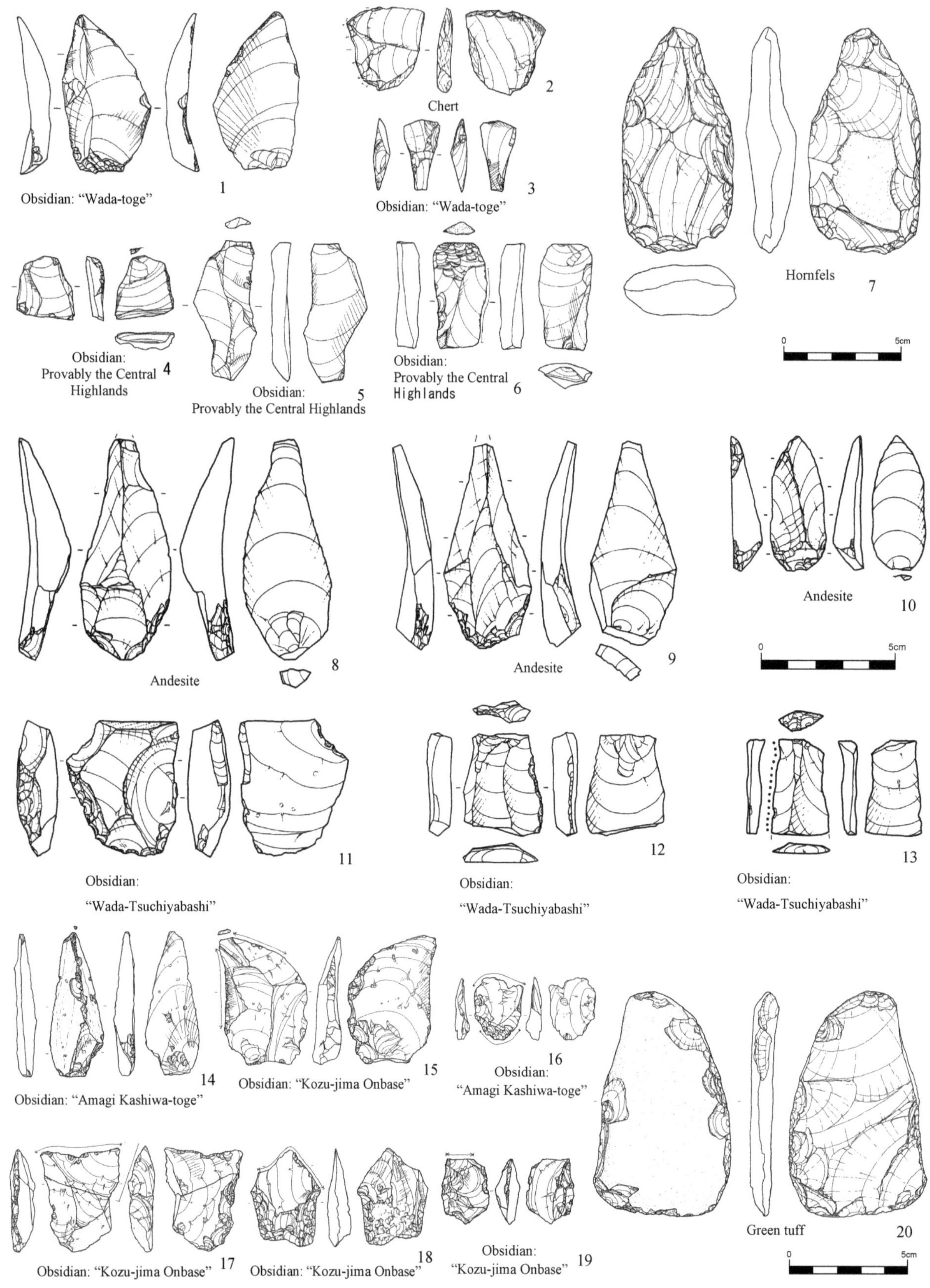

Fig. 2 The lithic assemblages in the Layer X Stage of the Tachikawa Loam sediments (the initial mode of obsidian use in the Kanto plain and its surroundings). 1-7: Nishidaigotoda Site (Fujinami et al. 1999); 8-13: Happusan II Site (Suto 1999); 14-20: Nishibora Site (Sasahara 1999); 1, 8-10, 14: Blades with retouched base; 2, 3, 15-19: Trapezoids; 7, 20: Bifacial stone axes; 11, 12: Side-scrapers; 4-6,13: Blades or elongated flakes.

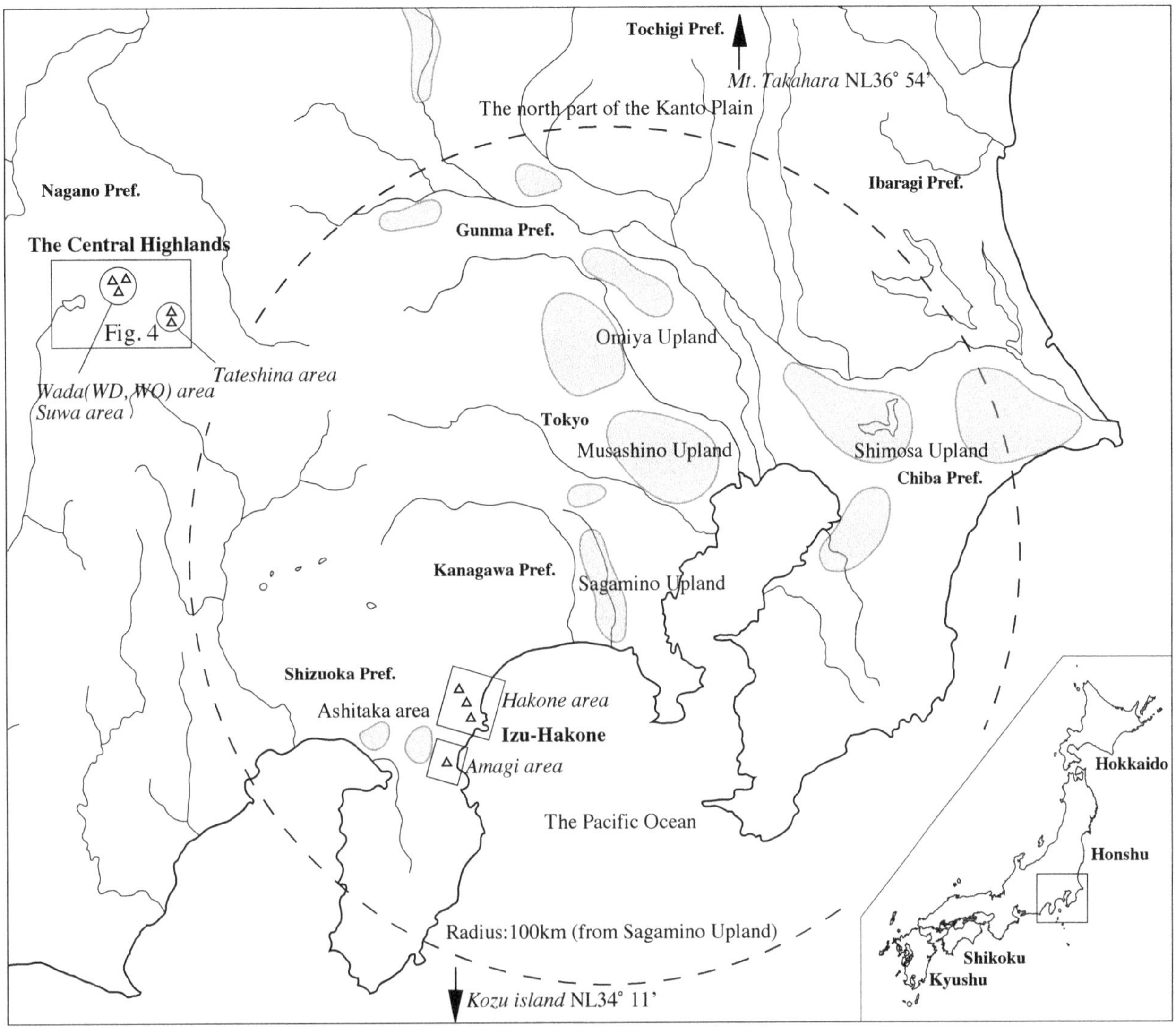

Fig3. Distribution of major Upper Palaeolithic site clusters and obsidian sources in the central part of Japan.: obsidian sources, : site clusters.

Cultural Layer III of Kanedori seems to be most likely parallel to the Layer X Stage. These sites demonstrate a distinctively wider range of distribution within each region than those dating prior to 40 ka.

The Japanese Islands are situated in the easternmost periphery of Eurasia. From the viewpoint of the dispersal of modern *Homo sapiens* from Africa to East Asia, the Japanese Islands can be recognized as an important area for the search for the emergence of modern human behavior (Kaifu 2005:156). For modern humans who reached the Japanese Islands, it might well have been the endpoint of their journey. The earliest anatomically modern human fossils, infant bones, were discovered in Yamashitacho 1 Cave in Okinawa Island, part of the Ryukyu Islands (Takamiya *et al.* 1975, Matsu'ura and Kondo 2000). The radiocarbon date obtained from a charcoal layer close to the cultural layer was estimated to be 31,200±1000 ^{14}C BP (TK-78) (Takamiya *et al.* ibd.). Although radiocarbon dates are roughly parallel, the relationship between these findings in Okinawa and the lithic industries of Layer X or Layer IX stages in the mainland are not obvious. There still has been no human fossil evidence directly indicating that anatomically modern humans were responsible for the Layer X Stage.

Two fundamental questions arose on the Middle and the initial Upper Palaeolithic in Japan. First, when did traces of modern human behavioral patterns appear in the archeological record? Second, how can we recognize them? For answers to these questions, I examine human behavior related to the beginning of obsidian use in the following sections. The exploitation of obsidian and distribution of obsidian artifacts over 100km from the sources have already been observed in Central Japan in the initial Upper Paleolithic dating to ca. 30-33 ka ^{14}C BP.

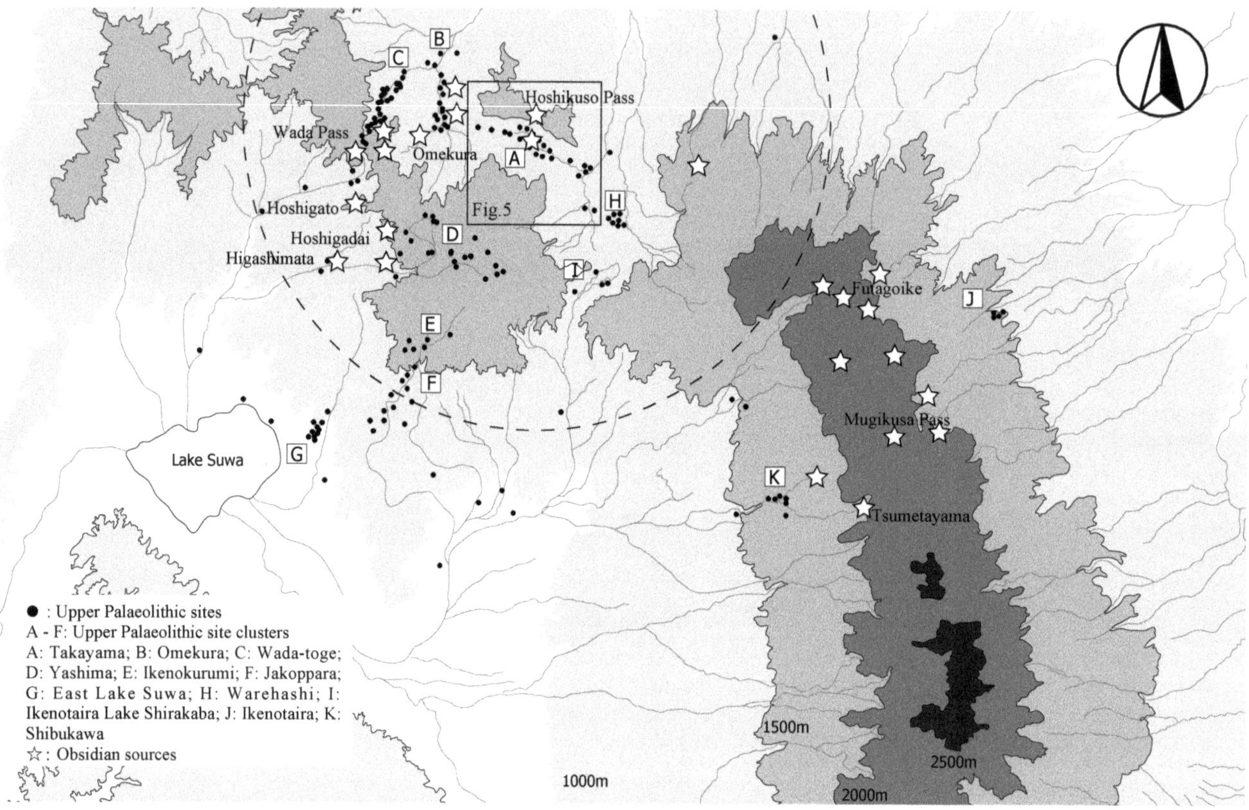

Fig. 4 Distribution of Upper Palaeolithic sites and obsidian sources in the Central Highlands.

Obsidian Sources in Central Japan

The high-density distribution of Upper Palaeolithic site clusters on the Kanto Plain and surrounding area (Fig. 3) demonstrates that this region was suitable for habitation in the last glacial period. Obsidian sources (Fig. 3, Tab. 1) are located in the Central Highlands of the Nagano Prefecture, as well as on Mt. Takahara in the Tochigi Prefecture, which is located more than 100 km from the Kanto Plain. In addition, a source of obsidian on Kozu Island in the Pacific Ocean is also known. Kozu Island is located 50 km from the tip of the Izu Peninsula, while sources in Izu-Hakone are relatively close to the Kanto Plain. Obsidian rocks transported from all of these sources were utilized throughout Central Japan for ca. 30,000 years throughout the Upper Paleolithic and Jomon Periods.

Particularly in the area of the Central Highlands (Fig. 4), there are numerous large-scale, high-density Upper Palaeolithic sites where obsidian procurement and lithic production took place. Mochizuki *et al.* (2006) presented a classification of obsidian sources using X-ray fluorescence analysis (XRF) (Tab.1). Thirty-two geological "sources" could be classified into 14 "groups" based on the chemical composition of the obsidian nodules, and several groups which are chemically close to each other comprise "areas", namely the "Suwa area", the "Wada (WD) area", the "Wada (WO) area" and the "Tateshina area". These obsidian sources and sites are situated in the mountainous area ca. 1,200m-2,000m above sea level.

Beginning and Changes in Obsidian Use

In this section, changes in the use of obsidian observable in the cultural sequence between the stages of Layer X and Layer IX in Central Japan are reviewed. In addition, some behavioral aspects of Palaeolithic toolmakers are examined.

1) The Initial Mode of Obsidian Use

According to Ito (2006) and Nakamura (2006), there are 58 excavated assemblages assigned to the Layer X Stage on the Musashino Upland, of which nine have yielded obsidian artifacts. The number of obsidian artifacts in each site is as follows: 170 pieces out of 1,889 artifacts (8.9 percent) in the lithic assemblage of Layer Xa of the Musashidai Site (Hayakawa *et al.* eds. 1984: 29-64; Itoh ibd.: Table 2); 18 pieces out of 274 artifacts (6.5 percent) in Cultural Layer 1 of the Tamaranzaka Site, Loc. 5 (Kamishikiryo 1999: 40-66); 8 pieces out of 119 artifacts (6.7 percent) in the lithic assemblage of Layer Xa of the Nishidaigotoda Site (Fujinami *et al.* 1999: 117-185) (Fig. 2); 2 pieces out of 15 artifacts (13 percent) in Cultural Layer 4 of the Shimayashiki Site (Kobayashi at al. 1998: 104-121); and 5 pieces out of 5 artifacts (100 percent) in Cultural Layer III of the Takainari Site (Maechi *et al.* eds 1989: 66-67).

The ratio of obsidian artifacts averages ca. 10 percent in each case. These obsidian assemblages also indicate that obsidian was usually transported to the sites as stone implements or flakes. Although the evidence for

Table 1 An example of the category of obsidian sources by X-ray fluorescence analysis (Mochizuki et al. 2006).

<table>
<tr><th>Prefectures</th><th>«Areas»</th><th>«Groups»</th><th>«Sources»</th></tr>
<tr><td rowspan="14">Nagano</td><td rowspan="7">Wada (WD)</td><td>Takayama</td><td rowspan="7">Takayama, Kobukazawa, Higasi-mochiya, Fuyo-raito, Furu-toge, Tsuchiyabashi-kita, Tsuchiyabashi-nishi, Tsuchiyabashi-minami, Chojigoryo</td></tr>
<tr><td>Kobukazawa</td></tr>
<tr><td>Tsuchiyabashi-kita</td></tr>
<tr><td>Tsuchiyabashi-nisi</td></tr>
<tr><td>Tsuchiyabashi-minami</td></tr>
<tr><td>Fuyo-raito</td></tr>
<tr><td>Furu-toge</td></tr>
<tr><td rowspan="3">Wada (WO)</td><td>Budozawa</td><td rowspan="3">Budozawa, Budozawa-ugan, Makigasawa-ue, Makigasawashita, Takamatsuzawa</td></tr>
<tr><td>Makigasawa</td></tr>
<tr><td>Takamatsuzawa</td></tr>
<tr><td>Suwa</td><td>Hoshigadai</td><td>Hoshigato-daiichi-koku, Hoshigato-daini-koku, Hoshigadai A, Hoshigadai B, Suigetsu-reien, Suigetsu-koen, Hoshigato-norikoshi</td></tr>
<tr><td rowspan="3">Tateshina</td><td>Tsumetayama</td><td>Tsumetayama, Mugikusa-toge, Mugikusa-toge-higashi, Shibunoyu, Utsukusimori, Yatsugatake7, Yatsugatake 9, Futago-ike</td></tr>
<tr><td>Futagoyama</td><td>Futago-ike</td></tr>
<tr><td>Suribachiyama</td><td>Suribachi-yama, Kiko-ike</td></tr>
<tr><td rowspan="4">Kanagawa</td><td rowspan="5">Hakone</td><td>Ashinoyu</td><td>Ashinoyu</td></tr>
<tr><td>Hatajuku</td><td>Hatajuku</td></tr>
<tr><td>Kuroiwabashi</td><td>Kuroiwabashi</td></tr>
<tr><td>Kajiya</td><td>Kajiya</td></tr>
<tr><td rowspan="2">Shizuoka</td><td>Kamitaga</td><td>Kamitaga</td></tr>
<tr><td>Amagi</td><td>Kashiwa-toge</td><td>Kashiwa-toge</td></tr>
<tr><td rowspan="2">Tokyo</td><td rowspan="2">Kozu-jima</td><td>Onbase-jima</td><td>Onbase-jima, Nagahama, Sawajiriwan</td></tr>
<tr><td>Sanukazaki</td><td>Sanukazaki, Nagahama</td></tr>
</table>

tool production of obsidian is rare at given sites, it is observed at the Musashidai Site. Several source analyses by XRF revealed that two of the obsidian artifacts in the Nishidaigotoda site were identified as belonging to the "Wada-toge (Wada Pass)" group in the Central Highlands (Palynosurvey 1999) (Fig. 2). In addition, two obsidian samples from the Shimayashiki Site were identified as belonging to the "Kirigamine group" (probably from a source in Hoshigato) in the Central Highlands and the "Mount Takahara group 1" in Tochigi Prefecture (Fujine 1998).

Obsidian transported from Kozu Island on the Pacific Ocean has been identified in the sites located in the Ashitaka area (Shizuoka Prefecture). The Loam sediments called Layer BB (black-band)-VI and BB-VII in the Ashitaka area are thought to be parallel to Layer X in the Kanto Plain (Takao 2006). Layer BB-VI in the Nishibora Site yielded trapezoids and a stone axe (Sasahara 1999: 12-125) (Fig. 2). Three radiocarbon dates obtained from the artifact-bearing stratum are 30,390±230 ^{14}C BP (Beta-122045), 30,200±360 ^{14}C BP (Beta-122043), 29,690±210 ^{14}C BP (Beta-122044) (Palaeoenvironment Research Institute 1999b). These may indicate that the Nishibora assemblage belongs to the relatively late part of the Layer X Stage. The excavated obsidian artifacts represent 253 pieces out of a total of 831 artifacts. Of these, at least 86 obsidian artifacts were identified as originating from Kozu Island by XRF analysis (Sasahara ibd.: 119). In recent years, several assemblages yielded from BB-VII, which are assigned to the early part of the Layer X Stage, have also been excavated in the Ashitaka area. Hirose *et al.* (2006) preliminarily reported that in these sites, lithic artifacts made of obsidian transported from both the Central Highlands and Kozu Island were discovered together in a single site.

The Happusan II Site, which is situated at the midway point between the Kanto Plain and the Central Highlands, yielded two obsidian side scrapers and one obsidian flake in addition to 5,744 glassy andesite artifacts (Suto 1999: 5-281) (Figs. 2, 7). These obsidian artifacts were determined to belong to the "Wada-Tsuchiyabashi" group by XRF analysis (Mochizuki 1999). The Happusan II site is situated in a large-scale andesite source located ca. 1,000 m above present sea level. The radiocarbon dates of the artifact-bearing stratum are 32,240±260 ^{14}C BP (Beta-86230), 32,180±260 ^{14}C BP (Beta-86233), 32,190±260 ^{14}C BP (Beta-86232), 31,860±250 ^{14}C BP (Beta-86229) and 31,360±230 ^{14}C BP (Beta-86231) (Palaeoenviromental Research Institute 1999a). The evidence from this site indicates that toolmakers in the Layer X Stage explored for sources of andesite and obsidian at high altitudes.

Table 2 The composition of the lithic assemblage in Hinatabayashi B Site and Oiwake Site.

Site		*Axe*	*Trapezoid*	*Knife*	*Other Tool*	*Core*	*Flake and Chip*	*Raw material*	*Total*	*Reference*
Hinatabayashi B (N=8999)	Obsidian	0	37	2	510	154	5786	0	6489	Tsuchiya *et al.* 2000
	Agate	0	14	0	84	93	1727	0	1918	
	Serpentinite	71	0	0	2	0	84	1	158	
	Others	0	8	0	39	9	376	2	434	
Oiwake (N=936)	Obsidian	0	26	2	8	63	825	4	928	Otake *et al.* 2001
	Others	0	0	0	6	0	2	0	8	

Table 3 Obsidian sources in Hinatabayashi B Site and Oiwake Site.

Site	*Wada-Takayama*	*Wada-Kobukazawa*	*Wada-Tsuchiyabashikita*	*Suwa-Hoshigadai*	*Tateshina-Tsumetayama*	*Tateshina-Futagoyama*	*unmeasurable*	*Total*	*Reference*
Hinatabayashi B	2833	516	39	34	71	2	45	3540/6489	Mochizuki 2000 and Kobayashi 2001
Oiwake	32	5	-	2	-	-	-	39/928	

On the contrary, archaeological sites which belong to the Layer X Stage have not been discovered at sources of obsidian in the Central Highlands. It is therefore difficult to present a detailed mobility pattern between the source areas and the residential areas during this archaeological stage. As exemplified by the sites in the Musashino Uplands, the amount of obsidian consumed in the residential area was not large as a whole. Rather, the lithic production was substantially supported by local raw materials, which far outnumbered the production of obsidian implements (Fig. 6). This may imply that toolmakers had not established effective routes for obsidian procurement between the residential and source areas. Fig. 7 shows tentative links between them based on the above-mentioned XRF analyses.

2) The Established Mode of Obsidian Conservation

During the Layer IX Stage, circular settlements emerged in various parts of Japan (Hashimoto 2003), and the residential areas in which toolmakers utilized obsidian expanded. Several sites assigned to the Layer IX Stage and located at obsidian sources have been excavated in the Central Highlands. In this section, an archeological case study concerning the relationship between an obsidian source and a residential area in the Layer IX Stage, which used sites located in the Central Highlands and the lakeside of Nojiri-ko (Lake Nojiri) is examined.

The Hinatabayashi B Site (Tsuchiya and Tani 2000a: 23-104) is situated on the lakeside of Lake Nojiri, which is located ca. 80 km north of the Central Highlands in Nagano Prefecture. The spatial arrangement of the lithic clusters of Cultural Layer 1 displays a circular formation of ca. 30 m in diameter with high-density lithic clusters in the center of the circle (Fig. 8). Numerous refits indicate that the lithic clusters most likely existed simultaneously. These features are diagnostic of circular settlements. The radiocarbon dates of the artifact-bearing stratum are 31,420±280 ^{14}C BP (Beta-120859), 29,870±250 ^{14}C BP (Beta-120858), 29,820±250 ^{14}C BP (Beta-120864), 29,640±240 ^{14}C BP (Beta-120866) and 28,230±210 ^{14}C BP (Beta-120863) (Yamagata 2000). The total number of lithic artifacts consists of 8,999 pieces, over 70 percent of which is obsidian (Tab. 2). The list of obsidian sources revealed by XRF analysis for the obsidian assemblages clearly indicates that the overwhelming number of obsidian tools and debris are attributed to the "Wada-Takayama" and "Wada-Kobukazawa" groups of the Central Highlands (Mochizuki 2000).

The Takayama site cluster (Fig. 5) located in the Kirigamine mountains between 1,300-1,400 m above sea level includes the obsidian source of the "Wada-Takayama" group (Hoshikuso Pass) (Tozawa *et al.* eds. 1989). The Oiwake Site (Fig. 5), Cultural Layer 5, which belongs to the Takayama site cluster and is located close to the obsidian source, yielded a relatively small quantity of lithics including trapezoids (Otake *et al.* 2001: 280-319) (Fig. 8). The total number of obsidian artifacts represents 928 pieces

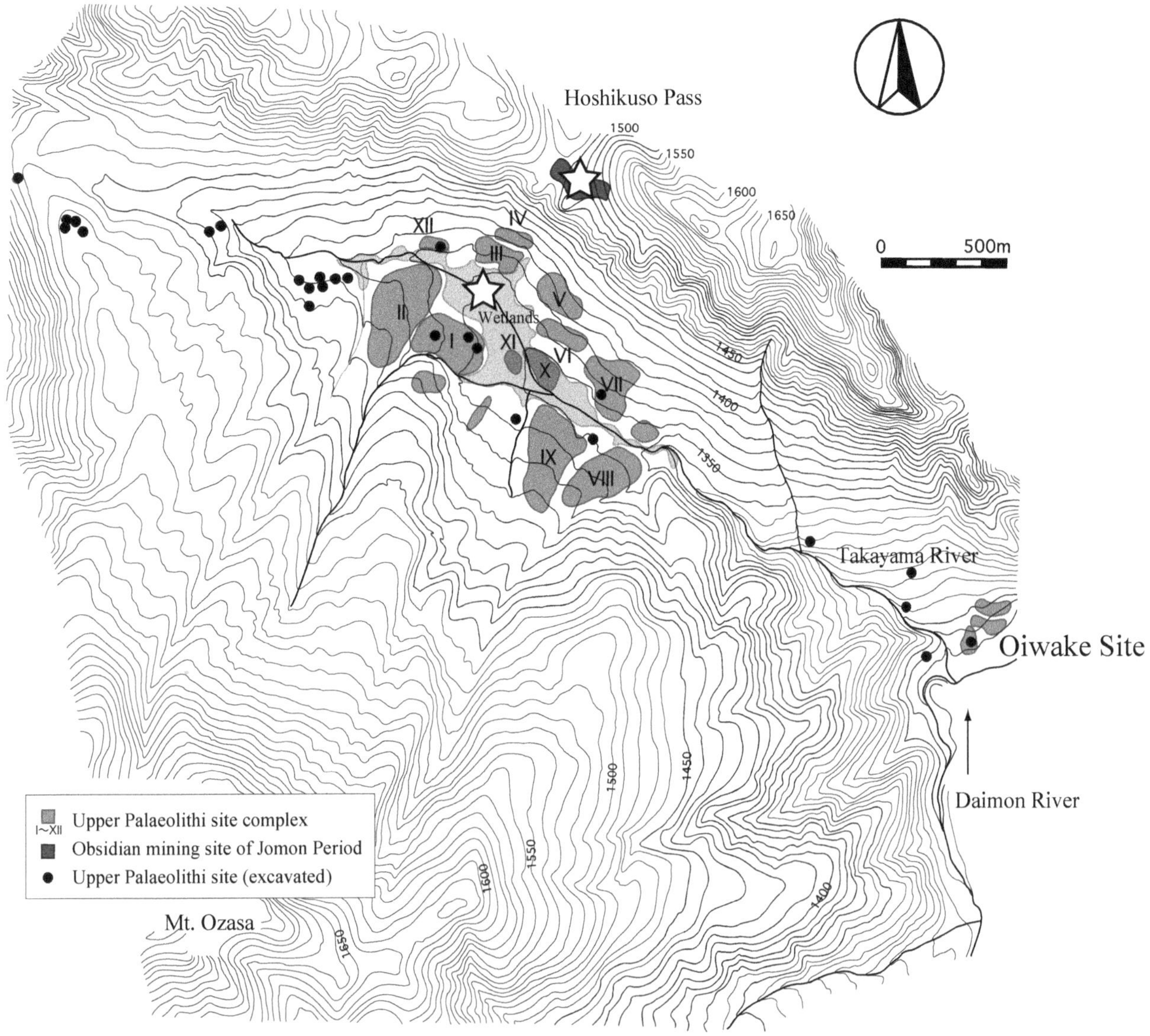

Fig. 5 Distribution of sites and obsidian sources in the Takayama Site Cluster. (Courtesy of Takayama Sites Research Group)

out of a total of 936 artifacts (Tab. 2). The radiocarbon dates of the artifact-bearing stratum are: 31,039±298 ^{14}C BP (TERRA-b030501c19), 30,635±296 ^{14}C BP (TERRA-b030501c23), and 29,306±248 ^{14}C BP (TERRA-b030501c20) (Yoneda 2001). The lithic assemblage shows the earliest trace of human activity in the Central Highlands.

When the obsidian sources were characterized by XRF analysis and compared, it was revealed that the Oiwake and Hinatabayashi B Sites have almost the same composition (Tab. 3). The trapezoids from these sites indicate similar typological features, in which bifacial retouch from both lateral sides of the flake is adopted to shape a trapezoidal lithic form. The radiocarbon dates are also very similar to one another. From these sites, an archaeological model for human activities between obsidian procurement in the source area and its consumption in the residential area can be reconstructed.

A combination of large- and small-sized settlements indicates that a small party dispatched from the circular settlement on the shore of Lake Nojiri in order to retrieve obsidian from the ground surface at Hosikuso Pass or the wetlands (Fig. 5). In the Hinatabayashi B Site, 844 pieces of obsidian flakes have cortex on their dorsal surface, which accounts for 42.5 percent of the 1,985 obsidian flakes. It is likely that unprocessed obsidian nodules were directly transported to the residential area near Lake Nojiri. From this treatment of obsidian, small-scale rather than large-scale lithic production would be expected to be seen in the source area. In agreement with this speculation, a relatively small lithic assemblage is present at the Oiwake Site. This site can be considered a temporal hunting camp located on the mountain containing the obsidian source. The relatively small-scale production of obsidian hunting gear at the Oiwake Site indicates that toolmakers exploited animal resources during their stay for obsidian procurement.

The XRF analyses identified source composition of obsidian in the circular settlements comprising the site cluster at

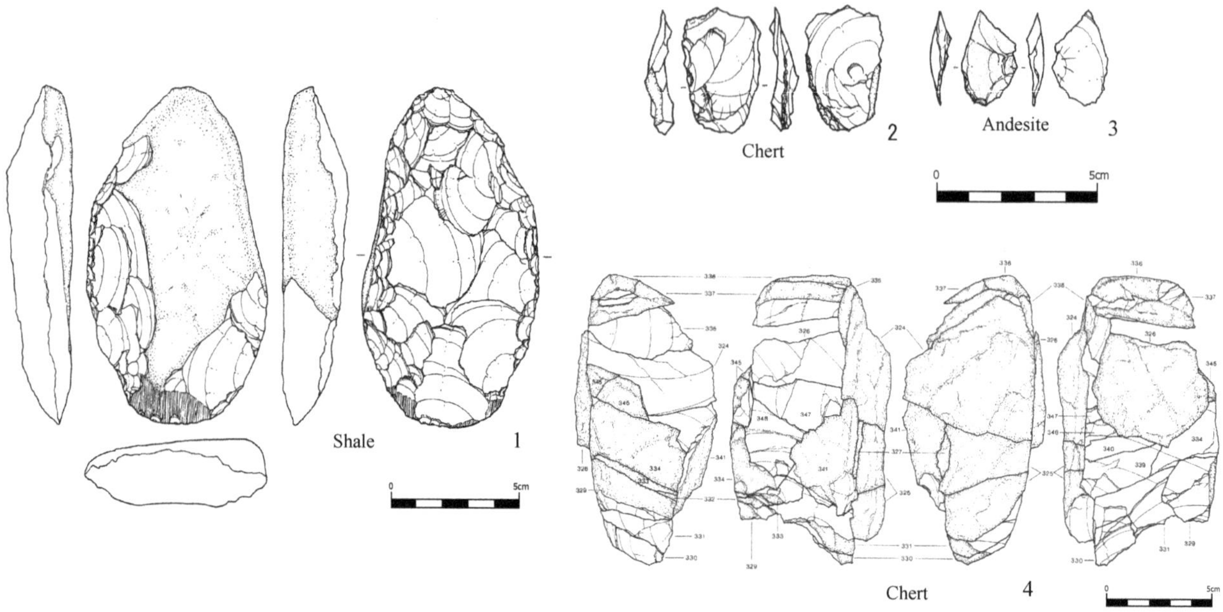

Fig. 6 The lithic assemblage of the Seta Site, Cultural Layer 8 utilizing local raw materials in the initial mode of obsidian use, the Musasino Uplands (Terada et al. 1997). 1: Bifacial stone axe with polished edge; 2, 3: Trapezoids; 4: Refits of lithic artifacts.

the Lake Nojiri site group. The results demonstrated that toolmakers made not only full use of the source at Hoshikuso Pass ("Wada-Takayama" group), but also obtained obsidian from other sources dispersed throughout the Central Highlands. For example, of the 2,819 obsidian artifacts from the circular settlement at the Kannoki Site, Loc. 3 (Tsuchiya and Otake 2000: 109-146), which is located close to the Hinatabayashi B Site, 1,051 pieces are attributed to the "Suwa-Hoshigadai" group and 483 pieces belong to the "Wada-Takayama" group (Mochizuki and Henmi 2000).

Of particular interest is what the type of human activity was undertaken in association with obsidian reduction at the Hinatabayashi B Site. Several characteristics of behavior can be interpreted from patterns of refits distribution in the site (Fig. 8). First, a notable concentration of refitted obsidian artifacts was demonstrated inside the circular periphery zone (N=130). This indicates that the lithic clusters in the central zone represent the center for the knapping obsidian nodules. Second, the refits were also frequently observed between the lithic clusters in the central zone and those in the circular periphery zone (N=65). This suggests that the central zone was strongly linked to the circular zone through the delivery and receiving of the cores and/or the lithic blanks. Specifically, the obsidian nodules were knapped in the central zone, and then the cores and/or the lithic blanks were shared amongst the members of the settlement. An important third detail concerns the location of the 54 refits among the lithic clusters in the circular periphery zone. If it is valid to assume that a lithic cluster represents a Palaeolithic hut-like feature and a group structure, this third type of refit pattern also indicates that toolmakers shared obsidian blanks amongst themselves. In this way, it can be postulated that a system of obsidian conservation by residential groups emerged in the Layer IX Stage, likely in conjunction with the emergence of circular settlements in Central Japan. It cannot be assumed, however, that the social role of circular settlements was only limited to the joint control of obsidian consumption. Previous studies have suggested that the settlement functioned as activity areas such as large mammal processing, rituals or meetings among residential groups (Kosuge 2005). It is reasonable to conclude that obsidian conservation constituted one part of complicated social activities that occurred in circular settlements.

In addition to the joint control of obsidian use in circular settlements, the complexity of toolmakers' behavior with respect to obsidian increased during this archaeological stage. The Kamibayashi Site, Cultural Layer 2, located in the northern part of the Kanto Plain in Tochigi Prefecture, yielded an obsidian assemblage (382 of 3,540 lithic artifacts) comprised of obsidian originating from almost all source areas in Central Japan along with the lithic artifacts of local raw materials such as chert (Idei *et al.* 2004). Also the Kamibayashi Site is one of the circular settlements (Fig. 9). Sources are attributed to "Suwa-Hoshigadai" group (314 pieces/375 pieces), "Tateshina-Tsumetayama" group (21/375), "Wada-Takayama" group (6/375), "Wada-Kobukazawa" group (3/375), "Tateshina-Futagoyama" group (1/375), "Hakone-Hatajuku" group (2/375), "Amagi-Kashiwatoge" group (1/375), "Takaharayama-Amayuzawa" group (17/375) and "Kozu jima-Onbase" group (10/375) (Idei *et al.* ibd.: 609-620). It is unreasonable to assume that the members of the settlement traveled long distances to amass all of the obsidian by themselves. Additionally, several sites indicate that the circular settlements that depend on obsidian from the Central Highlands or Mt. Takahara for stone tool production in proportion of around 50% or more occurred in the Kanto Plain. For example, the proportion of obsidian artifacts in the Seiganji-maehara

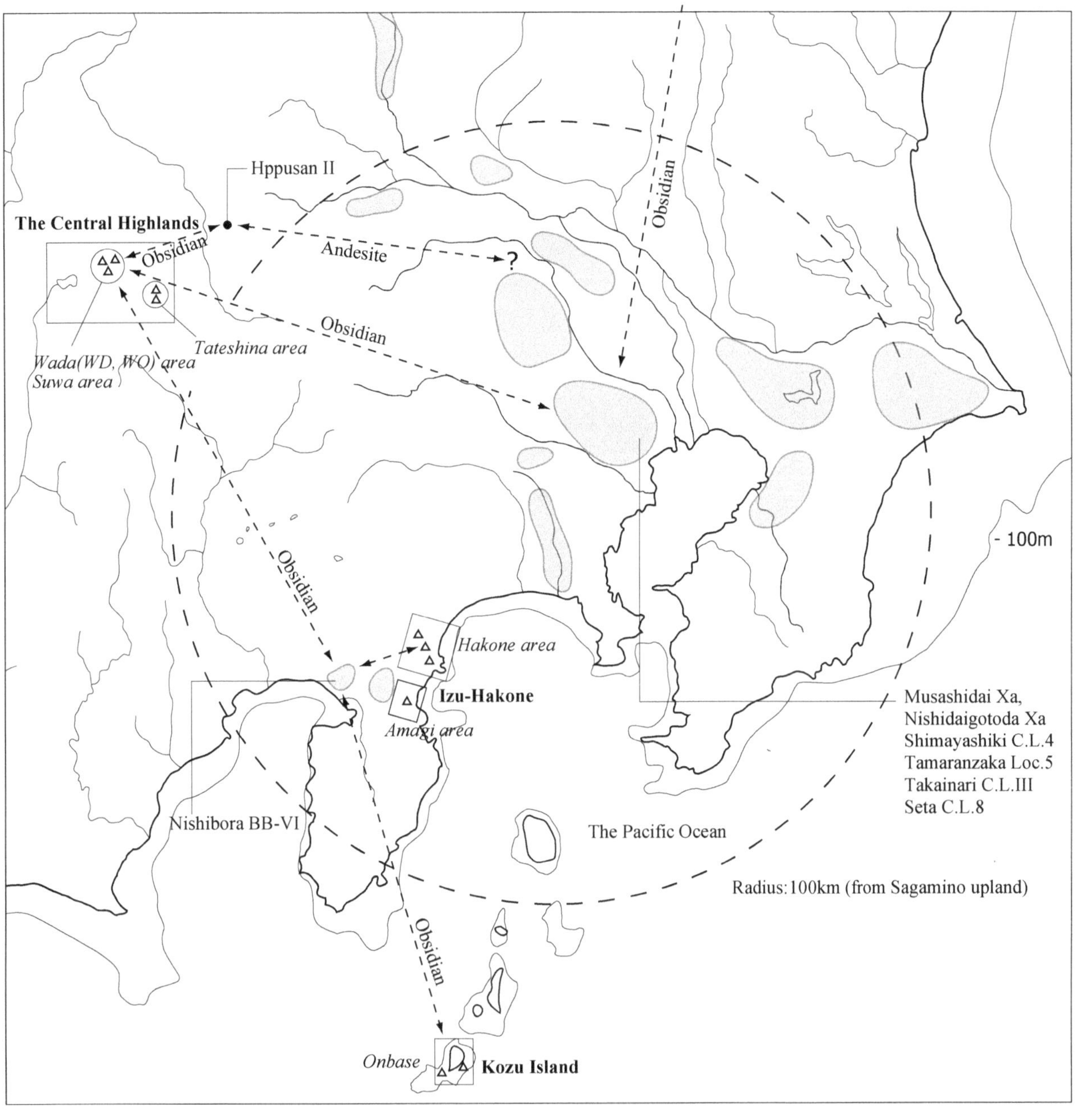

Fig. 7 The provisional links between obsidian sources and residential areas in the initial mode of obsidian use.

Site in the Saitama Prefecture (Nishii 2009), the Kojo 1C Site (Daikuhara 1988) and the Sanwa-kogyodanchi I Site (Tsushima 1999) in the Gunma Prefecture indicate 88% (1293 pieces/1472 pieces), 54% (283/523) and 46% (622/1340) respectively. These sites demonstrate that the toolmakers in this stage established an extended network for the transportation of obsidian gathered from various sources dispersed over a large area in Central Japan.

Pioneer Phase of Obsidian Use

From a regional perspective, layers below Layer X and corresponding layers in the Kanto Plain have not yielded any evidence of a lithic industry (Suwama 2006). As stated in the previous section, the existing data indicates that obsidian was obtained from various sources in the mountain ranges and the Pacific Ocean (Kozu Island) around the Kanto Plain (Fig. 7). The evidence also suggests that they were first discovered by the toolmakers of the Layer X Stage in approximately the same time period.

Specifically, toolmakers in the initial mode of obsidian use substantially exploited local raw materials for lithic production in the residential area of the Kanto Plain. At the same time they could access the sources of glassy andesite located ca. 1,000m above present sea level outside of the Kanto Plain. Moreover, they also detected obsidian sources located far distances from their principal areas of residence. It can be considered that such activities represented a comprehensive survey of both non-local and local raw material over a vast area. However, it is unlikely that the purpose of the surveys was only limited to exploiting these raw materials.

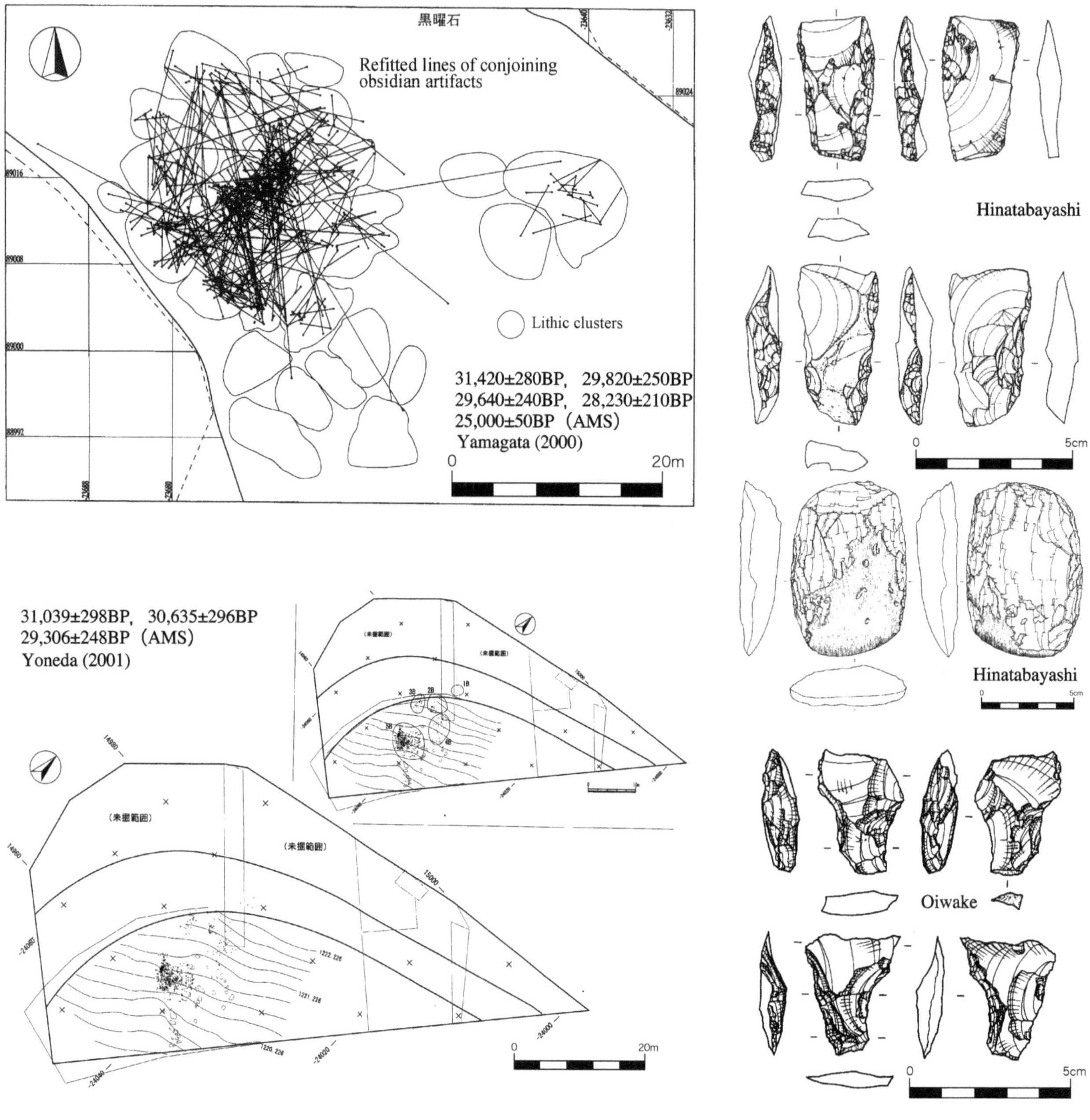

Fig. 8 Hinatabayashi B Site (upper: Tsuchiya et al. 2000) and Oiwake Site, Cultural Layer 5 (lower: Otake et al. 2001).

Based on this speculation, it is possible to formulate a scenario in which the population movement to the uninhabited Kanto Plain immediately resulted in intensive survey for natural resources in order for newcomers to secure their survival, focusing on a variety of foodstuffs or animal migration routes, as well as on lithic raw materials. Namely, the promotion of resource surveys may have led to the unexpected discovery of obsidian sources by toolmakers. The important thing is that the resource surveys covered high-altitude mountain ranges and the ocean. Obsidian in the initial mode of use might not have been treated as a significant lithic raw material, but rather as one indicator of potential habitability encountered by toolmakers in unpopulated lands.

In this way, the beginning of obsidian use in Central Japan coincided with a rapid increase in the number of Upper Palaeolithic sites covering a wide area of Japan. It is appropriate to refer to the archaeological stages of Layer X and Layer IX in this region as the "pioneer phase of obsidian use".

Emergence of Modern Human Behavior from the Perspective of Obsidian Archaeology

The initial mode of obsidian use in the Layer X Stage transitioned into the established mode of obsidian conservation in the Layer IX Stage. Although the development of blade technique gradually occurred, no discontinuity between both stages in terms of lithic technology producing trapezoids and stone axes is observable. Thus it is reasonable to assume that a

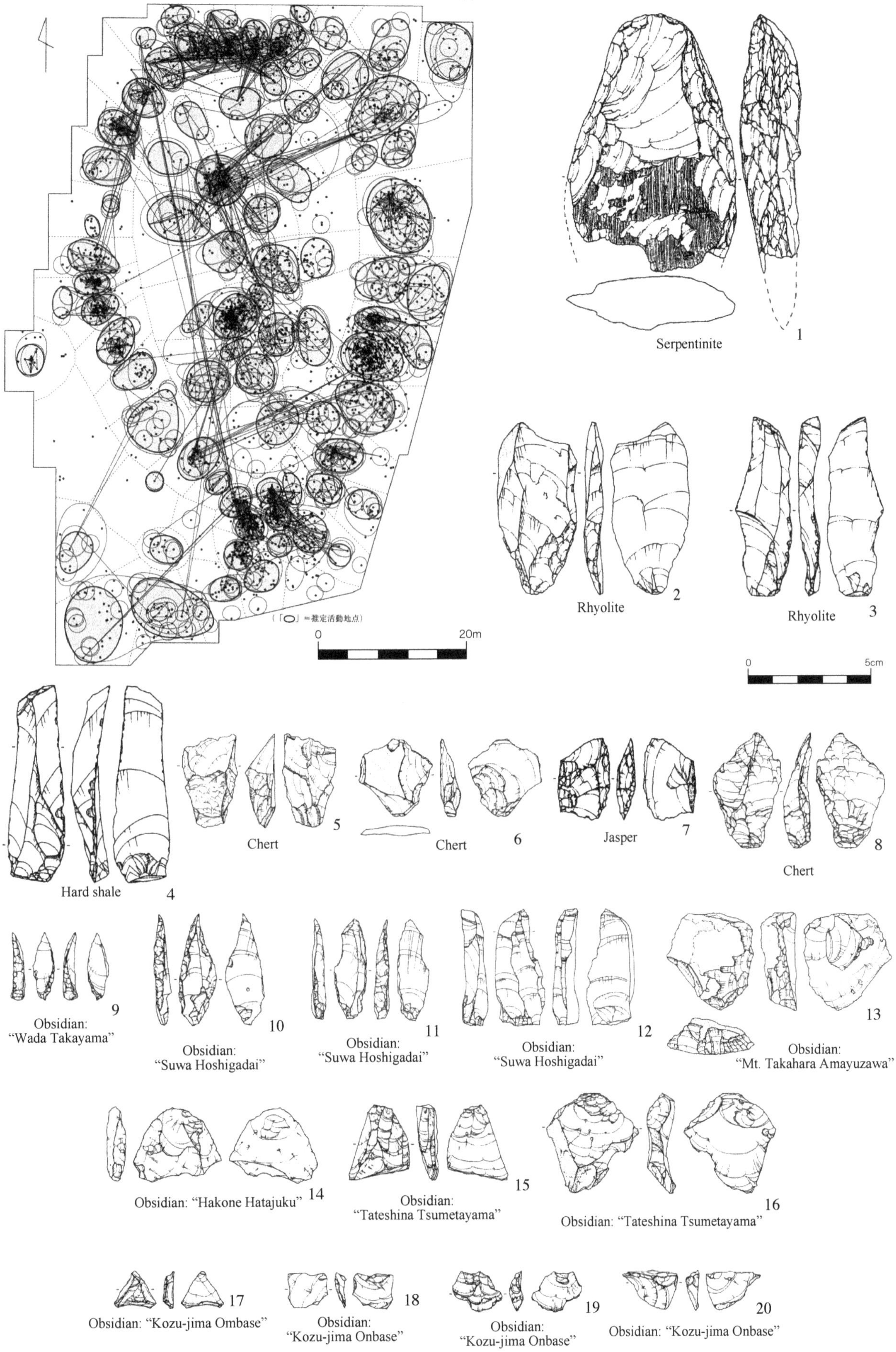

Fig. 9 The circular settlement of the Kamibayashi Site, Cultural Layer 2 in Tochigi Prefecture (after Idei et al. 2004).

population which continually utilized obsidian in successive generations was responsible for the changes relating to the conservation of obsidian as a non-local raw material.

Toolmakers in the pioneer phase of obsidian use possessed equipment and survival techniques which allowed them to travel to areas near or beyond the timberline ca. 1,500 m above sea level during the last glacial period. The transport of obsidian originating from Kozu Island to residential areas implies the possibility that toolmakers were not only equipped with a means of water transportation but also exploited marine resources. However, there is no archaeological evidence for such equipment or dietary remains relative to marine resources, as most artifacts other than lithics were completely decayed while buried in the volcanic ash soil of the Palaeolithic sites in Japan. If archaeological sites had existed on the Palaeolithic seashore, which was greatly extended by the regression of the Last Glacial Maximum, they would have disappeared by the transgression of the postglacial period.

Their social ability seems to have been sufficiently developed to allow the establishment of an organized conservation system for obsidian procurement, obsidian sharing within a settlement, and inter-settlement obsidian circulation. Moreover, the system for obsidian conservation established in the Layer IX Stage was achieved despite the difficulties associated with the long-distance transport of obsidian. The schedule of obsidian supply to the settlement would have been important, as shortages would have directly affected the production of hunting gear. This suggests that the toolmakers were capable of planning and managing a specific activity over a considerably long period of time. It is valid to consider that the archaeological records and their implications concerning behavior related to the beginning and change of obsidian use are the first archaeological manifestation of modern human behavior unique to *Homo sapiens* in the Japanese Islands, and that the pioneers of obsidian use were therefore behaviorally modern humans.

In the model outlined here of the initial mode of obsidian use, the first discovery of sources of non-local raw materials and their use resulted from the exploitation of a wide variety of natural resources in unpopulated lands. The successful migration of the pioneers in Central Japan must have fully depended on these activities. Taken into consideration with the sparsity of Middle Palaeolithic sites in Japan, it would be valid to question whether the people who belonged to these sites could have succeeded in migration. From the viewpoint of regional resource exploitation, it may be interpreted that most of the Middle Paleolithic sites represent traces of unsuccessful human migration in the Japanese Islands, which occurred intermittently before 40 ka.

It is reasonable to speculate that behaviorally modern humans, including the pioneers of obsidian use, first colonized the Japanese Islands from the continent of Asia. Based on a comparative study of the Middle and early Upper Paleolithic industries in southern Siberia, the Korean Peninsula, and China, Sato (2008) suggested that the migration route between the Korean Peninsula and northern Kyushu in Japan is the most important framework for archaeological investigations on this issue. Several Japanese authors have studied the initial phase of the Upper Palaeolithic from a regional perspective (Yamahara 1993, 1996; Tachibana 2000, 2002; Hitai 2004, Fujino 2006, Yoshikawa 2007). Hitai (2004) claimed that the dates for the emergence of the "small flake industry" (equal to the Layer X Stage) appear to be earlier in Central Japan than the emergence of another early industry in the northern part of Japan. This seems to imply that the dispersal of behaviorally modern humans proceeded from west to east, and is consistent with the sparsity of sites belonging to this period in the Hokkaido Region (the northernmost part of Japan). Thus, the archaeological evidence for the pioneer phase of obsidian use in Central Japan should be recognized as one of the forms of regional adaptation that constitutes a scenario of the first successful migration throughout the Japanese Islands.

Acknowledgements

I am grateful to the reviewers for many constructive comments on earlier draft. This research was supported by the Japan Society for the Promotion of Science (JSPS), Grant-in-Aid for Scientific Research (C) (20520664).

References

Akasaki, H., 2007, *Yamada-iseki* [*Excavation Report of the Yamada Site*]. Miyazaki Prefectural Center for Archaeological Operations, Miyazaki, 244p. (in Japanese)

Daikuhara, Y., 1988, *Kojo Iseki* [*Excavaton Report of the Kojo Site*]. Annaka City Board of Education, Gunma,132p. (in Japanese)

Fujinami, H. Hayashi, T. Nakamura, T. Nakamura, M. and Kaneko, Y., 1999, *Nishidaigotoda-iseki* [*Excavation Report of the Nishidaigotoda Site*]. The Second Research Group for the Sites within Metropolitan Area and the Nishidai Site Research Group, Tokyo, 604p. (in Japanese)

Fujine, H., 1998, Kokuyoseki no sanchi-suitei [Obsidian Source Analysis]. In *Shimayashiki-iseki* [*Excavation Report of the Shimayashiki Site*], pp. 136-138. Tokyo Metropolitan Center for Buried Cultural Properties, Tokyo. (in Japanese)

Fujino, T., 2006, Koki-kyusekki-jidai zenhan-ki (syogen-ki) no shuraku yoso [The aspect of colonial sites in the first half of the Late Palaeolithic Age]. In *Kyusekki Kenkyu* [*Palaeolithic Research*] 2, pp. 19-33. Japanese Palaeolithic Research Association, Aich. (in Japanese)

Habgood, P. J. and Franklin N. R., 2008, The revolution that didn't arrive: a review of Pleistocene Sahul. In *Journal of Human Evolution* 55, pp. 187-222.

Hagiwara, H., 2002, Nihon-retto saiko no kyusekki-bunka

[Japan's oldest Palaeolithic tools]. In *Hirado Shi-shi Kenkyu* [*The History of Hirado City*] 9, pp. 3-40. Hirado Historiography Committee, Nagasaki. (in Japanese)

Hashimoto, K., 2003, Kanjo-yunitto to sekihu no kankei ni tsuite [Relationships of circular unit and stone axe]. In *Kyusekki Kenkyu* [*Palaeolithic Research*] 4, pp. 35-46. Japanese Palaeolithic Research Association, Aichi. (in Japanese)

Hayakawa, I. and Kawachi, K., 1984, *Musashidai-iseki I* [*Excavation Report of the Musashidai Site I*]. Research Group for Sites in the Metropolitan Fuchu Hospital, Tokyo, 454p. (in Japanese)

Hirose, T., Iwana, K. and Takao, Y., 2006, Shizuoka-ken-nai no Iwajuku-jidai-I-ki shoto no sekki-gun [Lithic assemblages of the early part of the first stage of the Iwajuku Age in Shizuoka Prefecture]. In *The Proceedings of Iwajuku Forum 2006 "Iwajuku-jidai ha Dokomade Sakanoboreruka* [*How Far dose the Iwajuku Age Date Back to?*]*"*, pp. 14-21. Iwajuku Museum and the Organizing Committee for Iwajuku Forum, Gunma. (in Japanese)

Ikeda, T., 1999, *Ishinomoto-iseki-gun II* [*Excavation Report of the Ishinomoto Sites II*]. Kumamoto Prefectural Board of Education, Kumamoto, 226p. (in Japanese)

Idei, H., Matsu'ura, M., Kurihara, Y., 2004, *Kamibayashi-iseki* [*Excavation Report of the Kamibayashi Site*]. Sano City Board of Education, Tochigi, 948 p. (in Japanese)

Institute of Accelerator Analysis, 2009, Kanedori-iseki, Koyappata-iseki niokeru hosyaseitanso-nendai (AMS) [Radiocarbon dates of the Kanedori Site and the Koyappata Site (AMS)]. In *Tono-si Maizo-bunkazai Chosa Hokokusyo* [*Excavation Report of cultural properties in Tono City*] *6*, pp. 53-55. Tono City Board of Education, Iwate. (in Japanese)

Ito, T., 2006, Tamaranzaka, Musashi-kokubunji-ato-kanren, Musashidai-iseki no sekkigun [Lithic assemblages of the Tamaranzaka Sites, the Musashi-kokubunji-ato Sites and the Musashidai Sites]. In *The Proceedings of Iwajuku Forum 2006 "Iwajuku-jidai ha Dokomade Sakanoboreruka* [*How Far dose the Iwajuku Age Date Back to?*]*"*, pp. 40-50. Iwajuku Museum and the Organizing Committee for Iwajuku Forum, Gunma. (in Japanese)

Iwasaki, S., Miyata, E., Yoshidome, H., Ushinohama, O., Nagano, S., Higashi, K., Magome, R., Sugihara, T., Ogi, K., Miyata, T., Matsumoto, S. and Shiba, K., 2007, *Shinai-iseki (Uwaba-iseki hoka) Hakkutsu-chosa Hokokusyo* [*Excavation Report of the Uwaba Site and so forth*]. Izumi City Board of Education, Kagoshima, 197p. (in Japanese)

Japanese Archeological Association, 2003, *Zen-chuki-kyusekki-mondai no kensho* [*Inspection of the Early and Middle Palaeolithic Period in Japan*]. Japanese Archaeological Association, Tokyo, 625p. (in Japanese with English summary)

Japanese Palaeolithic Research Association, 2010, *Nihon-reto no Kyusekki-jidai Iseki* [*Palaeolithic Sites in the Japanese Islands: A Database*]. Japanese Palaeolithic Research Association, Tokyo, 337p. (in Japanese with English summary)

Kaifu, Y., 2005, *Jinrui ga Tadotte-kita Michi* [*Journey Followed by Humans*]. NHK books, Tokyo, 332p. (in Japanese)

Kikuchi, K., Takeda, Y., Komukai, H., Kumagaya, T., Nakagawa, S. and Takahashi, G., 2002, Iwate-ken Kanedori-iseki [The Kanedori Site in Iwate Prefecture]. In *The Archaeological Journal*, 495, pp. 6-10. Nyu-saiensu-sha, Tokyo. (in Japanese)

Kitamori, R., 2003, Kumamoto-ken hitoyoshi-shi Ono-iseki-gun-D-iseki [The Ono sites, Locality D in Hitoyoshi City, Kumamoto Prefecture]. In Anzai, M. (ed.) *Kokogaku* [*Archaeology*] I, pp. 129-166. Taira-denshi-insatsu, Chiba. (in Japanese)

Kisaki, Y., 2002, Kyusekki-jidai no iko to ibutsu [Palaeolithic Features and artifacts]. In *Shizume-iseki* [*Excavation Report of the Shizume Site*], pp. 15-62. Jonan Town Board of Education, Kumamoto. (in Japanese)

Kobayashi, K., 2001, Oiwake-iseki-gun syutudo kokuyoseki no gensanchi-suitei [Source analyses of the Obsidian assemblage in the Oiwake Sites]. In *Takayama-iseki-gun Dai-I-iseki oyobi Oiwake-isekigun Hakkutsu-chousa Hokoku-syo* [*Excavation Report of the Takayama Sites, Loc.1 and the Oiwake Sites*], pp. 439-454. Nagato Town Board of Education, Nagano. (in Japanese)

Kobayashi, S., Imai, E., Kogusuri, K. and Igarashi, A., 1998, *Shimayashiki-iseki* [*Excavation Report of the Shimayashiki Site*]. Tokyo Metropolitan Center for Buried Cultural Properties, Tokyo, 422p. (in Japanese)

Kosuge, M., 2005, *Kanjo-burokku-gun: san-mannen-mae no kyodai-Shuraku wo ou* [*In Search of the Circular Settlements in 30ky ago*]. Kasagake Town Museum of Iwajuku Culture, Gunma, 60p. (in Japanese)

Kuroda, A., 2005, Sokatu [Conclusion remarks]. In *Kanedori-iseki [The Kanedori Site: 2nd and 3rd Excavation Report]*, pp. 82-86. Miyamori Village Board of Education, Iwate. (in Japanese)

Kuroda, A., Yagi, K., Nagatomo, T., Terui, K., Soda, T. and Sugiyama S., 2005, *Kanedori Iseki* [*The Kanedori Site: 2nd and 3rd Excavation Report*]. Miyamori Village Board of Education, Iwate, 132p. (in Japanese)

Kunitake, S., 2004, Sekijin-seisan-gijutsu no tekio-ron-teki kosatsu: Boso-hanto IX-so no sekijin-seisan-gijutsu no hensen [An examination on the blade technology from a perspective of adaptive strategy: the transition of the blade technology in the Boso Peninsular]. In Anzai, M (ed.) *Kokogaku* [*Archaeology*] II, pp. 76-92. Taira-denshi-insatsu, Chiba. (in Japanese)

Machida, H., 2005, Nihon kyusekki-jidai no hennen: minami-kanto tachikawa romu-so no sai-kento [A tephrochronological review on the earlier Palaeolithic archaeological remains in South Kanto]. In *Kyusekki Kenkyu* [*Palaeolithic Research*] 1, pp. 7-16. Japanese Palaeolithic Research Association, Aichi. (in Japanese with English summary)

Maechi, H., Kimura, Y. and Saeki, H., 1989, *Takainari-iseki* [*Excavation Report of the Takainari Site*].

Archaeological Research Group of Nerima Ward, Tokyo, 147p. (in Japanese)

Matsufuji, K., 2007, Nihon-retto no kyusekki-jidai ha dokomade sakanoborunoka?: Iwateken Kanedori iseki dai IV bunka-so no nendai wo megutte [How far back can the Japanese Palaeolithic be traced?: A study on the date of the Kanedori Site, Cultural Layer IV, Iwate Prefecture]. In *Kokogaku ni Manabu [Learning from Archaeology]* III, pp. 1-16. Arcaeological Series of Doshisya University IX. The Committee for the Archaeological Series of Doshisya University, Kyoto. (in Japanese)

Matsu'ura, S. and Kondo, M., 2000, Nihon-retto no kyusekki-jidai-jinkotsu ha dokomade sakanoboruka [How far can the Palaeolithic human fossil remains date back to?]. In *Kokogaku to Kagaku wo Musubu [Linking Archaeology and Chemistry]*, pp. 135-167. University of Tokyo Press, Tokyo. (in Japanese)

McBrearty, S. and Brooks, A. S., 2000, The Revolution that wasn't: A new interpretation of the origin of modern human behavior. In *Journal of Human Evolution* 39, pp. 453-563.

Mellars, P., 1991, Cognitive changes and the emergence of modern humans in Europe. In *Cambridge Archaeological Journal* 1(1), pp. 63-76. Cambridge University Press, Cambridge.

Mellars, P., 2006, Going east: new genetic and archaeological perspectives on the modern human colonization of Eurasia. In *Science* 313, pp. 796-800.

Mochizuki, A., 1999, Kokuyoseki sanchi sutei hokoku [Report of obsidian source analyses]. In *Happusan-iseki-gun [Excavation Report of the Happusan Sites]*, pp. 610-613. Saku City Board of Education, Nagano. (in Japanese)

Mochizuki, A., 2000, Nojiriko-iseki-gun shutsudo kokuyoseki no sanchi-suitei (I) [Obsidian source analyses of the Lake Nojiri site cluster (I)]. In *Hinatabayashi-B-iseki, Hinatabayashi-A-iseki, Nanatsukuri-iseki, Odaira-iseki [Excavation Report of the Hinatabayashi B Site, the Hinatabayashi A Site, the Nanatsukuri Site, the Odaira B Site]*, pp. 233-236. Ngano Prefectural Center for Buried Cultural Properties, Nagano. (in Japanese)

Mochizuki, A. Yoshida, M. and Ambiru, M., 2006, Tsukimino iseki-gun dai-III-iseki C-chiten kokuyoseki-sei-sekki-gun no sanchi-suitei-bunseki [Report on source analyses of the obsidian lithic assemblage of the Tsukimino III C Site]. In *Kokuyoseki bunka kenkyu [Obsidian Culture]* 4, pp. 71-90. Meiji University Museum, Tokyo. (in Japanese)

Murasaki, T., 1999, *Mimikiri-iseki [Excavation Report of the Mimikiri Site]*. Kumamoto Prefectural Board of Education, Kumamoto, 256p. (in Japanese)

Mochizuki, A. and Henmi, T., 2000, Nojiri-ko-iseki-gun shutsudo kokuyoseki no sanchi-suitei (II). In *Kannoki Iseki, Nishioka A Isek [Excavation Report of the Kannoki Site, the Nishioka Site]*, pp. 269-280. Ngano Prefectural Center for Buried Cultural Properties, Nagano. (in Japanese)

Nakamura, M., 2006, Musashino-daichi chuo-bu no koki-kyusekki-jidai-syoto no sekkigun [Lithic assemblages in the initial Upper Palaeolithic in the central part of the Musashino Upland]. In *Proceedings of Iwajuku Forum 2006 "Iwajuku jidai ha dokomade sakanoboreruka [How Far Back can the Iwajuku Age be Traceed?]"*, pp. 51-60. Iwajuku Museum and the Organizing Committee for Iwajuku Forum, Gunma. (in Japanese)

Nagatomo, T. and Shitaoka, Y., 2004, Iriguchi iseki no IRSL nendai-sokutei [IRSL dating of the Iriguchi site]. In *Hirado shi-shi kenkyu [The History of Hirado City]* 9, pp. 110-114. Hirado Historiography Committee, Nagasaki. (in Japanese)

Nishii, Y., 2009, *Seiganji-maehara Iseki [Excavation Report of the Seiganji-maehara Site]*. Saitama Cultural Deposits Research Corporation, Gunma, 278p. (in Japanese)

Okamura, M., 2008, Nihon saiko no sekki to gisekki [The earliest lithics in Japan and pseudo-artifacts]. In the Committee for Publication of Essays in Memory of Professor Chosuke Serizawa (ed.) *Serizawa Chosuke sensei tsuito koko, minzoku, rekisi gaku ronso [Archaeology, Ethnology and History: Essays in memory of Professor Chosuke Serizawa]*, pp. 119-129. Rokuichi-shobo, Tokyo. (in Japanese)

Otake, N., 2008, Nihon-retto ni okeru kyusekki-jidai iseki-su [The numerical density of Palaeolithic sites in Japanese Islands]. In *Nihon-kyusekki-gakkai Dai-6-kai Kouen, Kenkyu-happyo, Sinpojiumu Yokousyu [Abstract Papers of the 6th Annual Meeting of Japanese Palaeolithic Research Association]*, pp. 32-35. Japanese Palaeolithic Research Association, Aichi. (in Japanese)

Otake, S. Katsumi, Y. Noguchi, A. Miki, Y. Kobayashi, K. Yoneda, M. and Nakajima, T., 2001, *Takayama-iseki-gun Dai-I-iseki oyobi Oiwake-iseki-gun Hakkutsu-chousa Houkoku-syo [Excavation Report of the Takayama Sites, Loc.1 and the Oiwake Sites]*. Nagato Town Board of Education, Nagano, 464p. (in Japanese)

Palaeoenvironmental Research Institute, 1999a, Hosya-sei tanso nendai-sokutei kekka [Results of the radiocarbon dating]. In *Happusan-iseki-gun [Excavation Report of the Happusan Sites]*, pp. 614-615. Saku City Board of Education, Nagano. (in Japanese)

Palaeoenvironmental Research Institute, 1999b, Numazu-shi Nishibora-iseki no shizen-kagaku-bunseki [Physical scientific analyses on the Nishibora Site, Numazu City]. In *Nishibora (b ku-1) Hakkutsu-chousa Houkoku-syo [Excavation Report of the Nishibora (b section -1) Site]*, pp. 297-300. Numazu City Board of Education, Shizuoka. (in Japanese)

Palaeoenvironmental Research Institute, 1999c, Shizen-kagaku-teki- bunseki [Physical scientific analyses] In *Ishinomoto-iseki-gun II [Excavation Report of the Ishinomoto Sites II]*, pp. 137-141. Kumamoto Prefectural Board of Education, Kumamoto. (in Japanese)

Palynosurvey, 1999, Nishidaigotoda iseki no shizen-kagaku-bunseki [Physical Scientific Analyses of the Nisidaigotoda Site]. In *Nishidaigotoda Iseki [Excavation Report of the Nishidaigotoda Site]*, pp. 561-604. The Second Research Group for the Sites within Metropolitan Area, Tokyo. (in Japanese)

Samukawa, T., Miyata, E., Somi, I., Uchimura, M., 2007, *Maeyama-iseki* [*Excavation Report of the Maeyama Site*]. Kagoshima Prefectural Center for Buried Cultural Properties, Kagoshima, 236p. (in Japanese)

Sato, H., 2002, Ushiromuta-iseki dai-III-bunkaso no hennen-teki igi to kodoron [Chronological meaning and behavioral pattern of the Ushiromuta Site, Cultural Layer III]. In *Ushiromuta-iseki* [*Excavation Report of Ushiromuta Site*], pp. 382-394. Ushiromuta site research group and Kawaminami Town Board of Education, Miyazaki. (in Japanese)

Sato, H., 2008, Higashi-ajia no koki-kyusekki-jidai no keisei [Transition to the Upper Palaeolithic in the East Asia]. In *Ibo* 26, pp. 2-15. Communities Research Group, Kanagawa. (in Japanese)

Serizawa, T., Yanagida, T., Akoshima, K. and Ono, S., 2006, Gunma-ken Kiryu-shi Tsurugaya-higashi-iseki no zenki-kyusekki [Lithic assemblage of the Early Palaeolithic in the Tsurugaya-higashi Site, Kiryu City, Gunma Prefecture]. In *Nihon kokogaku kyoukai dai 72kai sokai kenkyu happyo yoshi* [*Abstract Papers of the 72th Annual Meeting of Japanese Archaeological Association*], pp. 25-28. Japanese Archaeological Association, Tokyo. (in Japanese)

Sasahara, Y., 1999, *Nishibora (b ku-1) Hakkutsu-chousa Houkoku-syo* [*Excavation Report of the Nishibora (b section -1) Site*]. Numazu City Board of Education, Shizuoka, 300p. (in Japanese)

Shiotsuka, K., 2004, Iriguchi-iseki A, C chiten hakkutsu-chosa no gaiyo [Excavation Report of the Iriguchi Site, Locality A and C]. In *Hirado Shi-shi Kenkyu* [*The History of Hirado City*] 9, pp. 41-56. Hirado Historiography Committee, Nagasaki. (in Japanese)

Stern, N., 2009, The archeaological signature of behavioral modernity: a perspective from the southern periphery of the modern human range. In Shea, J. J. and Lieberman, E. D. (eds.) *Transitions in Prehistory: Essays in Honor of Ofer Bar-Yosef*, pp. 253-283. American School of Prehistoric Research, Harvard University.

Suto, T., 1990, *Happusan-iseki-gun* [*Excavation Report of the Happusan Sites*]. Saku City Board of Education, Nagano, 620p. (in Japanese)

Suwama, J., 1988, Sagamino-daichi ni okeru sekki-gun no hensen: soi-teki syutsudo-jirei no kento niyoru dankaiteki-haaku [Chronology of the lithic industries on the Sagamino Upland: the examination of stratigraphic orders of cultural layers]. In *Kanagawa Koko* [*Journal of the Kanagawa Archaeological Society*] 24, pp. 1-30. Kanagawa Archaeological Society, Kanagawa. (in Japanese)

Suwama, J., 2006, Kyusekki-jidai no saiko wo kangaeru [Rethinking the earliest Palaeolithic industry]. In *Proceedings of Iwajuku Forum 2006 "Iwajuku-jidai ha Dokomade Sakanoboreruka* [*How Far back can the Iwajuku Age be traced?*]*"*, pp. 2-12. Iwajuku Museum and the Organizing Committee for Iwajuku Forum, Gunma. (in Japanese)

Tachibana, M., 2000, Kyushu-to ni okeru chu-ki-kyusekki-jidai to koki-kyusekki-jidai seiritsu-ki zengo no sekki-gun [Middle and initial Upper Palaeolithic industries in the Kyushu Island]. In *Beppu-daigaku Hakubutsukan Kenkyu-hokoku* [*The Bulletin of Beppu University Museum*] 20, pp. 1-23. Beppu University Museum, Oita. (in Japanese)

Tabira, Y., 1999, *Tachikiri-iseki, Kyozuka-iseki* [*Excavation Report of the Tachikiri Site and the Kyozuka Site*]. Nakatane Town Board of Education, Kagoshima, 102 p. (in Japanese)

Tachibana, M., 2002, Ushiromuta-iseki AT-kai sekki-gun to Kyushu ni okeru koki-kyusekki-jidai-zenhan-ki no hensen [Ushiromuta lithic assemblages below AT tephra and the transition of the early Upper Palaeolithic of Kyushu]. In *Ushiromuta-iseki* [*Excavation Report of the Ushiromuta Site*], pp. 409-429. Ushiromuta Site Research Group and Kawaminami Town Board of Education, Miyazaki. (in Japanese)

Tachibana, M., Sato, H. and Yamada, T., 2002, *Ushiromuta-iseki* [*Excavation Report of the Ushiromuta Site*]. Ushiromuta Site Research Group and Kawaminami Town Board of Education, Miyazaki, 456p. (in Japanese)

Takamiya, H., Kinbu, M. and Suzuki, M., 1975, Naha Yamashita-cho doketsu hakkutsu-keika-hokoku [Excavation Report of the Yamashita cho cave site]. In *Jinrui-gaku zasshi* [*Journal of the Anthropological Society of Nippon*], pp. 125-130. The Anthropological Society of Nippon, Tokyo. (in Japanese)

Takao, Y., 2006, Tokai-chiho no chiiki-hennen [Chronology of the Tokai Region]. In Anzai, M. and Sato, H. (eds.) *Kyusekki-Jidai no Chiiki-hennen-teki Kenkyu* [*Regional Chronologies of the Palaeolithic in Japan*], pp. 61-102. Doseisya, Tokyo. (in Japanese)

Terada, Y., Takasugi, N., Morimoto, T., Ikuda, T., Hashizume, Y., Gondo, M. and Tanaka, K., 1997, *Seta-iseki* [*Excavation Report of the Seta Site*]. Setagaya Ward Board of Education, Tokyo, 278p. (in Japanese)

Tozawa, M., Ambiru, M., Yajima, K. and Kosuge, M., 1989, *Takayama-iseki-gun I* [*Excavation Report of Takayama Sites I*], Nagato Town Board of Education and Takayama Sites Research Group, Naganop, 135p. (in Japanese)

Tsuchiya, T. and Otake, N., 2000, *Kannoki-iseki, Nishioka-A-iseki* [*Excavation Report of the Kannoki Site, the Nishioka Site*]. Ngano Prefectural Center for Buried Cultural Properties, Nagano, 234p. (in Japanese)

Tsuchiya, T. and Tani, K., 2000, *Hinatabayashi-B-iseki, Hinatabayashi-A-iseki, Nanatsukuri-iseki, Odaira-iseki* [*Excavation Report of the Hinatabayashi B Site, the Hinatabayashi A Site, the Nanatsukuri Site, the Odaira B Site*]. Ngano Prefectural Center for Buried Cultural Properties, Nagano, 275p. (in Japanese)

Tsushima, H., 1999, *Sanwa-kogyodanchi I Iseki* [*Excavation Report of the Sanwa-kogyodanchi Site, volume 1*]. Gunma Archaeological Research Foundation, Gunma, 345p.

Yagi, K., 2005, Kanedori-iseki 4a-so no taiseki-nendai ni kansuru kazanbai-sojo-gaku-teki kenkyu [A tephrostratigraphical study on the sedimentary date of Layer 4a in the Kanedori Site]. In *Kanedori-iseki* [*The Kanedori Site: 2nd and 3rd Excavation Report*], pp. 61-

63. Miyamori Village Board of Education, Iwate. (in Japanese)

Yamada, S., 2008, Hoshino-iseki ni okeru hakuhen-hakuri no mondaiten [Re-examination of flaking techniques in the Hoshino Site]. In the Committee for Publication of Essays in Memory of Professor Chosuke Serizawa (ed.) *Serizawa Chosuke Sensei Tsuitoh Koko, Minzoku, Rekisigaku Ronso* [*Archaeology, Ethnology and History: Essays in memory of Professor Chosuke Serizawa*], pp.221-233. Rokuichi-shobo, Tokyo, (in Japanese)

Yamagata, H., 2000, Hoshasei-tanso-nendai-sokutei [Radiocarbon dating]. In *Hinatabayashi-B-iseki, Hinatabayashi-A-iseki, Nanatsukuri-iseki, Odaira-iseki* [*Excavation Report of the Hinatabayashi B Site, the Hinatabayashi A Site, the Nanatsukuri Site, the Odaira B Site*], p250. Ngano Prefectural Center for Buried Cultural Properties, Nagano. (in Japanese)

Yamahara, T., 1993, Hokkaido ni okeru daikeiyo-sekki wo tomonau sekki-gun ni tsuite [On the trapezoid lithic industry in the Hokkaido Region]. In the Committee for Publication of Essays Commemorating of Professor Hiroshi Shiomi's Retirement (ed.) *Koko-ronshu: Shiomi Sensei Taikan Kinen Ronbunshu* [*Essays Commemorating of Professor Hiroshi Shiomi's Retirement*], pp. 33-44. Hiroshima. (in Japanese)

Yamahara, T., 1996, Hokkaido ni okeru saisekijin-bunnka-ki izen no sekki-gun ni tsuite: Tokachi-chiiki no Eniwa-a kazanbai koka izen no sekki-gun no bunseki kara [On the lithic assemblages prior to the microblade industry in the Hokkaido Region: from the analysis on the lithic assemblages below the Eniwa-a tephra]. In *Obihiro Hyakunen Kinen-kan Kiyo* [*The Bulletin of Obihiro City 100years Memorial Museum*] 14, pp. 1-28. Obihiro City 100years Memorial Museum, Hokkaido. (in Japanese)

Yanagida, T. and Ono, S., 2007, Oita-ken Sozudai-iseki dai-6,7-ji-hakkutsu-chosa no kenkyu-hokoku: Nihon zenki-kyusekki-jidai no hennen to chiiki-sei no kenkyu [Research on the Early Palaeolithic industry discovered at the Sozudai Site, Oita Prefecture, Kyusyu Japan]. In *Bulletin of the Tohoku University Museum* 7, pp. 1-118. Tohoku University Museum, Miyagi. (in Japanese with English summary)

Yoneda, M., 2001, Naganoken Oiwake-iseki-gun ni okeru hosyasei-tanso-nendai-kettei [Radiocarbon determinations in the Oiwake site cluster, Nagano Prefecture]. In *Takayama-iseki-gun Dai-I-iseki oyobi Oiwake-isekigun Hakkutsu-chosa Hokoku-sho* [*Excavation Report of the Takayama Sites, Loc. I and the Oiwake Sites*], pp. 455-463. Nagato Town Board of Education, Nagano. (in Japanese)

Yoshikawa, K., 2008, Sekki-genryo no kakutoku-shohi to ido-ryoiki no hensei: Koki-kyusekki-jidai-zenhanki ni okeru keishitsu-ketsugan-chitai karano ichi shiron [The procurement and consumption of raw materials and reorganization of nomadic sphere in the Japanese Upper Palaeolithic Age]. In *Kyusekki Kenkyu* [*Palaeolithic Research*] 3, pp. 35-58. Japanese Palaeolithic Research Association, Aichi. (in Japanese with English summary)

www.ingramcontent.com/pod-product-compliance
Lightning Source LLC
LaVergne TN
LVHW071217210926
848482LV00015B/832